Advanced Sensing and Safety Control for Connected and Automated Vehicles: Volume II

Advanced Sensing and Safety Control for Connected and Automated Vehicles: Volume II

Guest Editors

Chao Huang
Yafei Wang
Peng Hang
Zhiqiang Zuo
Bo Leng
Hailong Huang

MDPI

Basel • Beijing • Wuhan • Barcelona • Belgrade • Novi Sad • Cluj • Manchester

Guest Editors

Chao Huang
Department of Industrial and Systems Engineering
The Hong Kong Polytechnic University
Hong Kong
China

Yafei Wang
Institute of Intelligent Vehicle
Shanghai Jiao Tong University
Shanghai
China

Peng Hang
School of Transportation Engineering
Tongji University
Shanghai
China

Zhiqiang Zuo
School of Electrical Automation and Information Engineering
Tianjin University
Tianjin
China

Bo Leng
School of Automotive Studies
Tongji University
Shanghai
China

Hailong Huang
Department of Aeronautical and Aviation Engineering
The Hong Kong Polytechnic University
Hong Kong
China

Editorial Office
MDPI AG
Grosspeteranlage 5
4052 Basel, Switzerland

This is a reprint of the Special Issue, published open access by the journal *Sensors* (ISSN 1424-8220), freely accessible at: https://www.mdpi.com/journal/sensors/special_issues/JVH5F48WY2.

For citation purposes, cite each article independently as indicated on the article page online and as indicated below:

Lastname, A.A.; Lastname, B.B. Article Title. *Journal Name* **Year**, *Volume Number*, Page Range.

ISBN 978-3-7258-2659-9 (Hbk)
ISBN 978-3-7258-2660-5 (PDF)
https://doi.org/10.3390/books978-3-7258-2660-5

Contents

About the Editors

Chao Huang

Dr. Chao Huang is a Research Assistant Professor at the Department of Industrial and System Engineering, The Hong Kong Polytechnic University (PolyU). She received her Ph.D. degree from the University of Wollongong, Australia, in 2018. Dr. Huang is the author/co-author of more than 80 papers in conferences, book chapters, and journals, with a significant number of them falling under the areas of mechanical engineering, civil engineering, transportation science, and robotics. Dr. Huang is currently serving as an Associate Editor in Engineering Applications of Artificial Intelligence, IEEE Transactions on Consumer Electronics, IEEE Transactions on Transportation Electrification, IEEE Women in Engineering Magazine, International Journal of Control, Automation and Systems (IJCAS) and IET Intelligent Transport Systems. Her research interests are human–machine collaboration, fault-tolerant control, mobile robots (EV, UAV), and path planning and control.

Yafei Wang

Yafei Wang received a B.Sc. Degree from Jilin University, Changchun; an M.Sc. Degree from Shanghai Jiao Tong University, Shanghai, China; and a Ph.D. Degree from the University of Tokyo, Tokyo, Japan. He is currently an associate professor at the School of Mechanical Engineering, Shanghai Jiao Tong University, Shanghai. His research interests include state estimation and control for connected and automated vehicles.

Peng Hang

Peng Hang received a Ph.D. Degree from the School of Automotive Studies, Tongji University, Shanghai, China, in 2019. He is currently a research professor at the Department of Traffic Engineering, Tongji University, Shanghai, China. In 2018, he was a visiting researcher with the Department of Electrical and Computer Engineering, National University of Singapore, Singapore. From 2020 to 2022, he was a research fellow at the School of Mechanical and Aerospace Engineering, Nanyang Technological University, Singapore. His research interests include vehicle dynamics and control, decision-making, motion planning, and motion control for autonomous vehicles. He is an associate editor of IEEE Transactions on Vehicular Technology, Journal of Field Robotics, IET Smart Cities, and SAE International Journal of Vehicle Dynamics, Stability, and NVH.

Zhiqiang Zuo

Prof. Zhiqiang Zuo received an M.Sc. Degree in Control Theory and Control Engineering in 2001 from Yanshan University and a Ph.D. Degree in Control Theory in 2004 from Peking University, China. In 2004, he joined the School of Electrical and Information Engineering, Tianjin University, where he is a full professor. From 2008 to 2010, he was a research fellow in the Department of Mathematics, City University of Hong Kong. From 2013 to 2014, he was a visiting scholar at the University of California, Riverside. His research interests include nonlinear control, robust control and multi-agent systems. He serves as an associate editor of the *Journal of the Franklin Institute* (Elsevier).

Bo Leng

Bo Leng received a Ph.D. Degree in Vehicle Engineering from Tongji University, Shanghai, China. He is currently an associate professor at the School of Automotive Studies, Tongji University. His current research interests include the dynamic control of distributed drive electric vehicles and motion planning and the control of intelligent vehicles. He won First Prize at the China Automobile Industry Technology Invention Award and First Prize at the Shanghai Science and Technology Progress Awards in 2020 and 2022, respectively. He was accepted by the Young Elite Scientists Sponsorship Program of the China Association for Science and Technology in 2022.

Hailong Huang

Hailong Huang received his Ph.D. Degree in Systems and Control from the University of New South Wales, Sydney, Australia, in 2018. He is now an assistant professor at the Department of Aeronautical and Aviation Engineering, the Hong Kong Polytechnic University, Hong Kong. His current research interests include guidance, navigation, and control of UAVs and mobile robots. He is an associate editor of *IEEE Transactions on Vehicular Technology, IEEE Transactions on Intelligent Vehicles, Intelligent Service Robotics*, and the *International Journal of Advanced Robotic Systems.*

sensors

MDPI

Article

Development of a Sliding-Mode-Control-Based Path-Tracking Algorithm with Model-Free Adaptive Feedback Action for Autonomous Vehicles

Kwangseok Oh [1] and Jaho Seo [2,*]

[1] School of ICT, Robotics & Mechanical Engineering, Hankyong National University, Anseong-si 17579, Republic of Korea

[2] Department of Automotive and Mechatronics Engineering, Ontario Tech University, Oshawa, ON L1G 0C5, Canada

* Correspondence: jaho.seo@ontariotechu.ca; Tel.: +1-905-721-8668 (ext. 7341)

Abstract: This paper presents a sliding mode control (SMC)-based path-tracking algorithm for autonomous vehicles by considering model-free adaptive feedback actions. In autonomous vehicles, safe path tracking requires adaptive and robust control algorithms because driving environment and vehicle conditions vary in real time. In this study, the SMC was adopted as a robust control method to adjust the switching gain, taking into account the sliding surface and unknown uncertainty to make the control error zero. The sliding surface can be designed mathematically, but it is difficult to express the unknown uncertainty mathematically. Information of priori bounded uncertainties is needed to obtain closed-loop stability of the control system, and the unknown uncertainty can vary with changes in internal and external factors. In the literature, ongoing efforts have been made to overcome the limitation of losing control stability due to unknown uncertainty. This study proposes an integrated method of adaptive feedback control (AFC) and SMC that can adjust a bounded uncertainty. Some illustrative and representative examples, such as autonomous driving scenarios, are also provided to show the main properties of the designed integrated controller. The examples show superior control performance, and it is expected that the integrated controller could be widely used for the path-tracking algorithms of autonomous vehicles.

Keywords: model-free adaptive feedback; sliding mode control; path tracking; autonomous vehicle; recursive least squares; forgetting factor; Lyapunov stability

Citation: Oh, K.; Seo, J. Development of a Sliding-Mode-Control-Based Path-Tracking Algorithm with Model-Free Adaptive Feedback Action for Autonomous Vehicles. *Sensors* **2023**, *23*, 405. https://doi.org/10.3390/s23010405

Academic Editors: Yafei Wang, Chao Huang, Hailong Huang, Zhiqiang Zuo, Bo Leng and Peng Hang

Received: 11 December 2022
Revised: 26 December 2022
Accepted: 27 December 2022
Published: 30 December 2022

1. Introduction

In addition to advanced hardware components such as steering, braking, and driving components, autonomous driving technology is one of the most important mobility technologies for improving safety, efficiency, and convenience. Because an autonomous vehicle aims to drive under any driving conditions and environment by itself, it needs various sensors—such as cameras, LiDAR, radar, and ultrasonic sensors—that can replace human sensory organs. In addition, mechanical actuators such as electric or hydraulic motors that can replace human muscle are needed to produce the desired force or pressure. Moreover, a computing system that functions like a human brain is required for data processing and decision-making for autonomous driving. Consequently, the vehicle system is more complicated and nonlinear as a result of the necessity of these various components that allow it to perform various driving tasks—such as lane changing, automatic parking, car-following, etc.

For driving tasks, accurate path-tracking performance should be ensured with reasonable path planning. Because vehicle conditions and driving conditions/environments can change unexpectedly, the path-tracking performance of autonomous vehicles can be degraded, causing fatal accidents on the road. To overcome the aforementioned limitation, various control technologies for the path tracking of autonomous vehicles have been developed, as follows.

1.1. Literature Review

Sun, C. et al. presented a model predictive control (MPC) path-tracking controller with switched tracking errors that can reduce the lateral tracking deviation and maintain vehicle stability for both normal and high-speed conditions [1]. They compared the performance of three MPC controllers with different tracking errors and analyzed their results. Baca, T. et al. proposed a linear MPC-based novel approach for optimal trajectory tracking for unmanned aerial vehicles (UAVs) using nonlinear state feedback [2]. They demonstrated the usability of the proposed approach through statistical and experimental evaluations of the platform in both simulated and real-world examples. Suh, J. et al. developed motion-planning algorithms for lane changing with a combination of probabilistic and deterministic prediction methods for automated driving under complex driving circumstances [3]. A collision probability and a safe driving envelope were defined by the authors using a reachable set and behavioral prediction of surrounding vehicles for safe lane changing. The developed model was evaluated based on simulations and experiments using an actual test vehicle under a lane change scenario. Xu, S. and Peng, H. presented a preview steering control algorithm for accurate, smooth, and computationally inexpensive path tracking for automated vehicles, along with an analysis of the closed-loop system [4]. In the study, the future road curvature as a dynamic disturbance was considered for the preview controller design, and its performance was evaluated based on simulations and experimental tests. Chowdhri, N. et al. developed a nonlinear MPC algorithm to perform evasive maneuvers and avoid a rear-end collision, with constraints [5] that are needed for ensuring vehicle stability and accounting for actuator limitations. Li, S. et al. proposed an obstacle avoidance controller based on nonlinear MPC for autonomous vehicle navigation [6]. It was designed so that the reference trajectory is adjusted when obstacles suddenly appear and the risk index is computed online for collision avoidance. Cao, J et al. developed a trajectory-tracking control algorithm for autonomous vehicles considering cornering characteristics with simplified vehicle dynamics and tire models [7]. Wang, Y et al. developed an MPC algorithm to optimize the reference trajectory with consideration of the motion prediction of other traffic participants using Monte Carlo simulations [8]. Quirynen, R. et al. studied the real-time feasibility of nonlinear MPC-based steering control on an embedded computer for autonomous vehicles [9]. In addition, Shen, C and Shi, Y investigated the nonlinear model predictive control (NMPC) method, looking for possible approaches to alleviate the heavy computational burden, and developed novel distributed NMPC algorithms by exploiting the dynamic properties of the autonomous underwater vehicle motion for trajectory-tracking control [10]. Chu, D. et al. presented a trajectory planning and tracking framework to obtain target trajectory and MPC with PID feedback to effectively track planned trajectory [11]. In [12], an improved MPC algorithm with fuzzy adaptive weight control was proposed for autonomous vehicles to ensure tracking accuracy and dynamic stability during path tracking. To implement trace planning and tracking for obstacle avoidance, Zhang, C et al. integrated a trajectory planner and a tracking controller for autonomous vehicles [13]. The study of [14] proposed a scheme for implementing an MPC path-following controller that considers feasible road regions, vehicle shapes, and the model mismatch caused by varying road conditions and small-angle assumptions in measurable disturbances [14]. To maintain a collision-free path for autonomous vehicles, the authors of [15] proposed a hierarchical path-planning and trajectory-tracking framework by solving a constrained finite-time optimal problem. Yue, M et al. developed a time-based quantic polynomial function for trajectory planning that takes into account the vehicle system's safety, comfort, and traffic efficiency [16]. A robust MPC with a finite time horizon was proposed by Peng, H et al. to achieve coordinated path tracking and direct yaw moment control for autonomous four-in-wheel-motor independent-drive electric vehicles [17].

The previous studies mentioned above used mathematical vehicle models to design path-tracking control algorithms; however, there are model uncertainties that have a negative impact on the path-tracking control performance. Hence, studies on the adaptive path-tracking control of autonomous vehicles have been conducted to reduce model uncertainty and improve performance under various driving conditions and environments.

Londhe, P. and Patre, B. designed a robust and adaptive tracking control algorithm for a complete nonlinear model of an autonomous underwater vehicle based on adaptive fuzzy sliding mode control (SMC) [18]. The study derived fuzzy control rules using the Lyapunov energy function to minimize chattering. Taghavifar, H. and Rakheja, S. applied an exponential-like sliding mode fuzzy type-2 neural network approach to design a robust adaptive indirect controller that can enhance the path-tracking performance of autonomous road vehicles [19]. In this study, the authors used the Lyapunov stability theorem to derive the adaptation laws for a hierarchical controller design and ensure the stability of the closed-loop system.

Zhou, X. et al. proposed an adaptive inverse controller to offset the dynamics of the steering system's backlash, and adaptive control laws were robustified by means of sigma modification [20]. The authors presented hardware-in-the-loop experimental results to show the main contribution of the proposed control algorithm. Yuan, X. et al. developed a course-angle optimal referential model and MPC-based adaptive control system for more adaptive path tracking at different velocities [21].

To improve tracking accuracy and stability, Lin, F. et al. developed an adaptive MPC controller by applying a recursive least squares algorithm that can estimate cornering stiffness and road friction online [22]. Liu, S et al. proposed a novel model-free adaptive control algorithm based on a dual successive projection method and analyzed it using the introduced method with a symmetrically similar structure of the controller [23]. Guerrero, J et al. designed an adaptive high-order sliding mode controller that does not require knowledge of the upper bound of the disturbance for trajectory tracking with the Lyapunov concept [24]. Tran, V et al. proposed a new concept of an adaptive strictly negative imaginary controller that minimizes a certain performance index robustly for 3D tracking of drones in the face of wind gusts [25]. Tian, Y et al. developed an adaptive path-tracking control strategy that coordinates active front steering and direct yaw moment based on an MPC algorithm. The authors used the recursive least squares method with a forgetting factor to identify the rear tires' cornering stiffness and update the path-tracking system prediction model [26]. For robust adaptive path tracking of an underactuated unmanned surface vehicle, Fan, Y et al. proposed an improved line-of-sight guidance law using a reduced-order extended state observer to address the large sideslip angle that occurs in practical navigation. [27]. Pereida, K and Schoellig, A developed a novel adaptive MPC with an underlying L_1 adaptive controller to enhance the trajectory tracking of a system under unknown and changing disturbances [28]. Kebbati, Y et al. presented an improved particle-swarm-optimized PID to handle the task of speed tracking based on nonlinear longitudinal dynamics for the coordinated longitudinal and lateral control in autonomous driving [29]. By applying dynamic trajectory planning and a robust adaptive nonlinear fuzzy backstepping controller, a novel nonlinear trajectory-tracking control strategy was developed for lane-changing maneuvers [30]. A sliding mode control approach with enhanced state observers was proposed in [31] to control both lane-keeping errors and roll angles within the prescribed performance boundaries. Liang, Y et al. proposed a novel scheme that integrates local motion planning and control to determine motion behaviors, track global paths, and conduct local motion commands based on adaptive MPC and lateral MPC [32]. For autonomous vehicles with four independent in-wheel motors, an integrated autonomous driving (AD) control system was developed in [33], consisting of two parts: an AD controller and a chassis controller. He, H et al. presented a hierarchical path-tracking control framework for two-axle autonomous buses with two layers that can prevent sideslip and rollover and can acquire the steering angle with stability constraints [34]. In order to design adaptive control algorithms for path tracking, mode-based or model-free adaptation rules are needed for control input adaptation. However, it is difficult to design adaptation rules ensuring robust stability of control systems while taking constraints into account. To tackle this issue, data-driven or learning-based path-tracking control algorithms have been developed.r

Chen, I. and Chan, C. developed deep reinforcement learning algorithms using proximal policy optimization that were combined with the conventional pure pursuit method to structure the controller's architecture [35]. Zhang, K. et al. proposed an adaptive learning MPC scheme for the trajectory tracking of perturbed autonomous ground vehicles based on unknown system parameter estimation [36]. The authors designed a set-membership-based parameter estimator using the recursive least squares technique. Jiang, Y. et al. investigated the path tracking control strategy of variable-configuration unmanned ground vehicle and proposed an improved model free predictive control scheme [37]. Li, X. et al. developed a novel robust adaptive learning control algorithm that can estimate the system uncertainties through the iterative learning method [38]. In this design, a two-degree-of-freedom vehicle model was reformulated into a parametric form. Wang, Z and Wang, J incorporated model-free strategies for control and direct data-driven control into a predictive control framework for trajectory tracking of automated vehicles [39]. For unmanned surface vehicles, Wang, N et al. developed an innovative self-learning system using only input–output signals [40]. They developed a data-driven performance-prescribed reinforcement learning control scheme to pursue control optimality and prescribe tracking accuracy simultaneously. Jiang, Y et al. studied the heading tracking problem of six-wheel independent-drive and four-wheel independent-steering unmanned ground vehicles under the influence of uncertainties based on the model-free adaptive control method and particle swarm optimization [41]. Parseh, M et al. proposed a data-driven motion planning method to minimize injury severity for vehicle occupants in unavoidable collisions by establishing a metric that models the relationship between impact location and injury severity using real accident data [42]. Wu, Q et al. developed a fuzzy-inference-based reinforcement learning approach for autonomous overtaking decision-making that was created using a multi-objective Markov decision process and a temporal difference learning method based on dynamic fuzzy inference [43]. By integrating model-free control and extreme-seeking control, Wang, Z et al. provided a new perspective on tuning model-free control gain while improving its performance [44]. Spielberg, N et al. designed a neural network MPC using vehicle operation data to construct a neural network model that could be efficiently implemented in MPC [45]. Peng, Z et al. proposed reduced- and full-order data-driven adaptive disturbance observers for estimating unknown input gains, as well as total disturbances consisting of unknown internal dynamics and external disturbances [46]. To avoid collisions efficiently, Wang, H and Liu, B proposed a collision-avoidance framework based on road friction estimation and dynamic stability control [47]. The study of [48] aimed to develop a model-based feasibility enhancement technique of constrained reinforcement learning that can enhance the feasibility of policies using a generalized control barrier function that is defined based on the distance to the constraint boundary [48]. With an iterative single-critic learning framework, Zhang, K et al. proposed adaptive resilient event-triggered control for rear-wheel-drive autonomous vehicles [49]. This control can be effective in balancing frequency and changes when adjusting the vehicle's control during the running process. Combining the event-triggered sampling mechanism and the iterative single-critic learning framework, the authors developed an event-triggered condition for adaptive resilient control.

1.2. Summary of the Proposed Control Algorithm and Major Contributions

Suitable path-tracking performance is essential for the driving tasks of autonomous vehicles, such as lane changing, automatic parking, and vehicle following. However, path-tracking performance can be degraded by unexpected and abrupt changes in vehicle conditions and the driving environment. To deal with this issue and ensure robust control performance, our study designed a new path-tracking control algorithm by integrating adaptive feedback control (AFC) inputs with SMC. Specifically, the AFC algorithm was created using the recursive least squares and gradient descent methods to adjust feedback gains. It was designed so that the SMC algorithm was able to consider the error terms regulated by the AFC input with finite stability and Lyapunov stability conditions. Furthermore, the designed SMC algorithm is capable of considering the error terms regulated by the AFC input with finite stability and Lyapunov stability.

The performance evaluation of the proposed path-tracking control algorithm was conducted under two scenarios: curved path tracking, and lane change scenarios with constant velocity conditions.

The following is a summary of the major contributions of this study:

- The proposed control method is an attempt to develop an integrative control algorithm for path tracking of autonomous vehicles using adaptive feedback and SMC algorithms that can reject model uncertainties and ensure robust stability.
- The proposed control scheme allows for the design of controllers using a simple mathematical model that requires low computational costs.

Based on the literature review above, Table 1 summarizes the pros and cons of the proposed control method in comparison with other related existing approaches, which are classified into five categories.

Table 1. A comparison of the pros and cons of several control methods.

Control Method and Representative Studies	Main Features	Pros	Cons
Proposed	Integrative control using AFC and SMC; a simple model can be used	Adaptive feedback action and robust control considering adaptation are possible	Parameters such as adaptation rate and weighting factor need to be properly determined
Model-based control Refs. [1–3]	Optimal control using a system mathematical model	Optimal control allocation is possible	It is necessary to know the system parameters and uncertainty, as well as their rejection
Model-based adaptive control Refs. [20–22]	Optimal control with a mathematical model and the adaptation law	Adaptive optimal control is possible	A proper determination of the controller's adaptation rate is needed for stability
Model-free adaptive control Refs. [23,39,44]	Adaptive control without a system mathematical model	A system mathematical model is not needed	Optimal control allocation is difficult for multi-input systems
Data-driven control Refs. [37,42,46]	Adaptive control and observation using control and system data	Control and observation are possible using only data (without a model)	A stability analysis is required
Learning-based control Refs. [40,48,49]	Control using a learning framework such as reinforcement learning	Performance can be enhanced gradually	To maintain stability, a stability analysis and adaptation of learning rate are required

The remainder of this paper is outlined as follows: Section 2 presents a control algorithm for path tracking using SMC with adaptive feedback. Section 3 provides the results of the performance evaluation. Section 4 concludes with a discussion of the limitations of the current work and prospects for future research.

2. SMC-Based Path Tracking with Adaptive Feedback Action

This section provides the mathematical formulation of the SMC-based path-tracking algorithm with adaptive feedback action. In order to design the path-tracking control algorithm, a kinematic mathematical error model was used. Figure 1 shows defined control errors such as lateral error and yaw angle error for path tracking.

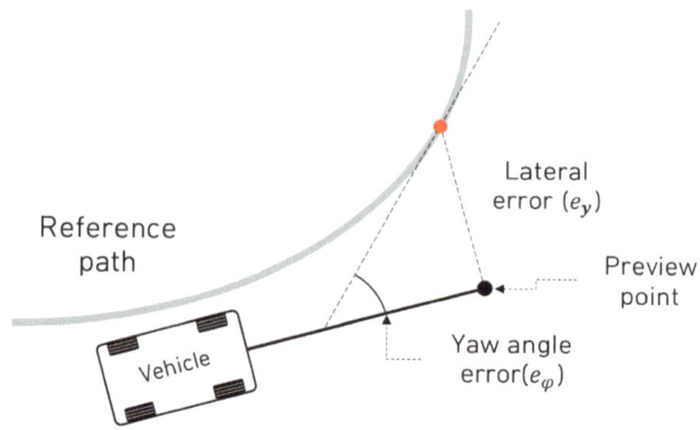

Figure 1. Defined control errors for path tracking.

Based on the defined path-tracking error, a kinematic-analysis-based mathematical error model was derived. The following equations represent the mathematical error model using kinematic parameters and its state-space representation:

$$\dot{e}_y = v_x e_\varphi \tag{1}$$

$$\dot{e}_\varphi = \frac{v_x}{L}\delta - \dot{\varphi}_d \tag{2}$$

$$\begin{bmatrix} \dot{e}_y \\ \dot{e}_\varphi \end{bmatrix} = \begin{bmatrix} 0 & v_x \\ 0 & 0 \end{bmatrix}\begin{bmatrix} e_y \\ e_\varphi \end{bmatrix} + \begin{bmatrix} 0 \\ v_x/L \end{bmatrix}\delta + \begin{bmatrix} 0 \\ -1 \end{bmatrix}\dot{\varphi}_d \tag{3}$$

where e_y and e_φ are the lateral error and yaw angle error with respect to the reference path for tracking of an autonomous vehicle, respectively, while v_x, $\dot{\varphi}_d$, δ, and L are the longitudinal velocity, desired yaw rate, front steering angle, and wheel base (i.e., the distance between the front-wheel axle and rear-wheel axle) of the vehicle, respectively. Figure 2 shows an overall block diagram for the model-free adaptive feedback action-based SMC algorithm.

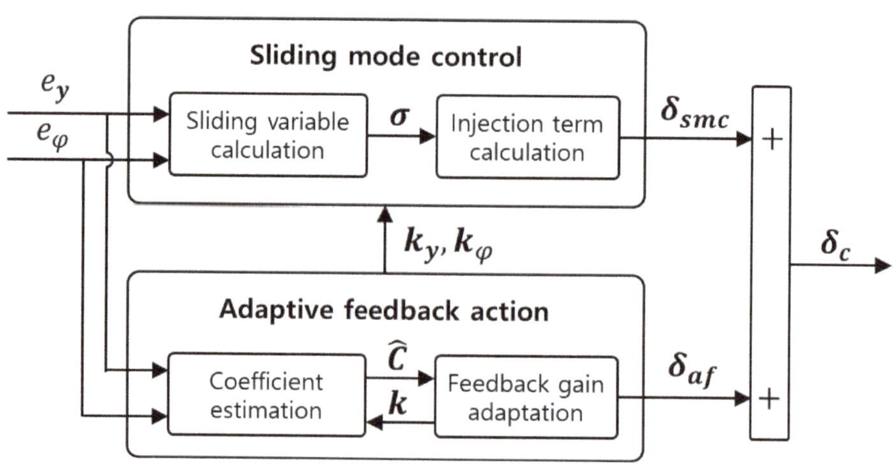

Figure 2. Block diagram for the adaptive feedback action-based sliding mode control.

The coefficient for feedback gain adaptation (the coefficient estimation block under the adaptive feedback action in Figure 2) can be estimated using the recursive least squares method with a forgetting factor. Using the estimated coefficient, a feedback gain is adapted based on the gradient descent method with a proper adaptation gain. The adaptive steering control input is calculated using the adapted feedback gain and the path-tracking control error. In this study, the SMC input for path tracking was computed with consideration of the adaptive steering control input to reduce the impact of the SMC input on the path-tracking control performance. The following equations were used to calculate the total steering control input using adaptive and sliding control inputs. In addition, mathematical definitions of the adaptive steering control and SMC inputs are presented below:

$$\delta_c = \delta_{af} + \delta_{smc} \tag{4}$$

$$\delta_{af} = k_y e_y + k_\varphi e_\varphi \tag{5}$$

$$\delta_{smc} = -\rho \mathrm{sign}(\sigma) \tag{6}$$

where δ_c is the total control input for the front steering wheel angle, δ_{af} and δ_{smc} are the adaptive feedback-based control input and SMC-based control input, respectively, k_y and k_φ are the feedback gains for the lateral and yaw angle errors, respectively, and ρ and σ are the magnitudes of the SMC input and sliding surface for controller design, respectively. Equation (3) can be rewritten by using the AFC input described in Equation (5). The following state-space-formed error mathematical model is the rewritten equation of Equation (3) using Equation (5):

$$\begin{bmatrix} \dot{e}_y \\ \dot{e}_\varphi \end{bmatrix} = \begin{bmatrix} 0 & v_x \\ k_y v_x/L & k_\varphi v_x/L \end{bmatrix} \begin{bmatrix} e_y \\ e_\varphi \end{bmatrix} + \begin{bmatrix} 0 \\ v_x/L \end{bmatrix} \delta_{smc} + \begin{bmatrix} 0 \\ -1 \end{bmatrix} \dot{\varphi}_d \tag{7}$$

In this study, the SMC input for path tracking was computed based on Equation (7). Calculating SMC inputs requires information about adaptive feedback gains, whose adaptation algorithms are explained in the next section.

2.1. Adaptive Feedback Action for Feedback Gain Adaptation

To estimate the coefficients for feedback gain adaptation, the two relationship functions shown in Equation (8) were designed and used for recursive least squares estimation with forgetting factors. This equation relates control errors to feedback gains for the derivation of coefficients $C_{ij}(i, j = 1, 2)$ [50].

$$\dot{e}_y = C_{11}\dot{k}_y + C_{12}\dot{k}_\varphi \dot{e}_\varphi = C_{21}\dot{k}_y + C_{22}\dot{k}_\varphi \tag{8}$$

The coefficients are estimated based on recursive least squares with properly determined forgetting factors, which are used for the feedback gain adaptation. The feedback gain is adapted by using the gradient descent method to minimize the control errors. The following equation is the cost function J_{af} defined for the gradient descent method:

$$J_{af} = \frac{1}{2}e_y^2 + \frac{1}{2}we_\varphi^2 \tag{9}$$

Based on the gradient descent method with the cost function defined above, the following feedback gain adaptation rules can be derived to reduce the control errors using the adaptation gain, weighting factor, and partial derivatives of path-tracking control errors with respect to feedback gains:

$$\dot{k}_y = -\gamma_y \frac{\partial J_{af}}{\partial k_y} = -\gamma_y (e_y + we_\varphi)\left(\frac{\partial e_y}{\partial k_y} + w\frac{\partial e_\varphi}{\partial k_y}\right) \tag{10}$$

$$\dot{k}_\varphi = -\gamma_\varphi \frac{\partial J_{af}}{\partial k_\varphi} = -\gamma_\varphi (e_y + we_\varphi)\left(\frac{\partial e_y}{\partial k_\varphi} + w\frac{\partial e_\varphi}{\partial k_\varphi}\right) \tag{11}$$

In this study, it was assumed that the estimated coefficients in Equation (8) were approximately equal to the partial derivatives of the path-tracking errors with respect to the feedback gains. Because this assumption may lead to unexpected control uncertainty, it was designed so that the SMC algorithm featured AFC inputs to ensure robustness. The following Equations (12) and (13) are rewritten versions of Equations (10) and (11) with this assumption; Equation (14) is the detailed AFC input obtained using the adapted feedback gains and adaptation gains:

$$\dot{k}_y = -\gamma_y \frac{\partial J_{af}}{\partial k_y} = -\gamma_y (e_y + w e_\varphi)(\hat{C}_{11} + w\hat{C}_{21}) \tag{12}$$

$$\dot{k}_\varphi = -\gamma_\varphi \frac{\partial J_{af}}{\partial k_\varphi} = -\gamma_\varphi (e_y + w e_\varphi)(\hat{C}_{12} + w\hat{C}_{22}) \tag{13}$$

$$\delta_{af} = -e_y \int \gamma_y (e_y + w e_\varphi)(\hat{C}_{11} + w\hat{C}_{21}) dt - e_\varphi \int \gamma_\varphi (e_y + w e_\varphi)(\hat{C}_{12} + w\hat{C}_{22}) dt \tag{14}$$

The next subsection explains the SMC algorithm that considers the designed AFC input for robust path-tracking performance of autonomous vehicles.

2.2. SMC with Adaptive Feedback Action

The AFC algorithm described in the previous subsection can adapt the feedback gain to reduce the path-tracking control, but it cannot guarantee the stability of the control algorithm if it is used alone. Therefore, an SMC algorithm that can consider the adaptation influence on the path-tracking performance is proposed in this study, based on the integration of two control algorithms (such as adaptive feedback and robust control algorithms).

By integrating the adaptive feedback and robust control algorithms, uncertainties can be reduced by the feedback gain adaptation, while stability can be ensured by the robustness of the sliding mode controller. In this study, a sliding surface (σ) was designed for path tracking using the following equation:

$$\sigma = e_y + w e_\varphi \tag{15}$$

where w is the weighting factor for the design of a sliding surface. The following equation is the cost function for the design of the SMC algorithm; the time derivative of the cost function is described in Equation (18) for the control input derivation:

$$J_{smc} = \frac{1}{2}\sigma^2 \tag{16}$$

$$\dot{J}_{smc} = \sigma\dot{\sigma} = \sigma(\dot{e}_y + w\dot{e}_\varphi) \tag{17}$$

Equation (17) above can be rewritten as follows by applying Equation (7) to derive the SMC input considering the adaptive steering control input:

$$\dot{J}_{smc} = \sigma\left(v_x e_\varphi + \frac{w k_y v_x}{L} e_y + \frac{w k_\varphi v_x}{L} e_\varphi + \frac{w v_x}{L}\delta_{smc} - w\dot{\varphi}_d \right) \tag{18}$$

All of the terms in the parentheses of Equation (18)—except for the control input term δ_{smc}—can be considered as disturbances, and an inequality condition using the disturbance boundary value L_b can be derived as follows:

$$L_b \geq \left| v_x e_\varphi + \frac{w k_y v_x}{L} e_y + \frac{w k_\varphi v_x}{L} e_\varphi - w\dot{\varphi}_d \right| \tag{19}$$

In order to design an asymptotically stable controller, the discrete injection term of SMC is defined as follows:

$$\frac{wv_x}{L}\delta_{smc} = -\rho \text{sign}(\sigma) \tag{20}$$

where ρ is the magnitude of the injection term, which was designed by considering the boundary value in Equation (19) for the stability of the controller. Equation (18) can be rewritten as follows using the boundary value and the definition in Equation (20):

$$\dot{J}_{smc} \leq \sigma(L_b - \rho \text{sign}(\sigma)) = -|\sigma|(\rho - L_b) \tag{21}$$

For the finite stability condition, the following inequality condition was derived based on the cost function condition, and the magnitude of the injection term can be determined with Equations (21) and (22):

$$\dot{J}_{smc} \leq -|\sigma|\alpha/\sqrt{2} \tag{22}$$

$$\rho = L_b + \alpha/\sqrt{2} \tag{23}$$

where α is a parameter for the finite stability condition. Based on the detailed disturbance boundary value, the magnitude of the injection term can be rewritten as follows:

$$\rho = \left| v_x e_\varphi + \frac{wk_y v_x}{L} e_y + \frac{wk_\varphi v_x}{L} e_\varphi - w\dot{\varphi}_d \right| + \alpha/\sqrt{2} \tag{24}$$

It is assumed in this study that the AFC input can reduce the control errors reasonably with the SMC input; therefore, the path-tracking control errors e_y and e_φ are taken to be zero. Equation (24) can be simplified based on this assumption, as shown in Equation (25).

$$\rho = |w\dot{\varphi}_d| + \alpha/\sqrt{2} \tag{25}$$

Using the magnitude of the injection term ρ in Equation (25), the SMC input can be computed using Equation (20) as follows:

$$\delta_{smc} = -\frac{L}{wv_x}\left(|w\dot{\varphi}_d| + \alpha/\sqrt{2}\right)\text{sign}(\sigma) \tag{26}$$

To reduce chattering of the SMC input, a sigmoid function was adopted and used in Equation (26) instead of a sign function. The following equation is the sigmoid-function-based SMC input:

$$\delta_{smc} = -\frac{L}{wv_x}\left(|w\dot{\varphi}_d| + \alpha/\sqrt{2}\right)\left(\frac{m\sigma}{1 + m|\sigma|}\right) \tag{27}$$

where m is a coefficient that is used to adjust the gradient of the sigmoid function near zero.

Using Equations (4), (14) and (27), the total steering control input that requires the adaptation gain, weighting factor, and other parameters (α, m) can be derived as follows:

$$\delta_c = -e_y \int \gamma_y (e_y + we_\varphi)(\hat{C}_{11} + w\hat{C}_{21})dt - e_\varphi \int \gamma_\varphi (e_y + we_\varphi)(\hat{C}_{12} + w\hat{C}_{22})dt - \frac{L}{wv_x}\left(|w\dot{\varphi}_d| + \alpha/\sqrt{2}\right)\left(\frac{m\sigma}{1 + m|\sigma|}\right) \tag{28}$$

The next section provides the performance evaluation results under various evaluation scenarios (i.e., curved path tracking and lane change).

3. Performance Evaluation

The performance evaluation was conducted using a planar vehicle model called the bicycle model under two path-tracking scenarios: curved path tracking, and lane change. The longitudinal velocities for the curved path tracking and lane change scenarios were kept constant at 30 kph and 60 kph, respectively.

For a comparative study, the performance of the different types of designed path-tracking controllers was evaluated four times for each scenario. The control algorithms proposed in this study were designed and evaluated using MATLAB/Simulink. Figures 3 and 4 illustrate the two scenarios and an overall block diagram for the performance evaluation of the designed control algorithm, respectively.

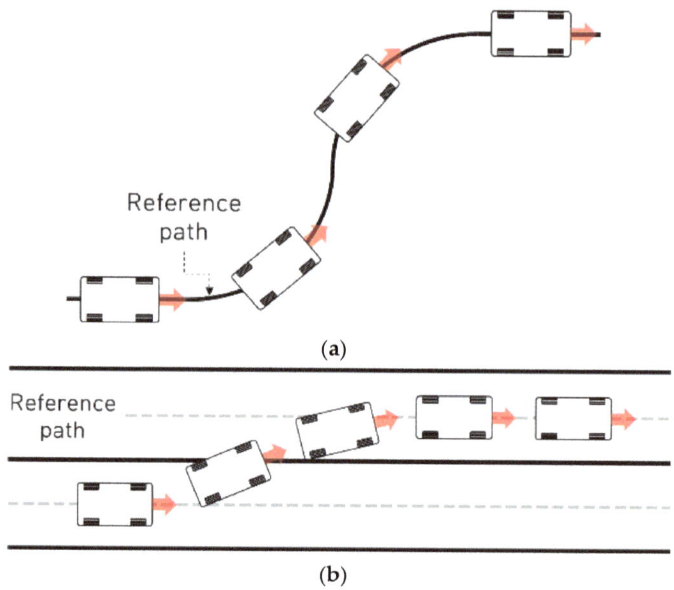

Figure 3. Two evaluation scenarios for performance evaluation: (**a**) Curved path-tracking scenario. (**b**) Lane change scenario.

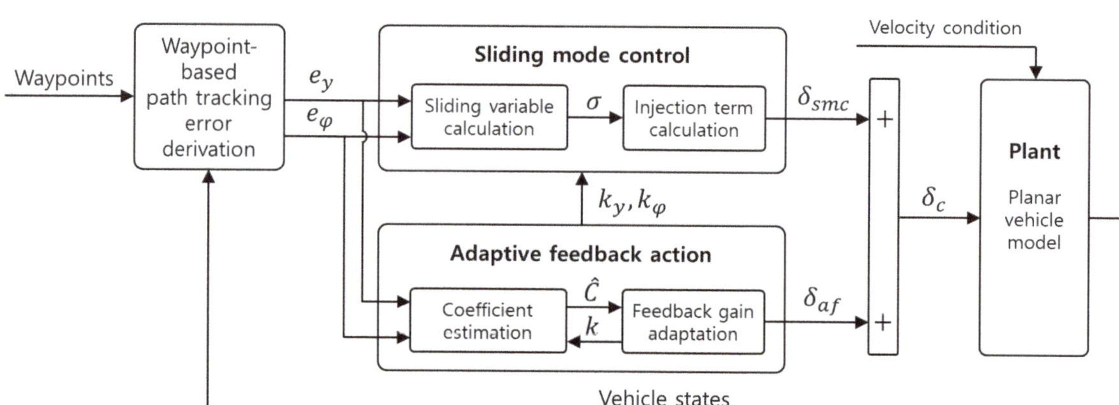

Figure 4. Overall block diagram for performance evaluation of the control algorithm.

In the waypoint-based path-tracking error derivation block, path-tracking control errors are computed using the designed waypoints and vehicle states in the block. The waypoints consist of x and y points of reference paths for curved and lane-change paths. Tables 2 and 3 show the vehicle specifications and the designed control parameters used for the performance evaluation.

Table 2. Vehicle specification.

Parameter	Unit	Value
Mass	kg	1600
Distance between the front axle and the center of mass	m	1.75
Distance between the rear axle and the center of mass	m	1.20
Wheel tread	m	1.65
Cornering stiffness, front	N/rad	74,000
Cornering stiffness, rear	N/rad	140,000

Table 3. Control parameters.

Parameter	Value (Curved Path)	Value (Lane Change)
Forgetting factor	0.999	0.999
Weighting factor (w)	5	5
Coefficient for sigmoid function (m)	1	1
Adaptation gain (γ_y)	1	0.001
Adaptation gain (γ_φ)	1	0.001
Parameter for stability condition (α)	1	1
Proportional gain (k_p)	0.05	0.008
Integral gain (k_i)	0.02	0.0001
Derivative gain (k_d)	0.001	0.00001

The next two subsections show the performance evaluation results for the curved path and lane change scenarios.

3.1. Path-Tracking Scenario: Curved Path Tracking (30 kph)

The results were compared between cases using AFC alone, SMC alone, SMC with AFC, and proportional–integral–derivative (PID) control.

The radius of curvature of the designed curved path was 100 m, and the longitudinal velocity of the vehicle was 30 kph. Figure 5 shows the steering control inputs for path tracking of all evaluation cases.

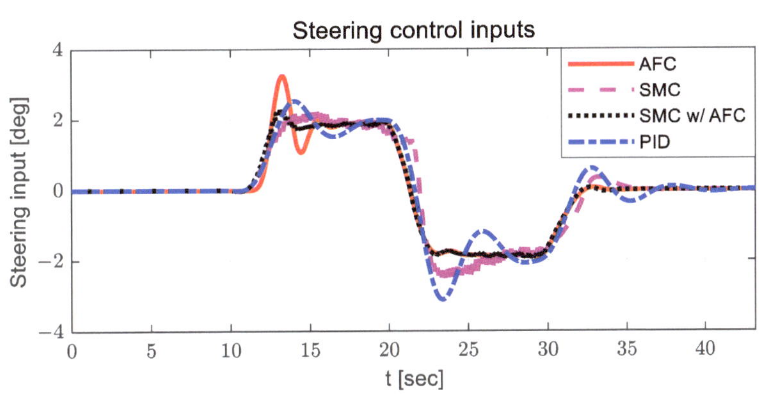

Figure 5. Results: steering control inputs for the curved path tracking.

For AFC, the steering control input is relatively large, and oscillation occurs between 10 and 15 s. The steering control input with SMC has a relatively large value around 23 s, with chattering. When using SMC with AFC, the steering control input is relatively stable compared to other steering control inputs. In the case of PID, the steering control input is relatively high after 23 s, with large oscillations. Figures 6 and 7 show the estimated coefficients for feedback gain adaptation in the cases of AFC and SMC with AFC, respectively.

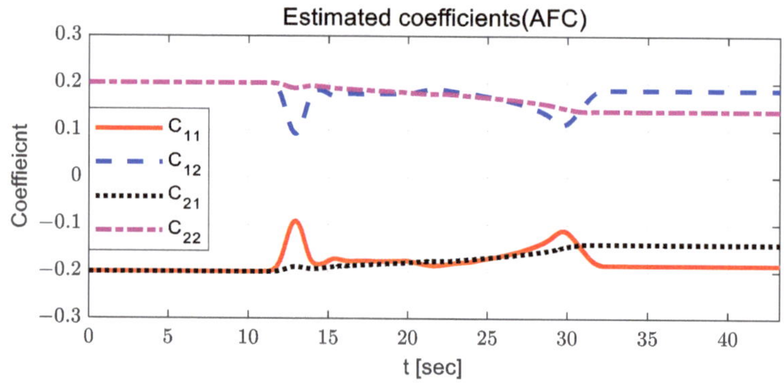

Figure 6. Results: estimated coefficients in the case of AFC for the curved path tracking.

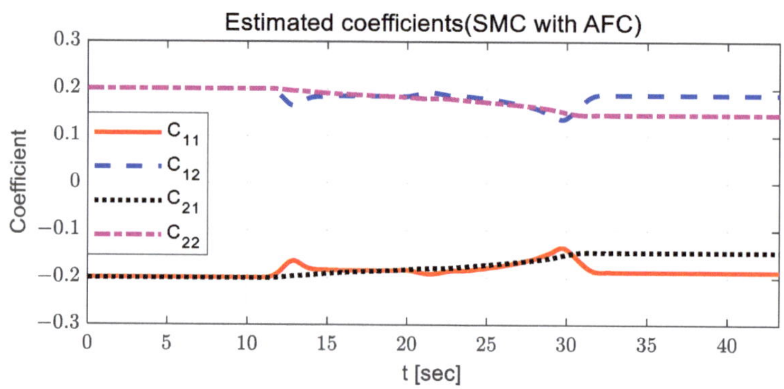

Figure 7. Results: estimated coefficients in the case of SMC with AFC for the curved path tracking.

It can be observed that there is no significant difference between AFC and SMC with AFC; however, the estimated coefficients for SMC with AFC have a relatively small variation around 13 and 30 s. Figures 8 and 9 show the adapted feedback gains and path-tracking control errors (i.e., preview lateral error and yaw angle error), respectively.

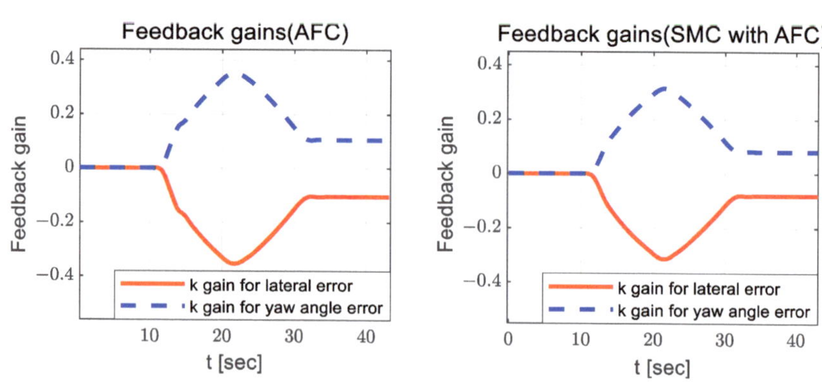

Figure 8. Results: adapted feedback gains (AFC—**left**; SMC with AFC—**right**) for the curved path tracking.

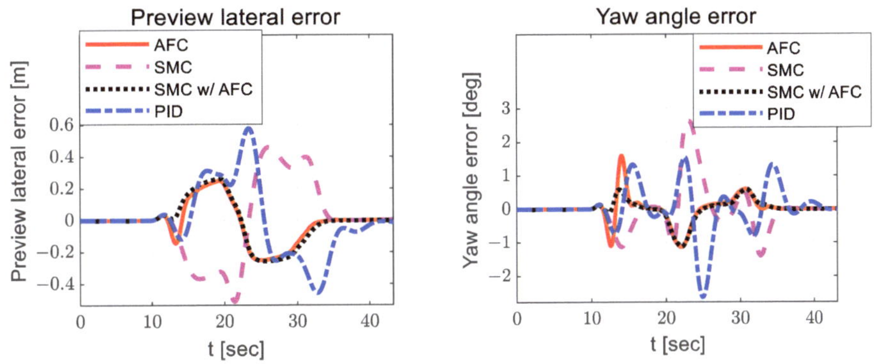

Figure 9. Results: path-tracking control errors (lateral—**left**; yaw angle—**right**) for the curved path tracking.

According to Figure 8, the adapted feedback gains between AFC and SMC with AFC do not differ significantly, but the feedback gains for SMC with AFC are slightly smaller than those for AFC. Additionally, SMC with AFC shows smaller preview yaw and lateral errors than AFC, SMC, and PID. Figures 10–12 show the dynamic behaviors, cost values for path tracking, and vehicle trajectories, respectively.

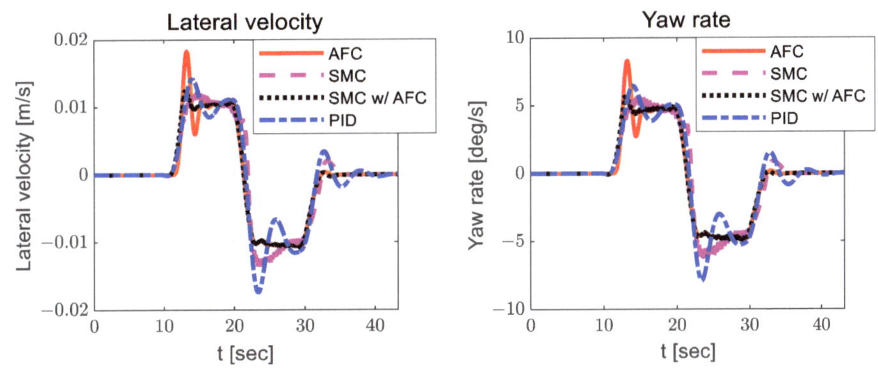

Figure 10. Results: dynamic behaviors (lateral velocity—**left**; yaw rate—**right**) for the curved path tracking.

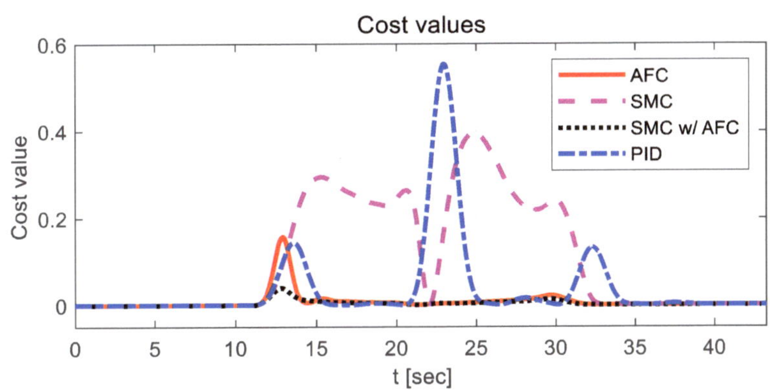

Figure 11. Results: cost value comparison for the curved path tracking.

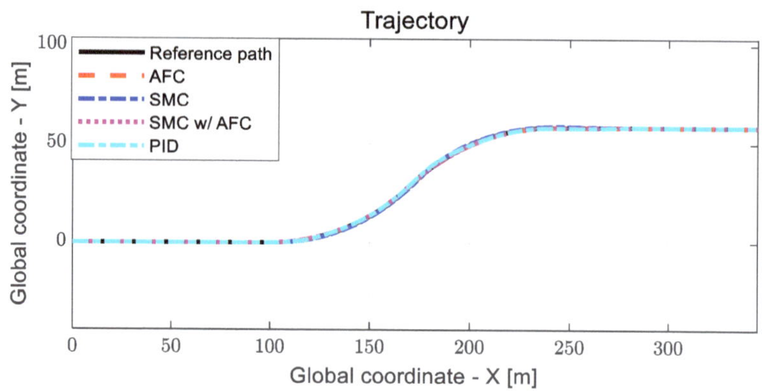

Figure 12. Results: trajectory comparison for the curved path tracking.

In Figure 11, PID has the highest cost value for path tracking. There is no significant difference between AFC and SMC with AFC in terms of cost value during the simulation, except for 13 s; however, SMC with AFC shows the smallest value among the three cases. Table 4 and Figure 13 compare the maximum and standard deviations of cost values in each case.

Table 4. Results of cost value comparison for the curved path tracking.

Division	Maximum	Standard Deviation
Adaptive feedback control (AFC)	0.1568	0.0231
Sliding mode control (SMC)	0.3964	0.1678
SMC with AFC	0.0395	0.0078
Proportional–integral–derivative (PID)	0.5535	0.1058

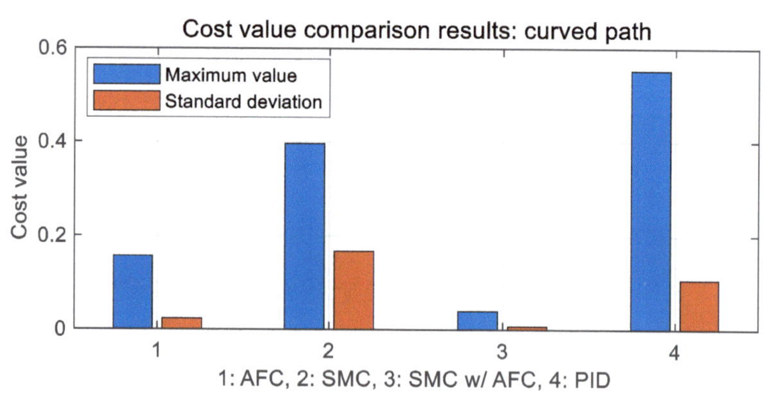

Figure 13. Results: cost value comparison in bar chart form for the curved path tracking.

We can note that the maximum and standard deviation values for SMC with AFC are the lowest of all cases. It can also be seen that the SMC-based path-tracking algorithm with adaptive feedback action shows better performance.

3.2. Path-Tracking Scenario: Lane Change (60 kph)

This section provides performance evaluation results for the lane change scenario with a constant velocity condition of 60 kph. The lane change scenario was designed by switching the desired straight paths so that the vehicle could perform the lane change task reasonably. The time delay function was also used to smooth the path-tracking control errors. Figure 14 illustrates the steering control inputs for the lane change scenario for all cases: AFC, SMC, SMC with AFC, and PID. It can also be observed that the steering control input in the case of SMC with AFC has relatively large values compared to the others. Finally, AFC and PID show some oscillations in the steering control input and slower responses.

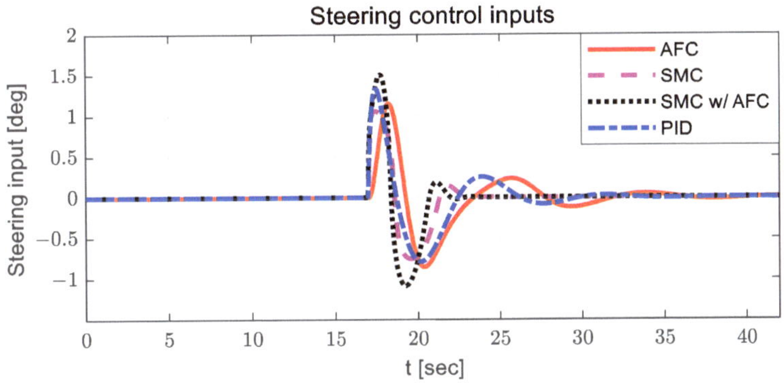

Figure 14. Results: steering control inputs for the lane change.

Figures 15 and 16 show the estimated coefficients for feedback gain adaptation in the cases of AFC and SMC with AFC, respectively.

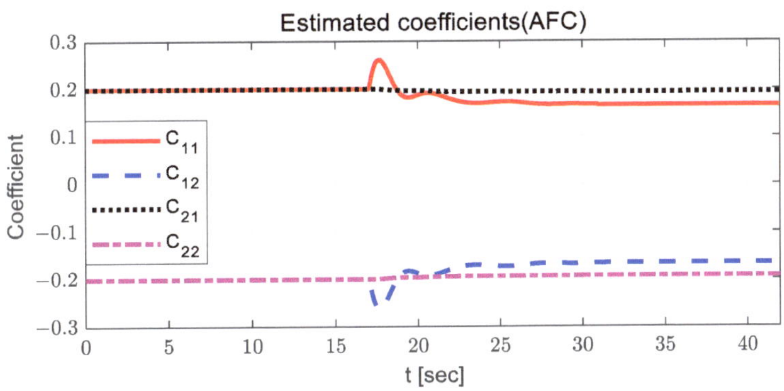

Figure 15. Results: estimated coefficients in the case of AFC for the lane change.

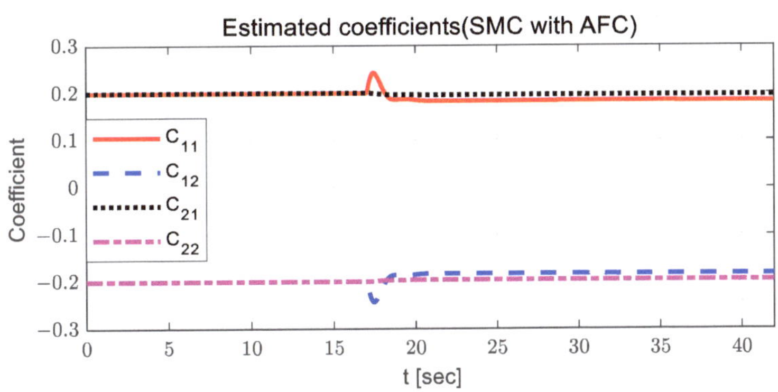

Figure 16. Results: estimated coefficients in the case of SMC with AFC for the lane change.

There are no significant differences between SMC with AFC and AFC in terms of their estimated coefficients and their variation patterns. In the case of using only AFC, there is a relatively larger change in the estimated coefficients because AFC produce steering control inputs for path tracking exclusively, without further assistance from the SMC input. Figures 17 and 18 show the adapted feedback gains and path-tracking control errors, respectively.

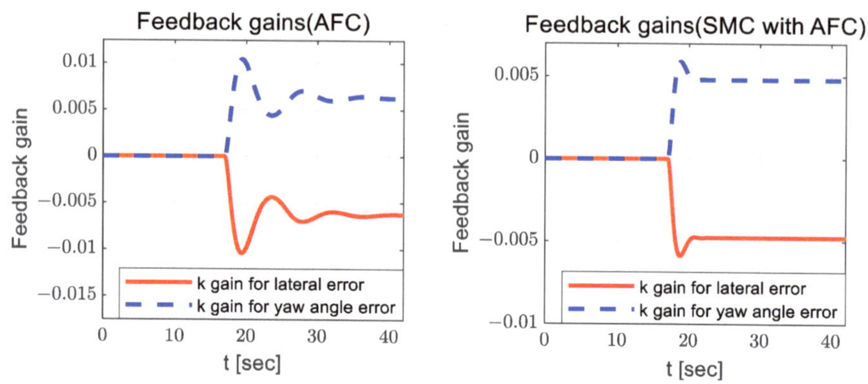

Figure 17. Results: adapted feedback gains (AFC—**left**; SMC with AFC—**right**) for the lane change.

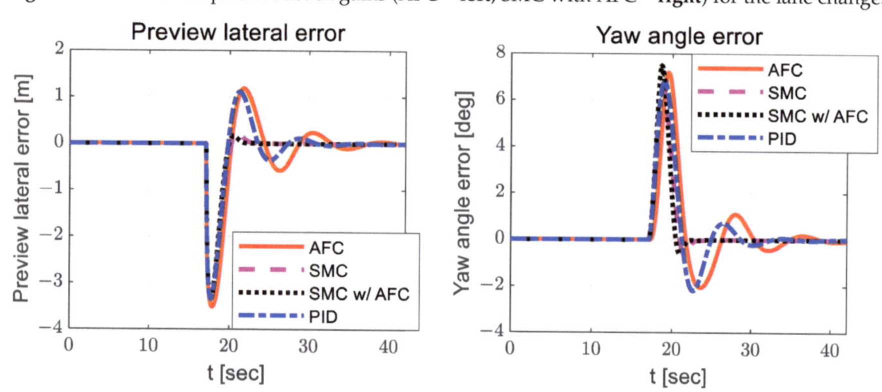

Figure 18. Results: path-tracking control errors (lateral—**left**; yaw angle—**right**) for the lane change.

In Figure 17, AFC and PID exhibit relatively larger oscillations than SMC with AFC. In Figure 18, there are also similar oscillations in path-tracking control error between SMC and SMC with AFC, but their values are not greatly different. In addition, SMC with AFC shows a higher convergence rate for the preview lateral error and yaw angle error than AFC, SMC, or PID.

With AFC and SMC, the preview lateral error and yaw angle error are more likely to converge than with AFC or SMC alone.

Figures 19–21 show the dynamic behaviors, cost values for path tracking, and vehicle trajectories, respectively.

As shown in Figure 20, AFC has the highest cost value with oscillations for a lane change, while SMC and SMC with AFC show similar variations in cost values. Figure 21 shows the vehicle trajectories for the same lane change. The results indicate that path-tracking control with AFC occurs a little later than the other cases, while showing relatively large overshoots and oscillations. Furthermore, the stabilization rates of the path-tracking controllers using SMC and SMC with AFC are higher than those of AFC alone and PID. In Table 5 and Figure 22, the maximum values and standard deviations of the cost values are compared for each control method.

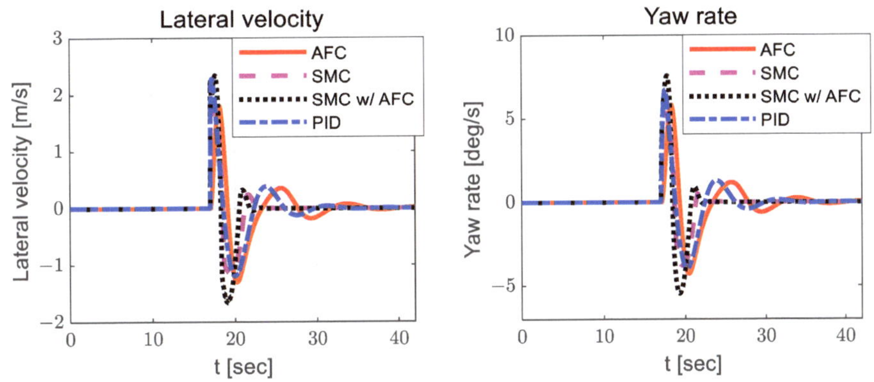

Figure 19. Results: dynamic behaviors (lateral velocity—**left**; yaw rate—**right**) for the lane change.

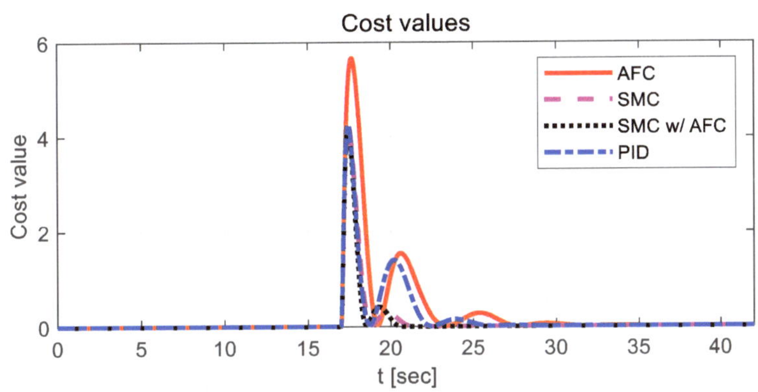

Figure 20. Results: cost value comparison for the lane change.

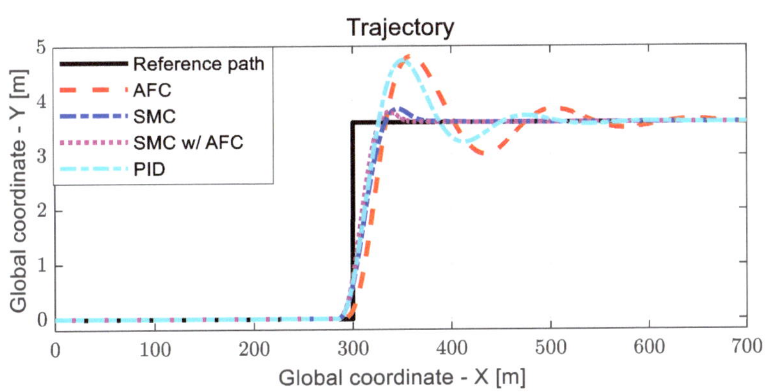

Figure 21. Results: trajectory comparison for the lane change.

Table 5. Results of cost value comparison for the lane change.

Division	Maximum	Standard Deviation
AFC	5.6590	0.8657
SMC	4.2110	0.5197
SMC with AFC	4.1395	0.4816
PID	4.2591	0.5635

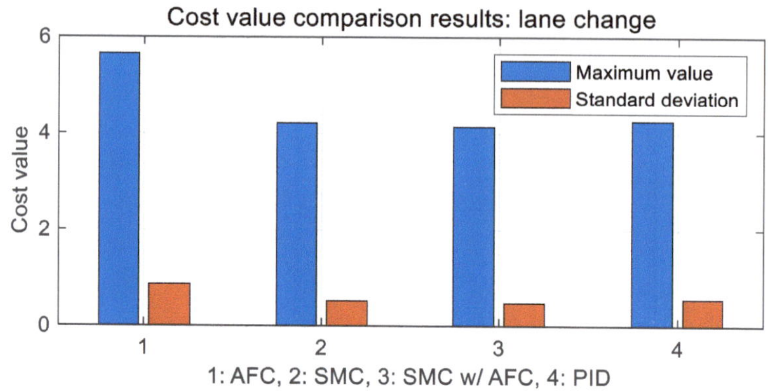

Figure 22. Results: cost value comparison in bar chart form for the lane change.

The above table shows that the maximum and standard deviation values in the case of SMC with AFC are the lowest among the four cases, while they differ slightly for SMC, SMC with AFC, and PID.

Based on the above results, it can be seen that the SMC-based path-tracking algorithm with adaptive feedback action shows reasonable tracking performance under the lane change scenario. In the next section, we discuss this study's conclusions, limitations, and prospects for future work.

4. Conclusions

This study proposes an SMC-based path-tracking control algorithm with adaptive feedback action for autonomous vehicles. The adaptive feedback and SMC algorithms were integrated to enhance the adaptiveness and robustness of the path-tracking control algorithm. The mathematical error model used for the controller design was based on the kinematic mathematical error model. The AFC algorithm was designed using recursive least squares with the forgetting factor and gradient descent methods based on a designed relationship function that uses a combination of path-tracking control errors and feedback gains. Based on the modification of the mathematical error model by the AFC input, the SMC algorithm was designed with finite stability conditions using the Lyapunov theorem. To avoid chattering phenomena and conflict of the SMC input with the AFC input, the sigmoid function was used with proper parameters for gradients. The performance evaluation was conducted under two scenarios (i.e., curved path tracking and lane changes) with constant velocity conditions. The evaluation results show that the control algorithm proposed in this study was able to track the designed reference path reasonably. However, some control parameters should be determined properly for reasonable performance. Therefore, future work will focus on improving the model-free adaptiveness and robustness of the control algorithm. Despite these limitations, it is expected that the developed control algorithm could be widely used for path-tracking algorithms for autonomous vehicles using a simple mathematical model with low computational costs.

Author Contributions: Writing—original draft preparation, K.O. and J.S.; writing—review and editing. All authors have read and agreed to the published version of the manuscript.

Funding: This research received no external funding.

Institutional Review Board Statement: Not applicable.

Informed Consent Statement: Not applicable.

Data Availability Statement: Not applicable.

Conflicts of Interest: The authors declare no conflict of interest.

References

1. Sun, C.; Zhang, X.; Zhou, Q.; Tian, Y. A Model Predictive Controller with Switched Tracking Error for Autonomous Vehicle Path Tracking. *IEEE Access* **2019**, *7*, 53103–53114. [CrossRef]
2. Baca, T.; Hert, D.; Loianno, G.; Saska, M.; Kumar, V. Model Predictive Trajectory Tracking and Collision Avoidance for Reliable Outdoor Deployment of Unmanned Aerial Vehicles. In Proceedings of the 2018 IEEE/RSJ International Conference on Intelligent Robots and Systems (IROS), Madrid, Spain, 1–5 October 2018. [CrossRef]
3. Suh, J.; Chae, H.; Yi, K. Stochastic Model-Predictive Control for Lane Change Decision of Automated Driving Vehicles. *IEEE Trans. Veh. Technol.* **2018**, *67*, 4771–4782. [CrossRef]
4. Xu, S.; Peng, H. Design, Analysis, and Experiments of Preview Path Tracking Control for Autonomous Vehicles. *IEEE Trans. Intell. Transp. Syst.* **2020**, *21*, 48–58. [CrossRef]
5. Chowdhri, N.; Ferranti, L.; Iribarren, F.; Shyrokau, B. Integrated Nonlinear Model Predictive Control for Automated Driving. *Control Eng. Pract.* **2021**, *106*, 104654. [CrossRef]
6. Li, S.; Li, Z.; Yu, Z.; Zhang, B.; Zhang, N. Dynamic Trajectory Planning and Tracking for Autonomous Vehicle with Obstacle Avoidance Based on Model Predictive Control. *IEEE Access* **2019**, *7*, 132074–132086. [CrossRef]
7. Cao, J.; Song, C.; Peng, S.; Song, S.; Zhang, X.; Xiao, F. Trajectory Tracking Control Algorithm for Autonomous Vehicle Considering Cornering Characteristics. *IEEE Access* **2020**, *8*, 59470–59484. [CrossRef]
8. Wang, Y.; Liu, Z.; Zuo, Z.; Li, Z.; Wang, L.; Luo, X. Trajectory Planning and Safety Assessment of Autonomous Vehicles Based on Motion Prediction and Model Predictive Control. *IEEE Trans. Veh. Technol.* **2019**, *68*, 8546–8556. [CrossRef]
9. Quirynen, R.; Berntorp, K.; Cairano, S. Embedded Optimization Algorithms for Steering in Autonomous Vehicles based on Nonlinear Model Predictive Control. In Proceedings of the 2018 Annual American Control Conference(ACC), Milwaukee, WI, USA, 27–29 June 2018. [CrossRef]
10. Shen, C.; Shi, Y. Distributed Implementation of Nonlinear Model Predictive Control for AUV Trajectory Tracking. *Automatica* **2020**, *115*, 118863. [CrossRef]
11. Chu, D.; Li, H.; Zho, C.; Zhou, T. Trajectory Tracking of Autonomous Vehicle Based on Model Predictive Control with PID Feedback. *IEEE Trans. Intell. Transp. Syst.* **2022**, 1–12. [CrossRef]
12. Wang, H.; Liu, B.; Ping, X.; An, Q. Path Tracking Control for Autonomous Vehicles Based on an Improved MPC. *IEEE Access* **2019**, *7*, 161064–161073. [CrossRef]
13. Zhang, C.; Chu, D.; Liu, S.; Deng, Z.; Wu, C.; Su, X. Trajectory Planning and Tracking for Autonomous Vehicle Based on State Lattice and Model Predictive Control. *IEEE Intell. Transp. Syst. Mag.* **2019**, *11*, 29–40. [CrossRef]
14. Guo, H.; Cao, D.; Chen, H.; Sun, Z.; Hu, Y. Model Predictive Path Following Control for Autonomous Cars Considering a Measurable Disturbance: Implementation, Testing, and Verification. *Mech. Syst. Signal Process.* **2019**, *118*, 41–60. [CrossRef]
15. Xu, Y.; Zheng, H.; Xu, W.; Wu, J. Robust Hierarchical Model Predictive Control for Trajectory Tracking with Obstacle Avoidance. *IFAC-PapersOnLine* **2020**, *53*, 15745–15750. [CrossRef]
16. Yue, M.; Hou, X.; Zhao, X.; Xu, X. Robust Tube-Based Model Predictive Control for Lane Change Maneuver of Tractor-Trailer Vehicles Based on a Polynomial Trajectory. *IEEE Trans. Syst. Man Cybern. Syst.* **2020**, *50*, 5180–5188. [CrossRef]
17. Peng, H.; Wang, W.; An, Q.; Xiang, C.; Li, L. Path Tracking and Direct Yaw Moment Coordinated Control Based on Robust MPC With the Finite Time Horizon for Autonomous Independent-Drive Vehicles. *IEEE Trans. Veh. Technol.* **2020**, *69*, 6053–6066. [CrossRef]
18. Londhe, P.; Patre, B. Adaptive Fuzzy Sliding Mode Control for Robust Trajectory Tracking Control of an Autonomous Underwater Vehicle. *Intell. Serv. Robot.* **2019**, *12*, 87–102. [CrossRef]
19. Taghavifar, H.; Rakheja, S. Path tracking of Autonomous Vehicles Using a Novel Adaptive Robust Exponential-like-Sliding-Mode Fuzzy Type-2 Neural Network Controller. *Mech. Syst. Signal Process.* **2019**, *130*, 41–55. [CrossRef]
20. Zhou, X.; Wang, Z.; Shen, H.; Wang, J. Robust Adaptive Path tracking Control of Autonomous Ground Vehicles with Considerations of Steering System Backlash. *IEEE Trans. Intell. Veh.* **2022**, *7*, 315–325. [CrossRef]
21. Yuan, X.; Huang, G.; Shi, K. Improved Adaptive Path Following Control System for Autonomous Vehicle in Different Velocities. *IEEE Trans. Intell. Transp. Syst.* **2020**, *21*, 3247–3256.
22. Lin, F.; Chen, Y.; Zhao, Y.; Wang, S. Path Tracking of Autonomous Vehicle based on Adaptive Model Predictive Control. *Int. J. Adv. Robot. Syst.* **2019**, *16*, 1729881419880089. [CrossRef]
23. Liu, S.; Hou, Z.; Tian, T.; Deng, Z.; Li, Z. A Novel Dual Successive Projection-Based Model-Free Adaptive Control Method and Application to an Autonomous Car. *IEEE Trans. Neural Netw. Learn. Syst.* **2019**, *30*, 3444–3457. [CrossRef]
24. Guerrero, J.; Torres, J.; Creuze, V.; Chemori, A. Trajectory Tracking for Autonomous Underwater Vehicle: An Adaptive Approach. *Ocean Eng.* **2019**, *172*, 511–522. [CrossRef]
25. Tran, V.; Santoso, F.; Garratt, M. Adaptive Trajectory Tracking for Quadrotor Systems in Unknown Wind Environments Using Particle Swarm Optimization-Based Strictly Negative Imaginary Controllers. *IEEE Trans. Aerosp. Electron. Syst.* **2021**, *57*, 1742–1752. [CrossRef]
26. Tian, Y.; Hang, P.; Wang, S. Adaptive Coordinated Path Tracking Control Strategy for Autonomous Vehicles with Direct Yaw Moment Control. *Chin. J. Mech. Eng.* **2022**, *35*, 1. [CrossRef]
27. Fan, Y.; Liu, B.; Wang, G.; Mu, D. Adaptive Fast Non-Singular Terminal Sliding Mode Path Following Control for an Underactuated Unmanned Surface Vehicle with Uncertainties and Unknown Disturbances. *Sensors* **2021**, *21*, 7454. [CrossRef] [PubMed]

28. Pereida, K.; Schoellig, A. Adaptive Model Predictive Control for High-Accuracy Trajectory Tracking in Changing Conditions. In Proceedings of the 2018 IEEE/RSJ International Conference on Intelligent Robots and Systems (IROS), Madrid, Spain, 1–5 October 2018. [CrossRef]
29. Kebbati, Y.; Oufroukh, N.; Vigneron, V.; Ichalal, D. Coordinated PSO-PID based Longitudinal Control with LPV-MPC based Lateral Control for Autonomous Vehicles. In Proceedings of the 2022 European Control Conference (ECC), London, UK, 1–5 August 2022. [CrossRef]
30. Guo, J.; Luo, Y.; Li, K. Adaptive Non-linear Trajectory Tracking Control for Lane Change of Autonomous Four-wheel Independently Drive Electric Vehicles. *IET Intell. Transp. Syst.* **2018**, *12*, 712–720. [CrossRef]
31. Hu, C.; Wang, Z.; Qin, Y.; Huang, Y.; Wang, J.; Wang, R. Lane Keeping Control of Autonomous Vehicles with Prescribed Performance Considering the Rollover Prevention and Input Saturation. *IEEE Trans. Intell. Transp. Syst.* **2020**, *21*, 3091–3103. [CrossRef]
32. Liang, Y.; Li, Y.; Khajepiur, A.; Huang, Y.; Qin, Y.; Zheng, L. A Novel Combined Decision and Control Scheme for Autonomous Vehicle in Structured Road Based on Adaptive Model Predictive Control. *IEEE Trans. Intell. Transp. Syst.* **2022**, *23*, 16083–16097. [CrossRef]
33. Ahn, T.; Lee, Y.; Park, K. Design of Integrated Autonomous Driving Control System That Incorporates Chassis Controllers for Improving Path Tracking Performance and Vehicle Stability. *Electronics* **2021**, *10*, 144. [CrossRef]
34. He, H.; Shi, M.; Li, J.; Cao, J.; Han, M. Design and Experiential Test of a Model Predictive Path Following Control with Adaptive Preview for Autonomous Buses. *Mech. Syst. Signal Process.* **2021**, *157*, 107701. [CrossRef]
35. Chen, I.; Chan, C. Deep Reinforcement Learning based Path Tracking Controller for Autonomous Vehicle. *Proc. Inst. Mech. Eng. Part D J. Automob. Eng.* **2021**, *235*, 541–551. [CrossRef]
36. Zhang, K.; Sun, Q.; Shi, Y. Trajectory Tracking Control of Autonomous Ground Vehicles Using Adaptive Learning MPC. *IEEE Trans. Neural Netw. Learn. Syst.* **2021**, *32*, 5554–5564. [CrossRef] [PubMed]
37. Jiang, Y.; Xu, X.; Zhang, L.; Zou, T. Model free predictive path tracking control of variable-configuration unmanned ground vehicle. *ISA Trans.* **2022**, *129*, 485–494. [CrossRef] [PubMed]
38. Li, X.; Liu, C.; Chen, B.; Jiang, J. Robust Adaptive Learning-Based Path Tracking Control of Autonomous Vehicles under Uncertain Driving Environments. *IEEE Trans. Intell. Transp. Syst.* **2022**, *23*, 20798–20809. [CrossRef]
39. Wang, Z.; Wang, J. Ultra-Local Model Predictive Control: A Model-Free Approach and Its Application on Automated Vehicle Trajectory Tracking. *Control. Eng. Pract.* **2020**, *101*, 104482. [CrossRef]
40. Wang, N.; Gao, Y.; Zhang, X. Data-Driven Performance-Prescribed Reinforcement Learning Control of an Unmanned Surface Vehicle. *IEEE Trans. Neural Netw. Learn. Syst.* **2021**, *32*, 5456–5467. [CrossRef]
41. Jiang, Y.; Xu, X.; Zhang, L. Heading tracking of 6WID/4WIS unmanned ground vehicles with variable wheelbase based on model free adaptive control. *Mech. Syst. Signal Process.* **2021**, *159*, 107715. [CrossRef]
42. Parseh, M.; Asplund, F.; Svensson, L.; Sinz, W.; Tomasch, E.; Torngren, M. A Data-Driven Method towards Minimizing Collision Severity for Highly Automated Vehicles. *IEEE Trans. Intell. Veh.* **2021**, *6*, 723–735. [CrossRef]
43. Wu, Q.; Cheng, S.; Li, L.; Yang, F.; Meng, L.; Fan, Z.; Liang, H. A Fuzzy-Inference-based Reinforcement Learning Method of Overtaking Decision Making for Automated Vehicles. *Proc. Inst. Mech. Eng. Part D J. Automob. Eng.* **2022**, *236*, 75–83. [CrossRef]
44. Wang, Z.; Zhou, X.; Wang, J. Extremum-Seeking-Based Adaptive Model-Free Control and Its Application to Automated Vehicle Path Tracking. *IEEE/ASME Trans. Mechatron.* **2022**, *27*, 3874–3884. [CrossRef]
45. Spielberg, N.; Brown, M.; Gerdes, J. Neural Network Model Predictive Motion Control Applied to Automated Driving with Unknown Friction. *IEEE Trans. Control. Syst.* **2021**, *30*, 1934–1945. [CrossRef]
46. Peng, Z.; Wang, D.; Wang, J. Data-Driven Adaptive Disturbance Observers for Model-Free Trajectory Tracking Control of Maritime Autonomous Surface Ships. *IEEE Trans. Neural Netw. Learn. Syst.* **2021**, *32*, 5584–5594. [CrossRef]
47. Wang, H.; Liu, B. Path Planning and Path Tracking for Collision Avoidance of Autonomous Ground Vehicles. *IEEE Syst. J.* **2022**, *16*, 3658–3667. [CrossRef]
48. Ma, H.; Chen, J.; Eben, S.; Lin, Z.; Guan, Y.; Ren, Y.; Zhen, S. Model-based Constrained Reinforcement Learning using Generalized Control Barrier Function. In Proceedings of the 2021 IEEE/RSJ International Conference on Intelligent Robots and Systems (IROS), Prague, Czech Republic, 27 September 2021. [CrossRef]
49. Zhang, K.; Su, R.; Zhang, H.; Tian, Y. Adaptive Resilient Event-Triggered Control Design of Autonomous Vehicles with an Iterative Single Critic Learning Framework. *IEEE Trans. Neural Netw. Learn. Syst.* **2021**, *32*, 5502–5511. [CrossRef] [PubMed]
50. Oh, K.; Seo, J. Sensitivity-based model-free adaptive displacement and velocity control algorithms for unknown single-input multi-output systems with recursive least squares. *J. Mech. Sci. Technol.* **2021**, *35*, 5173–5185. [CrossRef]

sensors

Article

Autonomous Vehicle Dataset with Real Multi-Driver Scenes and Biometric Data

Francisca Rosique *, Pedro J. Navarro, Leanne Miller and Eduardo Salas

División de Sistemas e Ingeniería Electrónica (DSIE), Campus Muralla del Mar, s/n, Universidad Politécnica de Cartagena, 30202 Cartagena, Spain
* Correspondence: paqui.rosique@upct.es

Abstract: The development of autonomous vehicles is becoming increasingly popular and gathering real-world data is considered a valuable task. Many datasets have been published recently in the autonomous vehicle sector, with synthetic datasets gaining particular interest due to availability and cost. For a real implementation and correct evaluation of vehicles at higher levels of autonomy, it is also necessary to consider human interaction, which is precisely something that lacks in existing datasets. In this article the UPCT dataset is presented, a public dataset containing high quality, multimodal data obtained using state-of-the-art sensors and equipment installed onboard the UPCT's CICar autonomous vehicle. The dataset includes data from a variety of perception sensors including 3D LiDAR, cameras, IMU, GPS, encoders, as well as driver biometric data and driver behaviour questionnaires. In addition to the dataset, the software developed for data synchronisation and processing has been made available. The quality of the dataset was validated using an end-to-end neural network model with multiple inputs to obtain the speed and steering wheel angle and it obtained very promising results.

Keywords: autonomous vehicles; multimodal driving datasets; LiDAR; driver biometric data

Citation: Rosique, F.; Navarro, P.J.; Miller, L.; Salas, E. Autonomous Vehicle Dataset with Real Multi-Driver Scenes and Biometric Data. *Sensors* **2023**, *23*, 2009. https://doi.org/10.3390/s23042009

Academic Editors: Chao Huang, Yafei Wang, Peng Hang, Zhiqiang Zuo, Bo Leng and Hailong Huang

Received: 13 December 2022
Revised: 27 January 2023
Accepted: 30 January 2023
Published: 10 February 2023

1. Introduction

Rapid advances in artificial intelligence, electronics, information and communications technology (leading to miniaturisation and improved performance of computers, sensors and networks) has led to the development of new approaches to Autonomous Vehicle technologies [1]. This together with new consumption habits and environmental awareness, where technology is vital and allows us to be more efficient and sustainable, has led to a considerable increase in the amount of research carried out on autonomous vehicles, making it the latest trend in the automotive industry [2]. Evidently, there is plenty of motivation and enthusiasm for speeding up progress, especially with the recent success of Big Data, Machine Learning and Deep Neural Networks.

Given the growing popularity of the development of autonomous vehicles, the collection of real data is considered a valuable task, with it being necessary for this sector to provide high-quality, multimodal and real-world datasets which can be used for benchmarking, simulation development, algorithms testing and diverse computer vision training exercises, among others.

The vehicle used for the data collection is usually equipped with a variety of sensors, such as cameras, Light Detection and Ranging (LiDAR) sensors, RADAR, GPS and Inertial Measurement Units (IMU). The raw data obtained by these sensors is recorded on a disk while the vehicle is being driven manually. Subsequently, the recorded data can be used to train and test different algorithms for autonomous driving, e.g., vehicle/pedestrian detection and tracking, Simultaneous Localization and Mapping (SLAM) and motion estimation [3].

In this context, many datasets have been published, a summary of the most popular datasets and their features is presented in Table 1. These datasets vary greatly in terms of traffic conditions, sensor configuration, application focus, data format, size, tool support, as well as other aspects. The most sought-after datasets dedicated to autonomous vehicle systems (AVS) are the so-called multimodal datasets. These datasets have gained particular attention recently, as datasets containing data from an individual sensor are insufficient to provide a complete perception of the environment. Furthermore, the most exploited sensors in this field, such as cameras, LIDARs, radars, etc., offer complementary data and their collaboration can guarantee a better understanding of the surroundings [3].

Table 1. Comparison of the main Datasets.

Ref./Year	Samples	Image Type	LIDAR	RADAR	IMU/GPS	Control Actions	Raw Data	Driver Data	Real Data	Biometrics Data	Driver Behaviour
UPCT	78 K	RGB, Depth	Yes	No	Yes	Steering wheel, Speed	Yes	Yes	Yes	Yes	Yes
KITTI [4]/2012	15 K	RGB	Yes	No	Yes	-	Yes	No	Yes	No	No
Udacity [5]/2016	34 K	RGB	Yes	No	Yes	Steering wheel	Yes	No	No	No	No
Lyft L5 [6]/2019	323 K	RGB	Yes	No	Yes	-	Yes	No	Yes	No	No
nuScenes [7]/2019	1.4 M	RGB	Yes	Yes	Yes	-	Partial	No	Yes	No	No
Pandaset [8]/2019	48 K	RGB	Yes	No	Yes	-	Partial	No	Yes	No	No
Waymo [9]/2019	1 M	RGB	Yes	No	Yes	-	Yes	No	Yes	No	No
PreSIL [10]/2019	50 K	RGB	Yes	No	No	-	No	No	No	No	No
GAC [11]/2019	3.24 M	RGB	No	No	No	Steering wheel, Speed	N/A	No	Yes	No	No
A2D2 [12]/2020	392 K	RGB	Yes	No	Yes	Steering angle, brake, accelerator	Partial	No	Yes	No	No
IDDA [13]/2020	1 M	RGB, Depth	No	No	No	-	No	No	No	No	No
Appollo Scape [14]/2020	100 K	RGB	Yes	No	No	-	No	No	Yes	No	No
Cityscapes [15]/2020	25 K	RGB	No	No	Yes	-	No	No	Yes	No	No
OLIMP [16]/2020	47 K	RGB	No	Yes	No	-	Yes	No	Yes	No	No
PixSet [17]/2021	29 K	RGB	Yes	Yes	Yes	-	No	No	Yes	No	No
ONCE [18]/2021	1 M	RGB	Yes	No	No	-	No	No	Yes	No	No

Table 1 provides a comparison of the main existing datasets, at both an academic and professional level and consists of a brief survey of datasets relevant to the development of autonomous driving systems. We focus on the most comparable and recent datasets, which strongly emphasise multimodal sensor data. Although they are not recent, we also include the KITTI and Udacity datasets as we consider them to be two of the most significant early driving datasets. We present the datasets in chronological order.

Despite the large number of existing studies, most of these datasets do not provide raw data, but instead offer labelled data to support training and evaluation, in particular semantic segmentation techniques. Obtaining real labelled data in large quantities is far from trivial. To start with, it is arduous and expensive to deploy multiple vehicles to collect images and data in a wide range of environmental, weather and lighting conditions. Secondly, the task of manually classifying each image is extremely time-consuming. Lastly, the accuracy of manually produced labels may be inconsistent across the dataset. These reasons, along with the level of fidelity achieved by 3D graphics engines, have encouraged the creation of synthetic datasets of artificial data based on scenes recreated by simulators [5].

As stated in the work by [19], this method of offering already labelled and even segmented data often presents problems in data quality due to the methods or models used. Another disadvantage of those models trained using only synthetic datasets is that in real-world scenarios, these tend to perform poorly, suffering from domain shift [20,21].

On the other hand, for a real implementation and correct evolution of autonomous vehicles at levels 4–5, it is also necessary to consider human interaction. Whether to infer a pedestrian's intent to cross the road, identify a driver's intent to perform a certain manoeuvre or detect potentially reckless moves, autonomous vehicles must have a high-level understanding of human behaviour. In most existing datasets, it is precisely this human data factor which is lacking. As can be seen in Table 1, apart from our proposal, the UPCT dataset, existing datasets dedicated to autonomous vehicles do not include biometric data or driver behaviour data.

In this article, we present the UPCT dataset, a public dataset of high-quality, multimodal data, obtained using state-of-the-art sensors equipped by the CICar autonomous vehicle belonging to the UPCT. The CICar includes sensors such as cameras, LiDAR, IMU, GPS and encoders, as well as biometric data from the drivers and driver behaviour questionnaires. The UPCT dataset offers the data acquired during 20 manual driving tests carried out by different drivers on an urban circuit, which consists of a circular route in the Spanish town of Fuente Alamo. To facilitate the use of the dataset, three large subgroups of data have been differentiated: Perception, Positioning and Driver data (biometrics and Driver Behaviour Questionnaire) and both the pre-processed raw data and the code which facilitates its use have been made available for download.

2. Materials and Methods

2.1. Experimental Design

To obtain the data, we decided to carry out ad hoc driving tests with a group of 50 healthy subjects of different ages and sex from the Region of Murcia (Spain), following the distribution shown in Table 2. The subjects were in possession of a valid type B driving licence (for driving cars, vans and, in general, vehicles with a maximum authorised mass of 3500 kg) at the time of the test. After performing the tests, the results of some subjects were excluded due to technical problems during the performance of the test or during the recording of the results, leaving a total of n = 20 subjects (11 male/9 female) with valid raw data to make up the final dataset.

Table 2. Demographic distribution of subjects by gender and age.

	Categories	n Initial	% Initial	n Final	% Final
Gender	Male	26	52	11	55
	Female	24	48	9	45
Age	18–24	6	12	3	15
	25–44	22	44	10	50
	45–64	17	34	5	25
	>=65	5	10	2	10

2.1.1. Driver Test Design

Before starting the experiment, in addition to the informed consent, each subject filled in two questionnaires: (1) the Biographic Questionnaire and (2) the Driver Behaviour Questionnaire.

- The Biographic Questionnaire identifies key facts about the subject, such as gender, age and driving record.
- The Driver Behaviour Questionnaire (DBQ) collects self-reported data from the drivers, as there are no objective records of driving behaviour and previous traffic violations. The original DBQ consists of 50 items and is used to score the following three underlying factors: errors, violations, and lapses.

For this experiment, we have chosen to use the Spanish Driver Behaviour Questionnaire (SDBQ) [22], a shorter version adapted to Spanish drivers consisting of 28 items adapted to the peculiarities of the Spanish population. The version used consists of four factors, composed as follows: 6 traffic law violation items, 6 violation/aggressive manifestation items, 8 error items, and 8 lapse items. Participants were asked to indicate, on a 5-point scale, how often they had been involved in the behaviours or situations mentioned in the questionnaire.

2.1.2. Driving Test Design

The driving test consists of one of the participating drivers, who has been equipped with a non-invasive smart band device, manually driving the UPCT-CICar vehicle (the equipment onboard and its characteristics will be explained in more detail in the following platform setup subsection) and following a previously established and identical route which is the same for all tests. Each driver had to complete one lap of the circuit, which included a parking exercise situated approximately halfway along the circuit.

The selected route is an urban circuit in the town of Fuente Álamo in the Region of Murcia, Spain, with the tests performed by multiple drivers manually driving the UPCT-CICar in real traffic situations (see Figure 1). This route provides a significant Point of Interest (POI) of typical urban driving situations: (a) intersections with priority and with "Give way"; (b) joining a roundabout, internal circulation and leaving the roundabout; (c) circulation in streets with "green wave" traffic lights; (d) traffic jams; (e) rapid incorporation to a high-density road through a side lane; and (f) pedestrian traffic on public roads.

Figure 1. Urban route selected for the driving tests.

In order to contemplate a variety of environmental and driving conditions, the tests were carried out at different times of the day (morning, afternoon or night). Figure 2 shows some images from the dataset, where different situations captured during the tests are shown.

(**a**) (**b**) (**c**) (**d**)

Figure 2. Images from dataset. (**a**) Pedestrian crossing; (**b**) saturation due to reflections on the road; (**c**) car braking; (**d**) complex shadows on the road.

After each driving test, the data acquired from the vehicle's perception systems (LiDARs and cameras), positioning systems (IMU, GPS, rotation angle, acceleration, etc.) and biometric data from the driver are transferred to the central server.

2.1.3. Platform Setup

For this work, the UPCT autonomous vehicle (UPCT-CICar [23]), was driven by a human pilot in manual mode. CICar is a real-world prototype, based on a commercial electric vehicle, the Renault Twizy, which has undergone a series of modifications to provide it with the required functionality. The CICar has been equipped with multiple sensors, including LiDAR, cameras, IMU, GPS, encoders, etc., necessary for the vehicle to perform autonomous driving tasks. This platform setup integrates a perception system, a control system, and a processing system on board the vehicle.

Perception System: The purpose of a sensor system is to collect data from the surrounding environment of the AV and send that data to the control system. These sensors measure different physical quantities, which are typically selected to overlap each other, providing the redundant information needed to correctly merge and correlate the information. In our autonomous vehicle, two types of sensors are used to measure the environment: short-range sensors (up to 10 m) and long-range sensors. Installed short-range sensors include a Sick 2D laser ranging scanner and time-of-flight camera. The long-range sensors are a 3D LIDAR scanner and a camera in the visible spectrum. Table 3 and Figure 3 show the different devices involved in data acquisition during the tests, as well as the details of the variables involved in obtaining them.

Driver Biometric System: The drivers' biometric signal collection system has been carried out using a non-invasive wearable device, bracelet type, called Empatica E4. The Empatica E4 is a wrist-worn top-quality sensor device considered a Class IIa Medical Device according to 93/42/EEC Directive. Empatica E4 device measures the acceleration data (ACC), as well as other physiological parameters, namely the Blood Volume Pulse (BVP), from which the Heart Rate Variability (HRV) and the Inter-Beat Interval (IBI) are derived as well, skin temperature (TEMP) and also changes in certain electrical properties of the skin such as the Electrodermal Activity (EDA). For the creation of our dataset, among the several measurements recorded by the Empatica E4, this signal was considered, since it provides information better suited for activity recognition. A summary of the technical specifications of the accelerometer sensor is detailed in Table 4.

Table 3. Sensor data in CICar.

Device	Variable	Details
LiDAR 3D	Scene	Long-range sensors 3D High-Definition LIDAR (HDL64SE supplied by Velodyne) Its 64 laser beams spin at 800 rpm and can detect objects up to 120 m away with an accuracy of 2 cm 1.3 Million Points per Second Vertical FOV: 26.9°
2 × LiDAR 2D	Scene	Short-range sensors Sick laser 2D TIM551 Operating range 0.05 m–10 m Horizontal FOV 270° Frequency 15 Hz Angular resolution 1° Range 10% of reflectance 8 m
2 × ToF	Scene	Short-range sensors ToF Sentis3D-M420Kit cam Range: Indoor: 7 m, Outdoor: 4 m Horizontal FOV: 90°
RGB-D	Scene	Short-range sensors Depth Camera D435 Intel RealSense range 3 m Up to 90 fps Depth FOV: 87° × 58° RGB FOV: 69° × 42°
IMU	Localisation, longitudinal and transversal Acceleration	NAV440CA-202 Inertial Measurement Unit (IMU) 3-axis accelerometer Bandwidth: 25 Hz Pitch and roll accuracy of <0.4°, Position Accuracy < 0.3 m
GPS	Localisation	EMLID RTK GNSS Receiver 7 mm positioning precision
Encoder	Distance	
Biometric sensors	Driver Biometric signals	Empatica E4 EDA Sensor (GSR Sensor), PPG Sensor, Infrared Thermopile 3-axis Accelerometer

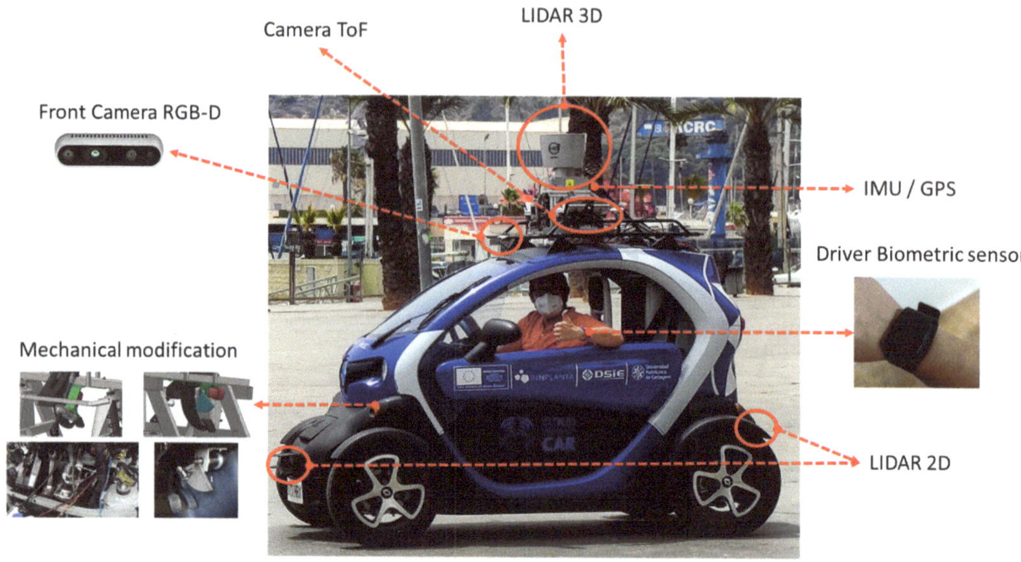

Figure 3. Autonomous vehicle UPCT-CICar and its different sensors and devices.

Table 4. Biometric variables details.

Variable	Sampling Frequency	Signal Range [Min, Max]	Details
ACC	32 Hz	[−2 g, 2 g]	Accelerometer 3 axes data (x, y, z).
EDA	4 Hz	[0.01 µS, 100 µS]	Electrodermal activity by capturing electrical conductance (inverse of resistance) across the skin.
BVP	64 Hz	n/a	Blood Volume Pulse.
IBI	64 Hz	n/a	Inter-beat interval (obtained from the BVP signal)
HR	1 Hz	n/a	Average Heart Rate (obtained from the BVP signal). Values are calculated at 10-s intervals.
TEMP	4 Hz	[−40 °C, 115 °C]	Skin Temperature.

Control System: The main control systems of the Renault Twizy have been automated in order to allow the vehicle to be autonomously controlled. The modified systems are the steering wheel, the brake pedal and the accelerator pedal (see mechanical modification in Figure 3). Despite the fact that all driving will be manual and not autonomous, the system will record the data with two controller drives through a CAN bus. The Compact Rio cRIO 9082 controls the accelerator, brake and steering wheel movements with the CAN-Open communication protocol, as well as I/O signals.

Processing System: Each sensor works with its own sample rate, and in most cases, this is different between devices. The achieve the synchronisation of the data and accurately reconstruct the temporal sequence, time stamps have been generated to synchronise the operating start and finish times. All of this is controlled and synchronised by the on-board processing system.

3. Results

As a result of the different executions of the experiment with the participating subjects, a raw data set has been obtained that has been curated and published in a repository. The data in the repository is organized under three major directories: (1) Driver, (2) Perception and (3) Position. The distribution of the data in the different directories is detailed below.

3.1. Driver Directory

This directory contains information regarding the drivers, from the questionnaires completed before the test and the biometric data obtained during the test. The directory contains 20 Biometric_XX.csv files (one per driver, where XX is the driver identifier number) and a DBQ.csv file with the data collected from the Biographic Questionnaire and the Driver Behaviour Questionnaire forms.

For the composition of the Biometric_XX.csv files, a normalised sampling frequency of 4 Hz has been used and in the case of sensors with lower frequencies, the table has been completed with NaN fields, with the HR column being the only one affected as the sample rate of this field is 1 Hz. The Biometric_XX.csv files have the following table format, where each column contains the following information:

- (TIME): The first column corresponds to the time stamp expressed as a unix timestamp in UTC.
- (TEMP): Data from the temperature sensor expressed as degrees in Celsius (°C).
- (EDA) Measurement of electrodermal activity by capturing electrical conductance (inverse of resistance) across the skin. The data provided is raw data obtained directly from the sensor expressed in micro siemens (μS).
- (BVP) The BVP is the blood-volume pulse and the raw output of the PPG sensor. The PPG/BVP is the input signal to algorithms that calculate Inter beat Interval Times (IBI) and Heart Rate (HR) as outputs.
- (HR): This file contains the average heart rate values calculated at 10-s intervals. They are not derived from real-time readings but are processed after the data is loaded into a session.
- (ACC_X, ACC_Y, ACC_Z) Data from the three-axis accelerometer sensor. The accelerometer is configured to measure acceleration in the range [−2 g, 2 g]. Therefore, the unit in this file is 1/64 g. Data from x, y and z axis are displayed in the sixth, seventh and eighth columns, respectively.

The DBQ.csv file is made up of a total of 45 columns, where the first column contains the subject identifier. The rest of the columns correspond to each of the items from the Biographic Questionnaire and the Driver Behaviour Questionnaire forms, where the last 25 columns are the questions from the DBQ form.

3.2. Perception Directory

This directory contains:

- Twenty .bin type files, called perceptionXX.bin, where XX corresponds to the identifier number assigned to each driver at the time of the test.
- Twenty images from the RGB-D camera (front view).

In the .bin file, the data from the 3D LiDAR sensor, 2D LiDAR sensors and TOF cameras, obtained by the CICar perception system is saved. The data from these sensors were recorded continuously during the driving test, with data packets being written by the different sensors one after the other and at the exact moment in which they arrived at the system, without contemplating an established order of sensor reading and recording.

Each data packet consists of a header made up of two 32-bit integers, which identify the source of the data followed by the data as it was received from the sensors. The header format is as follows:

uint32_t head [2];

- head [0]: indicates the size in bytes of the packet received by the sensor.
- head [1]: contains a sensor identifier which shows the source of the data packet received.

The sensor identifiers are the following:

//Packet identifiers in the data file
static const uint32_t LIDAR_PACKET_ID = 0 × 4C494452;   // 3D LiDAR

static const uint32_t GPS_PACKET_ID = 0 × 475053; //GPS
static const uint32_t NMEA_STRING_ID = 0 × 4E4D4541;//
static const uint32_t M420_FRAME_ID = 0 × 4D343230; //Camera ToF
static const uint32_t T551_FRONT_ID = 0 × 54354652; //Front 2D LiDAR
static const uint32_t T551_BACK_ID = 0 × 5435424B; //Rear 2D LiDAR

The following is a .bin file example:

(uint32_t) 1206 // Data packet size 1206 bytes
(uint32_t) 0 × 4C494452 // Data source: 3D LiDAR
(char [1206]) { ... } // 1206-byte vector with 3D LiDAR data
(uint32_t) 230,524 // Size of data packet 230,524 bytes
(uint32_t) 0 × 4D343230 // Data source: ToF camera
(char [230524]) { .. } // 230524-byte vector with ToF data.
(uint32_t) 1206 // Size of data packet 1206 bytes
(uint32_t) 0 × 4C494452 //Data source: 3D LiDAR
(char [1206]) { .... } //1206-byte vector with 3D LiDAR data.
(uint32_t) 1206 //Size of data packet 1206 bytes
(uint32_t) 0 × 4C494452 //Data source: 3D LiDAR
(char [1206]) { .... } //1206-byte vector with 3D LiDAR data.
(uint32_t) 921 //Size of data packet 921 bytes
(uint32_t) 0 × 54354652 //Data source: Front 2D LiDAR
(char [921]) { ..... } //921-byte vector with Front 2D LiDAR data.

To facilitate the use and processing of the data, a programme has been developed that allows the data from each sensor to be extracted separately and independently into an additional .bin file. In this case, by separating the data into different files, the data packet identifier is not necessary, but synchronisation with the system is lost. Therefore, to avoid loss of synchronisation between the data packet of each sensor, the time stamp of the exact moment of capture must be included. The .bin file format for each independent sensor is as follows:

uint32_t segundos // Capture timestamp seconds
uint32_t microseg // Capture timestamp microseconds
uint32_t numbytes // Number of bytes in the data packet
char datos[numbytes] // 'raw' data packet from the sensor

This structure is repeated continuously for each data packet until the end of the file.

Furthermore, the data has been pre-processed and the 3D LiDAR, 2D LIDAR and ToF camera data from each test carried out They have been merged into a single point cloud and extracted to a .csv file called POINTCLOUD_XX.csv, where XX is the identifier assigned to each driver at the start of the test.

3.3. Position Directory

This directory contains information regarding the position system, obtained during the driving tests. The directory contains 20 Position_XX.csv files, one for each driver where XX is the driver identifier number. Each of these systems collects information from the GPS, IMU and Encoder sensors.

The Position_XX.csv files are saved in the following table format, where each column contains the following information:

- (TIME) This first column contains the timestamp of the session expressed as a unix UTC timestamp.
- (LATITUDE) latitude values obtained by the GPS.
- (LONGITUD) longitude values obtained by the GPS.
- (ALTITUDE) altitude values obtained by the GPS.
- (STERING_ANGLE): Steering wheel angle.
- (SPEED): Speed/(m/s).
- (DISTANCE_TRAVELLED).
- (LIN_ACEL_X): acceleration obtained around the x-axis, obtained in g.

- (LIN_ACEL_Y): acceleration obtained around the *y*-axis, obtained in g.
- (LIN_ACEL_Z): acceleration obtained around the *z*-axis, obtained in g.
- (ANG_VEL_X): angular velocity obtained around the *x*-axis, in degrees/second.
- (ANG_VEL_Y): angular velocity obtained around the *y*-axis, in degrees/second.
- (ANG_VEL_Z): angular velocity obtained around the *x*-axis, in degrees/second.

The acceleration or angular velocity values are given by four bytes. These bytes correspond to a real number according to the IEEE-754 standard. The IEEE-754 standard is the most widely used for the representation of floating-point numbers.

4. Technical Validation

4.1. Driver Test Validation

To validate the data regarding the drivers, the following actions were carried out:

- A first validation is carried out by measuring the reliability of the data obtained from the DBQS tests carried out on the drivers. The reliability of the questionnaires was obtained with the entire sample, finding Cronbach's alpha indices and the two Guttman halves. The values to interpret the reliability were: <0.50 unacceptable; 0.50 $\geq$ poor < 0.60; 0.60 $\geq$ questionable/doubtful < 0.70; 0.70 $\geq$ acceptable < 0.80; 0.90 $\geq$ good < 0.90; and $\geq$0.90 excellent. The Cronbach Alpha coefficient is 0.796, which shows that the DBQ data set in this experiment has credibility [24].
- Missing data E4 data of seven participants (driver 1, driver 5, driver 17, driver 21, driver 30, driver 42, driver 45) were excluded due to a device malfunction during data collection. While physiological signals in the dataset are mostly error-free with most of the files complete above 95%, a portion of data is missing due to issues inherent to devices or a human error.

Raw data from the Empatica device was downloaded in csv format and analysed with the Kubios tool [25]. Kubios offers five artefact correction options based on very low to very high thresholds. No correction of the artefacts analysed by Kubios was necessary This is not surprising since the Empatica E4 already uses an algorithm that removes wrong IBIs or other wrong signals [26].

4.2. Driving Test Validation

Once the driving tests have been completed, a manual verification phase has been carried out on the data obtained (see Figure 4), where the data from those tests where reading or writing failures occurred, or failures in the test itself (routes, drivers, etc.) has been discarded.

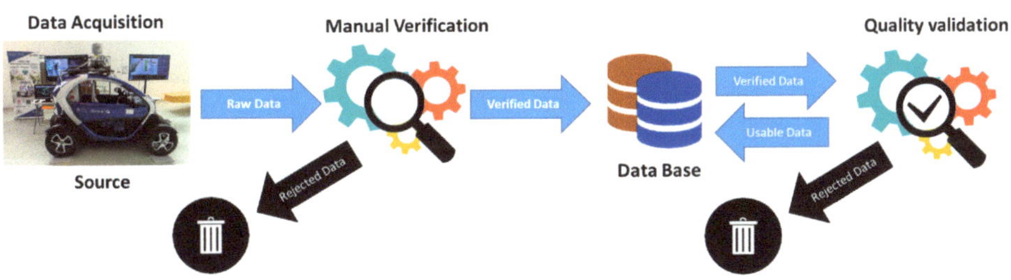

Figure 4. Verification process.

- Checking for abnormalities during the test. The time elapsed for the completion of each test has been checked, passing a filter, and discarding those tests in which the time has been either very short or too long. Data of five participants (driver 4, driver 17, driver 23, driver 29, driver 40) were excluded.
- Checking for errors in reading the sensors or writing to the disk. For each of the tests, the correct sending of information by the sensors during the test is verified. Those

tests where a total or partial failure has been detected have been discarded. To detect these failures, the following aspects were checked:

a. All files exist on the disk. At the end of each test, the number of files generated has been checked. The absence of any of the files implies a failure to read or write the data occurred, therefore this test was discarded completely. Data of four participants (driver 1, driver 10, driver 31, driver 34) were excluded.
b. Empty files. It has been verified that the files generated all contain data, discarding those tests where empty files have been detected. Data of two participants (driver 35, driver 36) were excluded.
c. Exploratory data analysis. Considering the different types of data processed, different types of descriptive analytics have been chosen: (1) Analysis of data deviation. A standard deviation analysis has been applied to those data with discrete values (average speed, time travelled, etc.), discarding those data with a sharp deviation. Data of two participants (driver 11, driver 38) were excluded (2) Time series analysis: most of the data correspond to time series of data, with a certain variation of speed, for this reason, it has been decided to use the Dynamic Time Warping (DTW) technique.

- Checking for driving route failures. For each of the tests carried out, the route taken by the driver during the test has been verified, to make sure the driver stuck to the route initially stipulated. The test where a small deviation from the track occurred has been discarded. To verify this, the following checks were made: (1) steering wheel rotation pattern during the test, given that for the same trajectory the steering wheel rotation pattern must be similar for all the tests. (2) GPS trajectory, the trajectory has been painted and the tests that do not comply with the marked route have been eliminated.

After this first screening process, a quality validation of the resulting data is performed to guarantee the quality of the data (see Figure 4). Our validation method comprised three steps: (1) Quality control of variables. (2) Quality control of support media. (3) Experimental validation.

4.2.1. Quality Control of Variables

An analysis of the internal structure of the set of circumstances of the DBQ form (content validity) has been performed. To be able to apply a factorial analysis correctly, those items with a declaration frequency of less than 5% were eliminated. Subsequently, and since the items on the form are dichotomous variables, the tetrachoric correlation coefficient was applied to obtain the correlation matrix between the 28 items.

The reliability of the questionnaires was obtained with the entire sample, finding Cronbach's alpha indices and the two Guttman halves. The values to interpret the reliability were: <0.50 unacceptable; 0.50 $\geq$ poor < 0.60; 0.60 $\geq$ questionable/doubtful < 0.70; 0.70 $\geq$ acceptable < 0.80; 0.90 $\geq$ good < 0.90; and $\geq$0.90 excellent. The Cronbach Alpha coefficient is 0.796.

Secondly, outliers in the acquired data, those values notably different compared to the patterns present in the rest of the data, may be due to errors in reading and writing the data from the sensors. Certain deviations were detected in the data from the GPS, due to momentary loss of signal or where the position has been calculated with a fewer number of satellites. In those cases in which the losses are less than two consecutive time intervals, a prediction of the vehicle's position is made. For cases where the loss is greater, the tests have been discarded. To apply this prediction a constant acceleration Kalman filter has been used.

4.2.2. Quality Control of Support Media

The clocks of all the sensors and devices were synchronised at the start of each experimental test session. All devices are controlled by the control unit on board the vehicle, which provides a perfect temporal and spatial synchronisation of the data obtained by the sensors (see Figure 5).

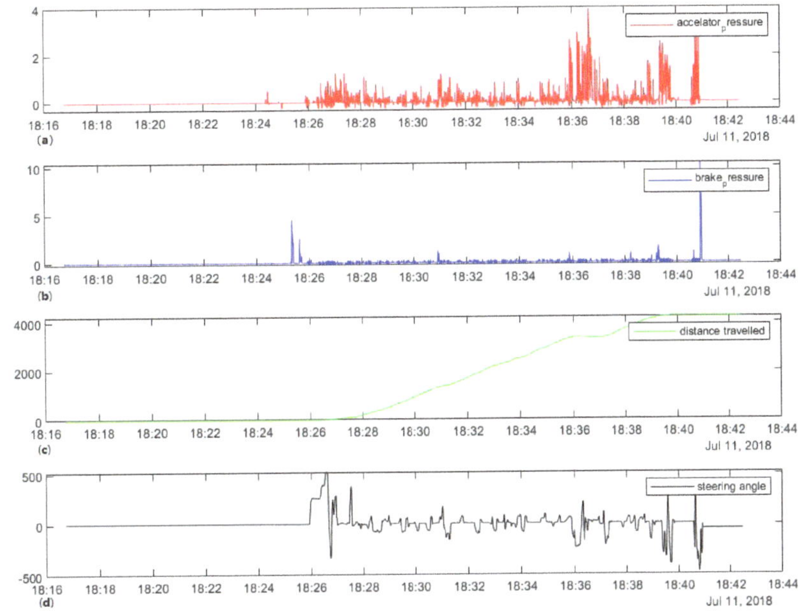

Figure 5. Data synchronisation with timestamp. (**a**) accelerator; (**b**) break; (**c**) distance; (**d**) steering wheel angle.

It has been verified that the data obtained by the encoder is synchronised with the rest of the sensor data. This was achieved by checking the distance indicated by the encoder coincides with the distance calculated between two consecutive GPS timestamps (GNSS). This was done using the Havershine expression shown in Equation (1), where d is the distance in metres between two points on the Earth's surface; r is the Earth's radius (6378 km); φ_1 and φ_2 are the latitudes in radians; Ψ_1 and Ψ_2 are the longitudes in radians of two consecutive timestamps.

$$d = 2r\sin^{-1}\left(\sqrt{\sin^2\left(\frac{\varphi_2 - \varphi_1}{2}\right) + \cos\varphi_1\cos\varphi_2\sin^2\left(\frac{\Psi_2 - \Psi_1}{2}\right)}\right) \quad (1)$$

The indirect data provided by the encoder has also been verified, for example, that the speed matches the direct data measurements provided by the GPS.

4.2.3. Experimental Validation

Finally, the most conclusive validation was performed: the usability analysis of the data contained in the final dataset. The work by Navarro et al. [20] presents the implementation of six end-to-end deep learning models trained using the UPCT dataset. The different end-to-end models were tested using different data sources from the vehicle, including RGB images, linear accelerations and angular velocities. We trained two models using only RGB image data, two using both the image data and IMU data as input to the models, and the last two used sequences of images as an input.

The best results were obtained using a mixed data input type end-to-end deep neural network model which used the front images obtained by the vehicle camera and angular speeds from the IMU to predict the speed and steering wheel angle, obtaining a mean error of 1.06%. An exhaustive optimization process of the convolutional blocks has demonstrated that it is possible to design lightweight end-to-end architectures with a high performance more suitable for the final implementation in autonomous driving.

5. Conclusions

In this work, we have presented the UPCT dataset, a real-world public driving dataset with 20 sets of driving data from 20 drivers which performed a driving test on an urban circuit in real traffic situations. The dataset contains different types of data which we have divided into three categories: (1) Driver, (2) Perception and (3) Position.

The dataset has been validated and tested with six end-to-end deep neural network models, using the RGB image data and IMU data, obtaining very promising results considering the size of the dataset. The detailed results are published in the work by Navarro et al. [20]. We plan to continue this research by making use of the depth images and comparing the results to those obtained when using just RGB images, as well as performing data augmentation to increase the sample sizes.

The main novelty of this dataset is the collection of biometric driver data which allows the behaviour of autonomous driving models to be compared to human drivers. In future research, we plan to use biometric driver data to perform driver behaviour studies. An interesting approach would be to relate the stress levels of the driver to certain driving situations, such as entering a roundabout, entering a main road or parking, for example. As each of the 20 drivers completed the same circuit, it would also be possible to compare the different driving styles and relate these tendencies to certain age groups or to a particular sex. In addition, in the driving behaviour questionnaire, each driver was asked about their driving style, and with the driving test, this can be compared to their real-life performance to determine if drivers correctly perceive their attitudes whilst driving.

Author Contributions: Conceptualization, F.R., P.J.N., L.M. and E.S.; methodology, F.R., P.J.N., L.M. and E.S.; software, P.J.N. and F.R.; validation, F.R. and L.M.; formal analysis, F.R., P.J.N., L.M. and E.S.; investigation, F.R., P.J.N., L.M. and E.S.; resources, F.R. and P.J.N.; data curation, P.J.N., L.M. and F.R.; writing—original draft preparation, F.R., P.J.N. and L.M.; writing—review and editing, F.R., P.J.N., L.M. and E.S.; All authors have read and agreed to the published version of the manuscript.

Funding: This work was partially supported by AEI METROPOLIS (ref. PLEC2021-007609) Spanish Government projects.

Institutional Review Board Statement: Not applicable.

Informed Consent Statement: Informed consent was obtained from all subjects involved in the study.

Data Availability Statement: The data set presented here, which is available within Figshare and released under a CC-BY 4.0 license. https://figshare.com/s/4b9a25a958c3ec578362 (accessed on 1 December 2022).

Conflicts of Interest: The authors declare no conflict of interest.

References

1. Olaverri-Monreal, C. Autonomous vehicles and smart mobility related technologies. *Infocommun. J.* **2016**, *8*, 17–24.
2. Alawadhi, M.; Almazrouie, J.; Kamil, M.; Khalil, K.A. A systematic literature review of the factors influencing the adoption of autonomous driving. *Int. J. Syst. Assur. Eng. Manag.* **2020**, *11*, 1065–1082. [CrossRef]
3. Rosique, F.; Navarro, P.J.; Fernández, C.; Padilla, A. A Systematic Review of Perception System and Simulators for Autonomous Vehicles Research. *Sensors* **2019**, *19*, 648. [CrossRef] [PubMed]
4. Geiger, A.; Lenz, P.; Stiller, C.; Urtasun, R. Vision meets robotics: The KITTI dataset. *Int. J. Robot. Res.* **2013**, *32*, 1231–1237. [CrossRef]
5. Yang, Z.; Zhang, Y.; Yu, J.; Cai, J.; Luo, J. End-to-end Multi-Modal Multi-Task Vehicle Control for Self-Driving Cars with Visual Perceptions. In Proceedings of the Proceedings—International Conference on Pattern Recognition, Beijing, China, 20–24 August 2018; Institute of Electrical and Electronics Engineers Inc.: New York, NY, USA, 2018; Volume 2018-Augus, pp. 2289–2294.
6. Kesten, R.; Usman, M.; Houston, J.; Pandya, T.; Nadhamuni, K.; Ferreira, A.; Yuan, M.; Low, B.; Jain, A.; Ondruska, P.; et al. Lyft Level 5 AV Dataset. 2019. Available online: https://www.woven-planet.global/en/woven-alpha/future-automated-driving-applications (accessed on 1 December 2022).
7. Caesar, H.; Bankiti, V.; Lang, A.H.; Vora, S.; Liong, V.E.; Xu, Q.; Krishnan, A.; Pan, Y.; Baldan, G.; Beijbom, O. Nuscenes: A multimodal dataset for autonomous driving. In Proceedings of the IEEE Computer Society Conference on Computer Vision and Pattern Recognition, San Juan, PR, USA, 17–19 June 1997; IEEE Computer Society: Washington, DC, USA, 2020; pp. 11618–11628.

8. Scale, A.I. Hesai Pandaset Open Dataset 2019. Available online: https://scale.com/open-av-datasets/pandaset (accessed on 1 December 2022).
9. Sun, P.; Kretzschmar, H.; Dotiwalla, X.; Chouard, A.; Patnaik, V.; Tsui, P.; Gou, J.; Zhou, Y.; Chai, Y.; Caine, B.; et al. Scalability in Perception for Autonomous Driving: Waymo Open Dataset. 2019. Available online: https://waymo.com/open/ (accessed on 1 December 2022).
10. Hurl, B.; Czarnecki, K.; Waslander, S. Precise Synthetic Image and LiDAR (PreSIL) Dataset for Autonomous Vehicle Perception. In Proceedings of the 2019 IEEE Intelligent Vehicles Symposium (IV), Paris, France, 9–12 June 2019; pp. 2522–2529.
11. Wang, D.; Wen, J.; Wang, Y.; Huang, X.; Pei, F. End-to-End Self-Driving Using Deep Neural Networks with Multi-auxiliary Tasks. *Automot. Innov.* **2019**, *2*, 127–136. [CrossRef]
12. Geyer, J.; Kassahun, Y.; Mahmudi, M.; Ricou, X.; Durgesh, R.; Chung, A.S.; Hauswald, L.; Pham, V.H.; Mühlegg, M.; Dorn, S.; et al. A2D2: Audi Autonomous Driving Dataset. *arXiv* **2020**, arXiv:2004.06320.
13. Alberti, E.; Tavera, A.; Masone, C.; Caputo, B. IDDA: A Large-Scale Multi-Domain Dataset for Autonomous Driving. *IEEE Robot. Autom. Lett.* **2020**, *5*, 5526–5533. [CrossRef]
14. Huang, X.; Wang, P.; Cheng, X.; Zhou, D.; Geng, Q.; Yang, R. The ApolloScape Open Dataset for Autonomous Driving and Its Application. *IEEE Trans. Pattern Anal. Mach. Intell.* **2020**, *42*, 2702–2719. [CrossRef] [PubMed]
15. Dataset Overview—Cityscapes Dataset. Available online: https://www.cityscapes-dataset.com/dataset-overview/ (accessed on 1 December 2022).
16. Mimouna, A.; Alouani, I.; Ben Khalifa, A.; El Hillali, Y.; Taleb-Ahmed, A.; Menhaj, A.; Ouahabi, A.; Ben Amara, N.E. OLIMP: A Heterogeneous Multimodal Dataset for Advanced Environment Perception. *Electronics* **2020**, *9*, 560. [CrossRef]
17. Déziel, J.; Merriaux, P.; Tremblay, F.; Lessard, D.; Plourde, D.; Stanguennec, J.; Goulet, P.; Olivier, P. PixSet: An Opportunity for 3D Computer Vision to Go Beyond Point Clouds With a Full-Waveform LiDAR Dataset. In Proceedings of the 2021 IEEE International Intelligent Transportation Systems Conference (ITSC), Indianapolis, IN, USA, 19–22 September 2021; pp. 2987–2993.
18. Mao, J.; Niu, M.; Jiang, C.; Liang, H.; Chen, J.; Liang, X.; Li, Y.; Ye, C.; Zhang, W.; Li, Z.; et al. One Million Scenes for Autonomous Driving: ONCE Dataset. 2021. Available online: https://arxiv.org/abs/2106.11037 (accessed on 1 December 2022).
19. Koopman, P.; Wagner, M. Challenges in Autonomous Vehicle Testing and Validation. *SAE Int. J. Trans. Saf.* **2016**, *4*, 15–24. [CrossRef]
20. Navarro, P.J.; Miller, L.; Rosique, F.; Fernández-Isla, C.; Gila-Navarro, A. End-to-End Deep Neural Network Architectures for Speed and Steering Wheel Angle Prediction in Autonomous Driving. *Electronics* **2021**, *10*, 1266. [CrossRef]
21. Cai, J.; Deng, W.; Guang, H.; Wang, Y.; Li, J.; Ding, J. A Survey on Data-Driven Scenario Generation for Automated Vehicle Testing. *Machines* **2022**, *10*, 1101. [CrossRef]
22. de Cózar, E.L.; Sanmartín, J.; Molina, J.G.; Aragay, J.M.; Perona, A. Behaviour Questionnaire (SDBQ). p. 15. Available online: https://www.uv.es/metras/docs/2005_metodologia_SDBQ.pdf (accessed on 1 December 2022).
23. Borraz, R.; Navarro, P.J.; Fernández, C.; Alcover, P.M. Cloud Incubator Car: A Reliable Platform for Autonomous Driving. *Appl. Sci.* **2018**, *8*, 303. [CrossRef]
24. Deng, Z.; Chu, D.; Wu, C.; He, Y.; Cui, J. Curve safe speed model considering driving style based on driver behaviour questionnaire. *Transp. Res. Part F Traffic Psychol. Behav.* **2019**, *65*, 536–547. [CrossRef]
25. Kubios HRV—Heart Rate Variability. Available online: https://www.kubios.com/ (accessed on 7 December 2022).
26. How Is IBI.csv Obtained? Available online: https://support.empatica.com/hc/en-us/articles/201912319-How-is-IBI-csv-obtained- (accessed on 7 December 2022).

sensors

MDPI

Article

Correction of Error of Airborne Anemometers Caused by Self-Excited Air Turbulence

Jianqiang Liu [1,2], Zhan Zhao [1,2], Zhen Fang [1,2], Yong Li [3] and Lidong Du [1,2,*]

[1] State Key Laboratory of Transducer Technology, Aerospace Information Research Institute, Chinese Academy of Sciences, Beijing 100194, China
[2] University of Chinese Academy of Sciences, Beijing 100049, China
[3] Key Laboratory of Transportation Meteorology, China Meteorological Administration, Nanjing 210008, China
* Correspondence: lddu@mail.ie.ac.cn; Tel.: +86-10-5888-7593

Abstract: An airborne anemometer, which monitors wind on the basis of Meteorological Multi-rotor UAVs (Unmanned Aerial Vehicles), is important for the prevention of catastrophe. However, its performance will be affected by the self-excited air turbulence generated by UAV rotors. In this paper, for the purpose of the correction of an error, we developed a method for the elimination of the influence of air turbulence on wind speed measurement. The corresponding correction model is obtained according to the CFD (Computational Fluid Dynamics) simulation of a six-rotor UAV which is carried out with the sliding grid method and the S-A turbulence model. Then, the model is applied to the developed prototype by adding the angle of attack compensation model of the airborne anemometer. It is shown by the actual application that the airborne anemometer can maintain the original measurement accuracy at different ascent speeds.

Keywords: air turbulence error; CFD simulation; multi-rotor UAVs; meteorological observation

Citation: Liu, J.; Zhao, Z.; Fang, Z.; Li, Y.; Du, L. Correction of Error of Airborne Anemometers Caused by Self-Excited Air Turbulence. *Sensors* **2023**, *23*, 4288. https://doi.org/10.3390/s23094288

Academic Editors: Yafei Wang, Chao Huang, Hailong Huang, Zhiqiang Zuo, Bo Leng and Peng Hang

Received: 8 March 2023
Revised: 15 April 2023
Accepted: 24 April 2023
Published: 26 April 2023

1. Introduction

Multi-rotor UAVs have prevailed in many fields such as chemical [1], agricultural [2,3] and meteorological monitoring. By integrating miniaturized instruments, they have greatly promoted the development of scientific, industrial, and regulatory fields, especially in meteorological environment monitoring. It has great advantages over traditional automatic weather stations (AWS), satellites, remote sensing, and other measurement methods. As a platform for meteorological monitoring, multi-rotor UAVs can collect sensor data more sensitively and timely, and can obtain data with high spatial and temporal resolution [4]. A lot of research has been initiated in recent years. The US and Europe have begun to use UAVs as important instruments for disaster and environmental monitoring [5,6]. Brooke Potter et al. [7] made use of a UAV to collect data from a remote stream site. Zhewen Xing [8] and Ruisheng Ma [9] used multi-rotor UAVs to monitor meteorological disasters. Daniel Leuenberger et al. [10] used drones to improve the accuracy of weather forecasts.

Although multi-rotor UAVs have advantages in various measurement tasks, there is an urgent demand to resolve the effect of air turbulence generated by rotors. Many researchers have done a lot of meaningful work. Seokkwan Yoon et al. [11] calculated and simulated the airflow of rotors to study the best separation distance between the fuselage and the wings. Neal [12] solved the time-dependent Navier–Stokes equations for isolated rotors in hover and forward flight using detached eddy simulation and adaptive mesh refinement. Scott E. [13] used a fixed LBM grid and an adaptive refinement method to establish a simulation model for the four rotors of the drone. Qiwei Guo et al. [14] studied the formation process and flow distribution of the downwash airflow of the quadrotor UAV, and established the calculation model of the downwash airflow of the quadrotor agricultural UAV using CFD simulation. Hao Zhang et al. [15] studied the downwash airfield distribution of a six-rotor UAV when hovering at different flight speeds and altitudes, and performed numerical simulations on the airflow field. Most of the papers are about the simulation and modeling of the downwash airflow for multi-rotor UAVs. However, for many meteorological monitoring UAVs, the sensors are established on top of the multi-rotor UAVs. Upwash airflow excited by the multi-rotor UAVs will disturb the sensor even more.

Therefore, the research on the influence of the upwash airflow on the multi-rotor UAVs is more significant, especially the anemometer-involved application. When multi-rotor drones are used as airborne anemometers, the impact of rotor airflow should be compensated. The angle of attack (AOA) of the multi-rotor UAVs will also affect the performance of the anemometer as well. It is more urgent to resolve the union effects which come from these two weak points. Taro Nakai et al. [16,17] made a very prominent contribution to the correction of the AOA. They improved the accuracy of the correction method for ultrasonic wind sensors. In this paper, the differential pressure anemometer developed by Cheng Liu and Yichen Pan [18,19], is used. Although it can maintain its original measurement accuracy in the AOA range of 0–45 degrees, the union effects still need to be corrected when it is used in multi-rotor UAVs.

In this paper, for the purpose of the correction of error, we developed a method for the elimination of the influence of air turbulence on wind speed measurement. The corresponding correction model is obtained according to the CFD (Computational Fluid Dynamics) simulation of a six-rotor UAV which is carried out with the sliding grid method and the S-A turbulence model. Then, the model is applied in the developed prototype by adding the angle of attack compensation model of the airborne anemometer. The model has been verified in actual measurement, and it can make the airborne anemometer maintain the original wind speed measurement accuracy in the angle of attack range of 0–45° at various ascent speeds.

2. Methods

The UAV used in this work is a common six-rotor UAV, which has six propellers, and all its attitude and position control are achieved by adjusting the speed of the six driving motors. When the UAV is working normally, the three propellers are separated by 120 degrees rotate clockwise, and the other three propellers rotate counterclockwise, as shown in Figure 1. In general, the motion state of a six-rotor UAV is mainly divided into five types: hovering, vertical motion, rolling motion, pitching motion, and yaw motion. Only the hovering and vertical motions are simulated in this paper to study the impact on the anemometer because the two types often occur in measurement scenarios.

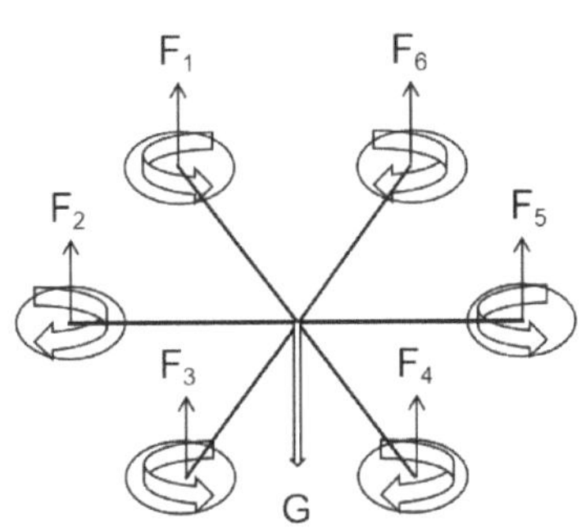

Figure 1. The working status of the six rotors.

2.1. Basic Control Equation

In the process of UAV flight, it is difficult to study the complex flow field and phenomenon generated by the rotation of the rotor using traditional aerodynamics. With the continuous development of computer technology and numerical methods, the use of computational fluid dynamics to calculate and simulate the rotor flow field has become one of the important methods for studying the characteristics of the rotor flow field.

The flying speed of meteorological UAVs is low, and the ascent speed is within 5 m/s normally. Therefore, the air medium in the external flow field can be regarded as incompressible. Navier–Stokes (NS) equations are the most suitable differential equation to express incompressible fluid. The NS equation reflects the basic laws of viscous fluids, and it relies on differential equations to describe fluid motion. The three-dimensional incompressible N-S equation is expressed as follows:

$$\begin{cases} \rho\frac{Du}{Dt} = \rho f_x - \frac{\partial p}{\partial x} + \mu\nabla^2 u \\ \rho\frac{Dv}{Dt} = \rho f_y - \frac{\partial p}{\partial y} + \mu\nabla^2 v \\ \rho\frac{Dw}{Dt} = \rho f_z - \frac{\partial p}{\partial z} + \mu\nabla^2 w \end{cases} \tag{1}$$

where u, v, and w are the components of the dimensionless velocity along the x, y, and z directions, p and t are the dimensionless pressure and time, f_x, f_y, and f_z denote the components of the external force per unit volume of fluid in the x, y, and z directions, respectively. Multiply the above equations by the unit vectors $\boldsymbol{i}, \boldsymbol{j}$, and $\boldsymbol{k}$ in the three directions and add them to obtain the simpler vector form of the N-S equation for incompressible viscous fluid:

$$\frac{D\vec{V}}{Dt} = \vec{f} - \frac{1}{\rho}\nabla p + \frac{\mu}{\rho}\nabla\vec{V} \tag{2}$$

where $\vec{V}$ is the velocity vector, ∇ is the Hamiltonian, and $\vec{f}$ is the total external force per unit volume of fluid.

2.2. Calculation Method

When using the CFD method to simulate the rotor flow field, there are two main methods. The first method is to use the Actuator Disk theory [20] to equate the rotating blade with an actuator disk. The momentum source method [21] is a kind of actuator disk method. Its basic idea is that the action of the blade on the airflow is added to the governing equations (Euler or N-S) equivalent to the time-averaged momentum source term. In this way, the effect of the blade on the airflow is characterized by the change of the airflow. The second method is the sliding grid method, which generates a body-fitted grid around each blade, and uses the entire rotor grid system as a motion-nested grid. In this grid, the rotor flow field is simulated by solving the Euler equation or N-S equation. Essentially, the rotor rotation of a multi-rotor UAV belongs to the mechanical rotation, so a simple and adaptable sliding grid can be used to complete the calculation of various states with a multi-reference (MRF) system model.

The MRF model is one of the multi-region calculation methods, which uses a steady-state approximation. Different rotation or movement speeds can be assumed in each region. The equations of the motion reference system are used to solve the flow problem in each motion area grid. On the interface of the computational domain, a local reference system is used to calculate the flux of the flow variables in one area and convert them to adjacent areas. The schematic diagram of a typical MRF system model is shown in Figure 2. It is a coordinate system that rotates at a stable angular velocity $\vec{w}$ for a stationary reference system. The origin of the rotating system is positioned by the position vector $\vec{r}$.

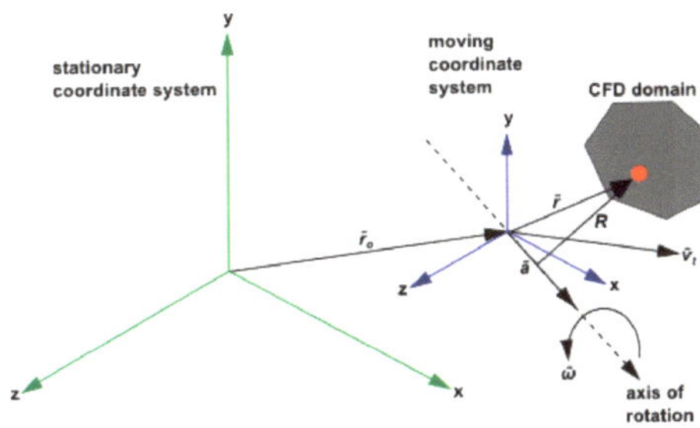

Figure 2. MRF model diagram.

The position of any point in the calculation domain of the rotation system can be determined by the position vector $\vec{r}$ and the origin of the rotation system. The implicated velocity can be expressed as follows:

$$\vec{u_r} = \vec{w} \times \vec{r} \tag{3}$$

The velocity $\vec{u_r}$ can be converted from a stationary system to a rotating system by the following equation:

$$\vec{v_r} = \vec{v} - \vec{u_r} \tag{4}$$

where $\vec{v_r}$ is the relative velocity and $\vec{v}$ is the absolute velocity. When solving the problem of multiple moving individuals in a rotating coordinate system, the additional term in the momentum equation will cause the fluid acceleration to increase. The fluid governing equations in the form of relative velocity are shown as follows:

$$\frac{\partial \rho}{\partial t} + \nabla \cdot \rho \vec{v_r} = 0 \tag{5}$$

$$\frac{\partial}{\partial t}\left(\rho \vec{v_r}\right) + \nabla \cdot \left(\rho \vec{v_r}\vec{v_r}\right) + \rho\left(2\vec{w} \times \vec{v_r} + \vec{w} \times \vec{w} \times \vec{v_r}\right) = -\nabla p + \nabla \cdot \vec{\tau} + \vec{F} \tag{6}$$

$$\frac{\partial}{\partial t}(\rho E_r) + \nabla \cdot \left(\rho \vec{v_r} H_r\right) = \nabla \cdot \left(k \nabla T + \overline{\overline{\tau}}_r \cdot v_r\right) + s_h \tag{7}$$

where Equation (5) is the continuity equation, Equation (6) is the momentum equation, and Equation (7) is the energy equation. The momentum equation contains two additional acceleration terms: Coriolis acceleration $2\vec{w} \times \vec{v_r}$ and centripetal acceleration $\vec{w} \times \vec{w} \times \vec{v_r}$. Compared with the original equation, the viscous stress $\overline{\overline{\tau}}_r$ uses the relative velocity derivative term. The energy equation uses relative internal energy E_r and relative total enthalpy H_r, and these variables are defined as:

$$E_r = h - \frac{p}{\rho} + \frac{1}{2}\left(v_r^2 - u_r^2\right) \tag{8}$$

$$H_r = E_r + \frac{p}{\rho} \tag{9}$$

In a sliding grid, the relative motion between the stationary and rotating parts causes transient interaction effects, which is a strong unsteady phenomenon, but these effects are ignored in the MRF system. The sliding grid technology uses two or more calculation areas, each area can generate a grid independently, which is extremely convenient for complex models. There is at least one interface between each area and adjacent areas. The interface of adjacent computing areas forms a "grid boundary", and the dynamic domain will move along the interface. The grid on the interface does not need to be aligned, and the flux is calculated by the information interpolation between the grid nodes. A virtual grid layer is generated on both sides of the slip surface, which overlaps the computational domain grid on both sides of the sliding surface. During calculation, the nodes on the virtual grid layer are interpolated to realize the flux transfer on the computational domains on both sides of the interface.

When using the sliding grid technology for numerical simulation, the model needs to be divided into two parts: the rotor part and the stator part, and these two parts have meshed separately. In this paper, the rotor part is the cylindrical area where the propeller rotates, and the stator part is the entire computational domain minus other areas of the rotor part. In the modeling, the connecting parts of the rotor part and the stator part are paired to form multiple interfaces.

2.3. Turbulence Model

In this paper, the method of numerical simulation calculation of the UAV flow field is the S-A turbulence model which is widely used in aviation. Compared with the k-ε turbulence model, the S-A turbulence model is more robust in simulating and calculating complex flows and consumes fewer computing resources. The S-A turbulence model is based on a transport equation of eddy viscosity containing the convection term, diffusion term, and source term. This application was proposed by Spalart and Allmaras [22]. Ashford and Powell [23] improved this to avoid negative values in the generated term. The fluctuating amount $\widetilde{v}$ of turbulent kinetic energy can be obtained from the transport equation:

$$\frac{\partial v}{\partial t} + \vec{V} \cdot \nabla \widetilde{v} = \frac{1}{\sigma}\left\{\nabla \cdot \left[v + (1 + c_{b2})\widetilde{v}\nabla\widetilde{v}\right] - c_{b2}\widetilde{v}\nabla\widetilde{v}\right\} + Q \tag{10}$$

where $\vec{V}$ is the mean velocity, Q is the source term, σ and c_{b2} are constant. Source term Q contains the generating term and dissipative term as follows:

$$Q = \widetilde{v}P\left(\widetilde{v}\right) - \widetilde{v}D\left(\widetilde{v}\right) \tag{11}$$

$$\widetilde{v}P(\widetilde{v}) = c_{b1}S\widetilde{v} \tag{12}$$

$$\widetilde{v}D(\widetilde{v}) = c_{w1}f_2\left(\frac{\widetilde{v}}{d}\right)^2 \tag{13}$$

The generating term can be obtained by Equations (14)–(16) in the following:

$$\widetilde{S} - Sf_{v3} + \frac{\widetilde{v}}{k^2d^2}f_{v2} \tag{14}$$

$$f_{v2} = \frac{1}{\left(1 + \frac{\chi}{c_{v2}}\right)^3} \tag{15}$$

$$f_{v3} = \frac{(1 + \chi f_{v1})(1 - \chi f_{v2})}{\chi} \tag{16}$$

where d is the minimum distance to the wall surface, S is the vorticity. f_m can be obtained by Equations (17)–(19) as follows:

$$f_{v2} = g\left(\frac{1 + c_w^6}{g^6 + c_w^6}\right)^6 \tag{17}$$

$$g = r + c_{w2}\left(r^6 - r\right) \tag{18}$$

$$r = \frac{\tilde{v}}{\tilde{S}k^2d^2} \tag{19}$$

The constant value in the S-A turbulence model is:

$$c_{w1} = 3.239, c_{w2} = 0.3, c_{w3} = 2, c_{v1} = 7.1, c_{v2} = 5, c_{b1} = 0.1355, c_{b2} = 0.622, k = 0.41, \sigma = 0.667 \tag{20}$$

2.4. Correction Model of the Angle of Attack

In this paper, a solid-state differential pressure anemometer is mounted on the multi-rotor UAV, and its structure is shown in Figure 3. The principle of the differential pressure anemometer in this work is that the differential pressure between the two ends of the cylinder varies with the wind speed. According to the variation in the differential pressure and distribution, the corresponding wind speed and wind direction can be calculated. Figure 4 is a schematic diagram of the anemometer measurement.

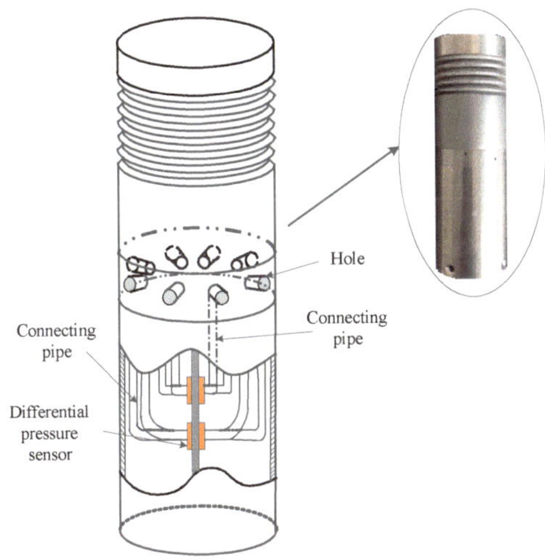

Figure 3. The internal structure of the differential pressure anemometer.

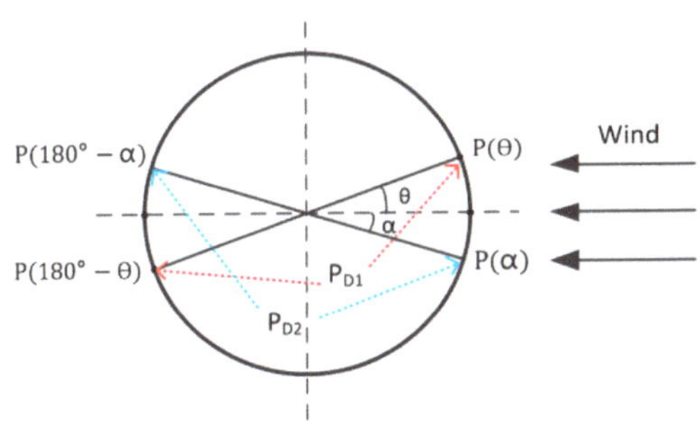

Figure 4. Schematic diagram of P_{D1} and P_{D2}.

The relationship between differential pressure and wind speed and direction can be expressed as follows [24]:

$$U_\infty = 2\sqrt{\frac{P_{D2}}{\rho(a\sin 2\theta + a + 2b)}} \tag{21}$$

$$\theta = \frac{1}{2}\left[\arccos\frac{(R_D - 1)\left(1 + \frac{2b}{a}\right)}{\sqrt{(R_D)^2 + 1}} - \arctan R_D\right] \tag{22}$$

where U_∞ is the wind speed, ρ is the air density, and a and b are the correction coefficients obtained by fitting the measured data. R_D is the ratio of the two largest differential pressures(P_{D1} *and* P_{D2}), which is expressed by:

$$R_D = \frac{P_{D1}}{P_{D2}} \tag{23}$$

When the angle of attack is greater than 15°, the measurement result is affected [25]. In this paper, the angle of attack is the angle between the wind speed vector and the anemometer measurement plane where the eight holes are located, as shown in Figure 5. The tilt angle α between the anemometer and the vertical axis is used to replace the angle of attack because it is equivalent to it and can easily be obtained by the accelerometer inside the anemometer in practical applications.

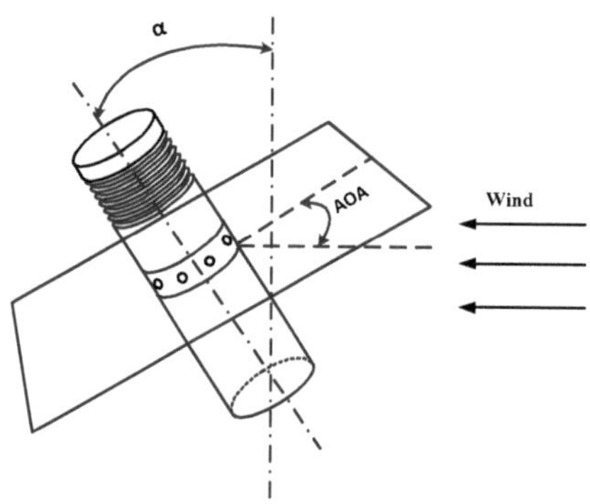

Figure 5. Display of AOA and α.

According to previous work, the angle of attack error of the anemometer can be corrected and compensated by the model as shown in Equations (24)–(27), so that the anemometer can maintain the original measurement accuracy and range [19].

$$U_T = 2\sqrt{\frac{P_{TD2}}{\rho(a\sin 2\theta + a + 2b)}} \tag{24}$$

$$P_{TD2} = \frac{P_{D2}}{T(g(\alpha, \theta))} \tag{25}$$

$$T(\alpha_r) = a_0 + a_1\cos\alpha_r + a_2\cos\alpha_r{}^2 \tag{26}$$

$$g(\alpha, \theta) = \alpha_r = \arcsin(\sin\alpha \cdot \cos\theta) \tag{27}$$

where U_T is the corrected wind speed under α, P_{TD2} is the second-largest differential pressure under the tilt angle α and $T(g(\alpha, \theta))$ represents the influence of the angle of attack on the pressure distribution. For Equations (24)–(27), there is a detailed derivation process and explanation in reference [19], which will not be introduced here.

3. Simulation and Modeling

3.1. Mesh and Boundary Conditions

In this paper, the geometric model is very complicated, there are small gaps between the rotating area and the static area. To better express as many detailed areas as possible, an unstructured grid method is used for numerical simulation calculations. For the calculation of the external flow domain of CFD, the larger the flow domain, the smaller the interference of the external flow field boundary on the flow field calculation. This requires the flow field to be set as large as possible during the calculation. However, a large computing domain needs to consume too many computing resources. When the calculated flow domain size reaches a certain range, the calculation accuracy remains stable. Before the formal simulation, we conducted a grid independence test to ensure the optimal grid size and distribution while maintaining calculation accuracy. The mesh size of the area where the airflow changes drastically and the area close to the surface is set to be smaller, and the mesh size of the area where the airflow is stable to be larger to keep the accuracy of calculation and save the calculation resources. We initially divided 5,376,248 mesh cells roughly according to the above rules, calculated the maximum wind speed error, and then refined the entire mesh four times. The final grid-independence test result is shown in Figure 6. The red triangle in the figure represents the number of grids divided in our five simulations and the corresponding maximum error.

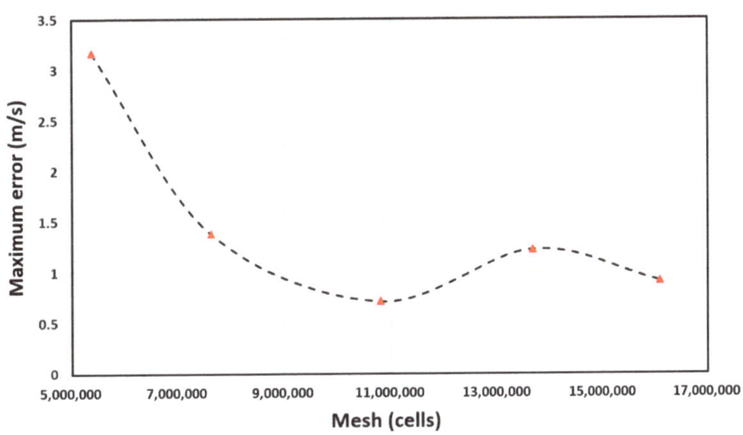

Figure 6. Grid-independence test.

According to the results of the grid-independence analysis, we ultimately used 10,803,973 grids for subsequent simulations. The dynamic calculation area is selected to cover the adjacent area of the propeller blades, as this area has the most significant impact on the air motion related to the rotation of the propeller. After further increasing the calculation area, the simulation results do not show significant changes, but the calculation time greatly increases. Therefore, after considering the calculation amount and simulation accuracy, we choose the adjacent area of the propeller blades as the dynamic calculation area. The external flow domain selected in this paper is shown in Figure 7. The "encryption area" in Figure 7 refers to the outer region of the "computing domain", which is usually used to avoid the influence of boundary effects on the calculation results and to improve the computational efficiency. Since the "computing domain" can be divided into multiple regions for parallel computing, adding an external "encryption area" can expand the range of the computing domain, thereby improving the accuracy and efficiency of numerical simulation. The size of the flow domain is about eight times the size of the UAV simulation model. Due to the symmetry of the UAV model, symmetrical boundary conditions are used in the calculation, and only half of the UAV model is calculated for the flow field, which saves computing resources without sacrificing calculation accuracy. The mesh diagram of the UAV is shown in Figure 8. The copter has a length of 1 m and a width of 1 m, and each blade has a length of 30 cm and a width of 5 cm. The anemometer is located at a height of 5 cm above the rotor. The number of mesh cells here is 3,689,216.

Figure 7. Computing domain and the encryption area.

Figure 8. The mesh diagram of the UAV.

For the sliding grid of the UAV, the rotors of the UAV should be wrapped in the rotation area. The static domain and each dynamic domain use interfaces to transfer data. The mesh diagram of the dynamic area of the UAV rotors is shown in Figure 9. The number of mesh cells here is 8,795,339.

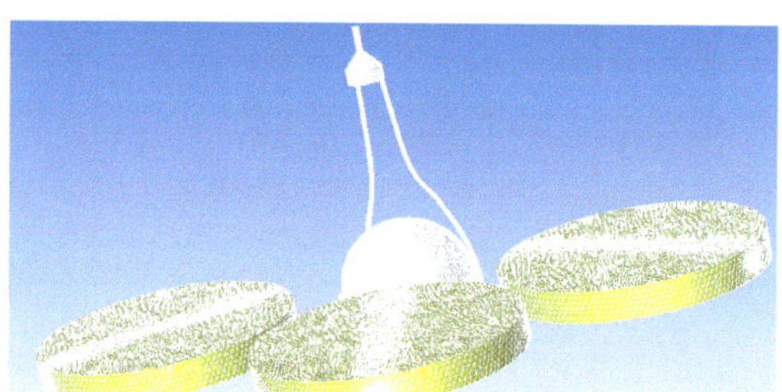

Figure 9. The mesh diagram of the dynamic area of the UAV rotors.

According to the calculation requirements in this work, the entire calculation domain is divided into two parts, the dynamic domain, and the static domain. The relevant boundary conditions include the object boundary conditions, the far-field boundary conditions, and the interface boundary conditions. The surface of the aircraft model is set with no slippage and no penetration. The contact surface between the flow domain of the UAV and the outer flow domain is set as interfaces. Similarly, the contact surfaces between the flow domain of the rotors and the overall flow domain of the UAV are set as interfaces, which allows the two-flow domain to exchange data during the calculation process. Except for the symmetry plane, the surface of the flow domain of the UAV is all set as a velocity inlet to simulate the realistic flow field of the UAV during flight.

3.2. Simulation and Results

The measurement accuracy of the wind sensor on the top of the UAV will be affected by the airflow generated by the rotors. It is necessary to compare the simulation value and the standard value of the wind speed of the UAV under different ascent speeds and different crosswind conditions through CFD simulation. The wind speed measurement error of the UAV will be corrected by comparing the two values. In the paper, the UAV velocity flow field diagram is obtained through CFD simulation at different ascent speeds of 0 m/s, 3 m/s, and 5 m/s and different crosswind speeds of 0 m/s, 3 m/s, 5 m/s, 7 m/s, 10 m/s, 13 m/s, 15 m/s, 17 m/s, and 20 m/s. Figures 10–12 are the velocity flow field diagrams of the UAV under different ascent speeds and crosswind speeds. The drone is hovering at the ascent speed of 0 m/s. When the crosswind speed is 0 m/s, the theoretical value of the wind speed measured by the UAV wind sensor should be 0 m/s. However, the flow field diagram shows a flow velocity exists at the wind sensor position, which indicates that the airflow driven by the rotation of the UAV rotors will affect the measurement results. When the UAV is rising at a constant speed, the UAV rotors have different effects on the flow velocity at the wind sensor position under different crosswind speeds.

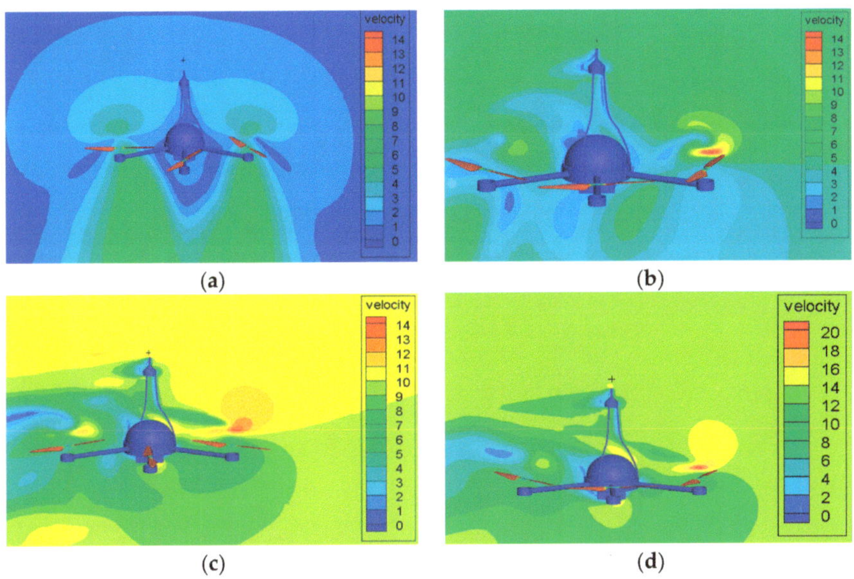

Figure 10. The UAV velocity flow field diagram at ascent speed of 0 m/s. when the crosswind speed is: (**a**) 0 m/s, (**b**) 5 m/s, (**c**) 10 m/s, (**d**) 13 m/s.

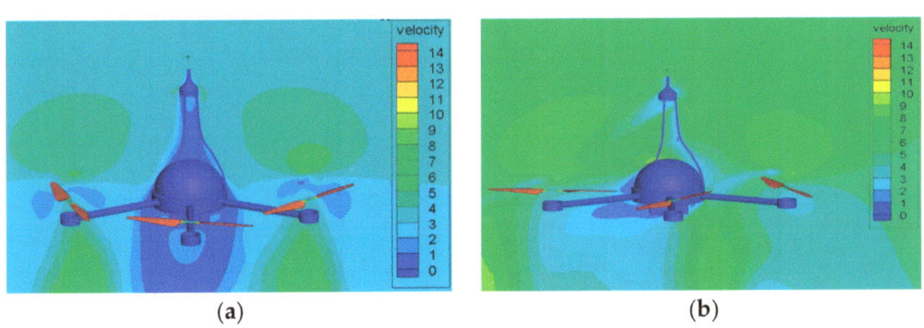

Figure 11. *Cont.*

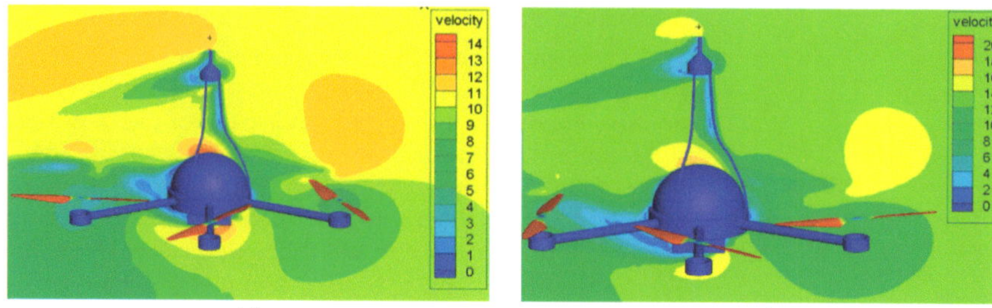

(**c**) (**d**)

Figure 11. The UAV velocity flow field diagram at ascent speed of 3 m/s. when the crosswind speed is: (**a**) 0 m/s, (**b**) 5 m/s, (**c**) 10 m/s, (**d**) 13 m/s.

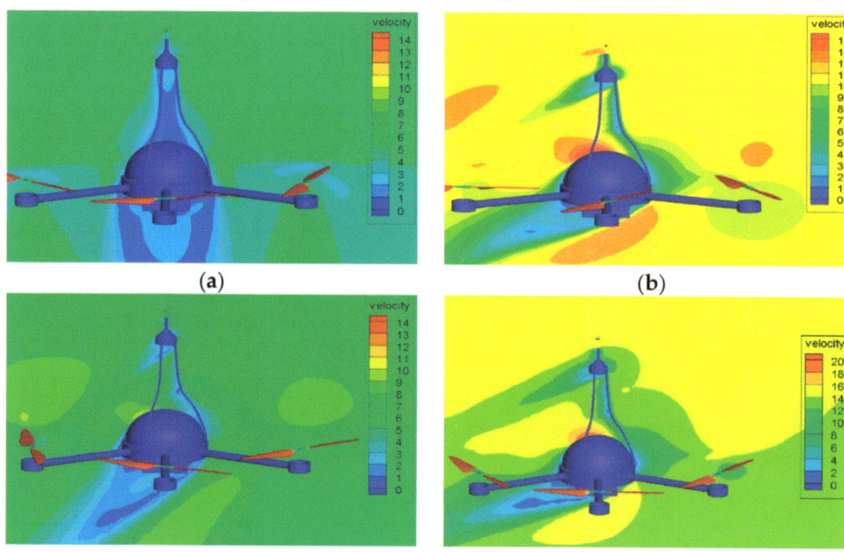

(**a**) (**b**)

(**c**) (**d**)

Figure 12. The UAV velocity flow field diagram at ascent speed of 5 m/s. when the crosswind speed is: (**a**) 0 m/s, (**b**) 5 m/s, (**c**) 10 m/s, (**d**) 13 m/s.

To clearly express the influence of the UAV rotors on wind speed measurement, specific simulation crosswind speed values under different standard crosswind speeds and ascent speeds are listed in Table 1. As shown in Table 1, when the crosswind speed of the drone is the same, the higher the ascent speed, the closer the simulation speed is to the standard crosswind speed. As the speed in the UAV flow field increases, the influence of the motion of the UAV rotors has a smaller effect on the flow field near the wind sensor.

Table 1. Simulation crosswind speed table under different standard crosswind speeds and ascent speeds.

Standard Crosswind Speed (m/s)	Simulation Crosswind Speed (m/s) at Ascent Speed of 0 m/s	Simulation Crosswind Speed (m/s) at Ascent Speed of 3 m/s	Simulation Crosswind Speed (m/s) at Ascent Speed of 5 m/s
0	0.168	0.015	0.009
3	3.344	3.333	3.222
5	5.928	5.758	5.419
7	7.612	7.552	7.438
10	10.970	10.670	10.661

Table 1. *Cont.*

Standard Crosswind Speed (m/s)	Simulation Crosswind Speed (m/s) at Ascent Speed of 0 m/s	Simulation Crosswind Speed (m/s) at Ascent Speed of 3 m/s	Simulation Crosswind Speed (m/s) at Ascent Speed of 5 m/s
13	13.807	13.794	13.682
15	16.148	15.989	15.871
17	18.282	18.099	17.979
20	21.483	21.263	21.142

3.3. Modeling

The wind speed measurement error of the airborne anemometer comes from the angle of attack of the UAV and the air turbulence of the rotors. Combining the two correction models can well correct the wind speed measurement error of the airborne anemometer. Figure 13 shows curves between simulation crosswind speed and standard crosswind speed under different ascent speeds based on the data in Table 1. It shows that the curves under the three ascent speeds are almost the same. In other words, although the airflow of the UAV rotors has an influence on the wind speed measurement, the speed of the rotor within 5 m/s is not related to it.

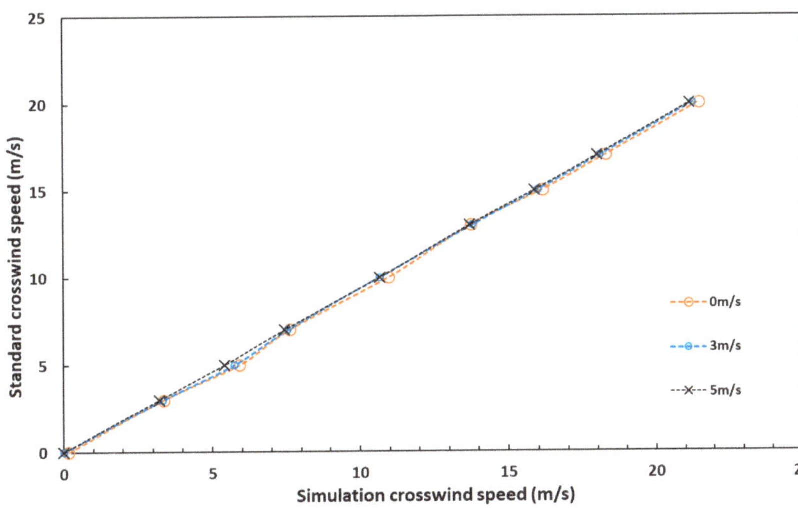

Figure 13. Curves between simulation crosswind speed and standard crosswind speed under different ascent speeds.

The three curves in Figure 13 are fitted by the least-squares method to obtain Equation (28) as expressed:

$$V_r = c_0 V_m + c_1 \tag{28}$$

where V_r is the real crosswind speed and V_m is the measured crosswind speed. c_0 and c_1 are the fitting coefficients. The values of c_0 and c_1 are the result of averaging the coefficients of three formulas which is expressed in Equation (29):

$$c_0 = 0.9456 \; c_1 = -0.1573 \tag{29}$$

According to Equations (24) and (28), the wind speed measurement correction model of the airborne anemometer can be expressed as:

$$U_R = 2c_0\sqrt{\frac{P_{TD2}}{\rho(a\sin 2\theta + a + 2b)}} + c_1 \tag{30}$$

4. Test and Results

The wind speed correction model of the airborne anemometer is obtained by combining the angle of attack correction model and the air turbulence correction model, and the model was verified through a UAV flight test. The drone flies at different ascent speeds near the meteorological tower at a height of 70 m and performs the wind speed measurement. The measurement results of the cup anemometer in the meteorological tower are used as standard data for comparison with the measurement results from the airborne anemometer before and after correction, which are shown in Figure 14. The wind speed measured by the drone after correction is dynamically changing and consistent with the measurement result of the cup anemometer, while the measurement results from the airborne anemometer before correction had a larger error. It can be seen from the figure that there are some deviations between the test points and the standard value. This is because the cup anemometer and the airborne anemometer are in close positions but not absolutely the same. In the boundary layer, the uneven airflow causes this deviation, but these deviations are within a reasonable range.

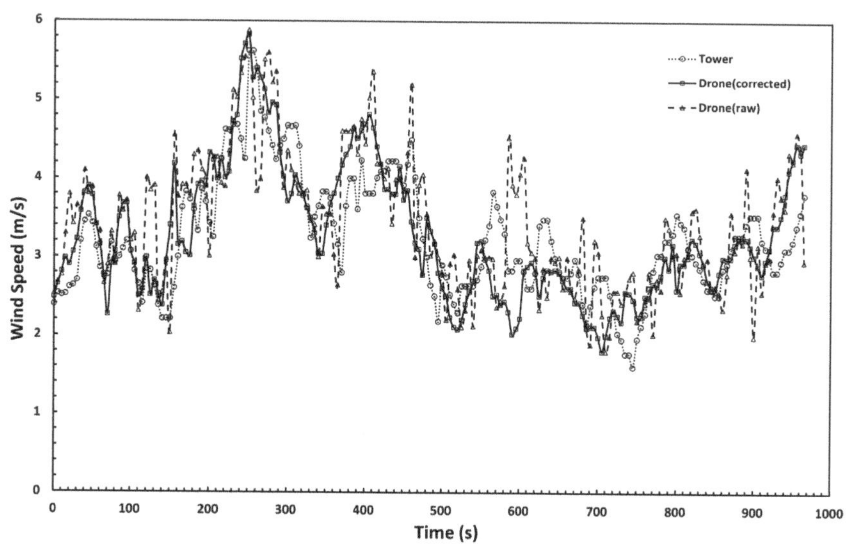

Figure 14. Wind speed measurement results between the meteorological tower and airborne anemometer.

To clearly verify the compensation model, the measurement error curve of the airborne anemometer is drawn, as shown in Figure 15. In this paper, the wind speed measurement error of the anemometer is $\pm(0.5 + 0.03\ V)$ m/s (V is the standard wind speed). The error bar in Figure 15 is obtained according to the standard value, which is the reason for its dynamic change. It can be seen from the figure that the wind speed measurement errors of the airborne anemometer are all within the error bar, which verifies that the model has a good correction effect.

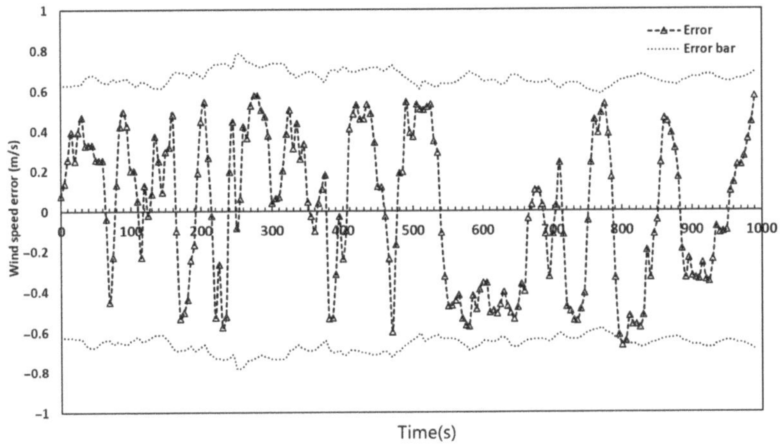

Figure 15. Wind speed measurement errors of the airborne anemometer.

5. Conclusions

In this paper, the influence of the air turbulence generated by the rotors of UAVs on the measurement of the airborne anemometer is studied. The CFD simulation of the UAV is carried out using the sliding grid method and the S-A turbulence model. The relationship between the measured wind speed and the standard wind speed was obtained, and an air turbulence correction model was established. The angle of attack compensation model of the differential pressure anemometer is added to the air turbulence correction model to make it more practical.

The model is verified in the actual measurement, and the result shows the model has a good correction effect. The airborne anemometer maintains the original measurement accuracy at different ascent speeds. This study proves that for a six-rotor UAV, the air turbulence generated by the rotors has an impact on the measurement, but it is not related to the speed of rotors within 5 m/s. Since the ascent speed of meteorological UAVs is low, this paper does not study speeds above 5 m/s. The effect of the high-speed rotating rotors on the airflow needs to be further explored. Whether this model has the same corrective effect on UAVs with other rotor numbers or UAVs with other structures requires further research and verification.

Author Contributions: J.L., Z.Z., Z.F., Y.L. and L.D. collaborated on the research for various aspects of the paper. L.D. and Z.Z. designed the mathematical model; J.L. and L.D. designed and performed the experiments; Z.F. and Y.L. analyzed the data; J.L. wrote the paper; Y.L. polished the paper. All authors have read and agreed to the published version of the manuscript.

Funding: This work was supported by the National Key Research and Development Project (2021YFC3000304).

Institutional Review Board Statement: Not applicable.

Informed Consent Statement: Not applicable.

Data Availability Statement: No new data was created or analyzed in this study. Data sharing is not applicable to this article.

Conflicts of Interest: The authors declare no conflict of interest.

References

1. Javier, B.; Marco, S. Environmental chemical sensing using small drones: A review. *Sci. Total Environ.* **2020**, *748*, 141172. [CrossRef]
2. Ahirwar, S.; Swarnkar, R.; Bhukya, S.; Namwade, G. Application of drone in agriculture. *Int. J. Curr. Microbiol. Appl. Sci.* **2019**, *8*, 2500–2505. [CrossRef]

3. Mogili, U.M.R.; Deepak, B. Review on application of drone systems in precision agriculture. *Procedia Comput. Sci.* **2018**, *133*, 502–509. [CrossRef]
4. Li, C.; Sun, X. A novel meteorological sensor data acquisition approach based on unmanned aerial vehicle. *Int. J. Sens. Netw.* **2018**, *28*, 80–88. [CrossRef]
5. Malaver Rojas, J.; Gonzalez, F.; Motta, N.; Depari, A.; Corke, P. Towards the development of a gas sensor system for monitoring pollutant gases in the low troposphere using small unmanned aerial vehicles. In Proceedings of the Workshop on Robotics for Environmental Monitoring, 11 July 2012; pp. 1–3. Available online: http://eprints.qut.edu.au/52042/ (accessed on 5 December 2022).
6. Zhang, J.; Tang, J.; Wang, T.; Chen, F. Energy-efficient data-gathering rendezvous algorithms with mobile sinks for wireless sensor networks. *Int. J. Sens. Netw.* **2017**, *23*, 248–257. [CrossRef]
7. Potter, B.; Valentino, G.; Yates, L.; Benzing, T.; Salman, A. Environmental Monitoring Using a Drone-Enabled Wireless Sensor Network. In Proceedings of the Systems and Information Engineering Design Symposium (SIEDS), Charlottesville, VA, USA, 26 April 2019. [CrossRef]
8. Xing, Z.; Zhang, Y.; Su, C.Y.; Qu, Y.; Yu, Z. Kalman filter-based wind estimation for forest fire monitoring with a quadrotor UAV. In Proceedings of the 2019 IEEE Conference on Control Technology and Applications (CCTA), Hong Kong, China, 19–21 August 2019; IEEE: Piscataway, NJ, USA, 2019; pp. 783–788. [CrossRef]
9. Ma, R.; Li, X.; Sun, M.; Kuang, Z. Experiment of Meteorological Disaster Monitoring on Unmanned Aerial Vehicle. In Proceedings of the 2018 7th International Conference on Agro-Geoinformatics (Agro-Geoinformatics), Hangzhou, China, 6–9 August 2018; IEEE: Piscataway, NJ, USA, 2018; pp. 1–6. [CrossRef]
10. Leuenberger, D.; Haefele, A.; Omanovic, N.; Fengler, M.; Martucci, G.; Calpini, B.; Fuhrer, O.; Rossa, A. Improving high-impact numerical weather prediction with lidar and drone observations. *Bull. Am. Meteorol. Soc.* **2020**, *101*, E1036–E1051. [CrossRef]
11. Yoon, S.; Lee, H.C.; Pulliam, T.H. Computational analysis of multi-rotor flows. In Proceedings of the 54th AIAA Aerospace Sciences Meeting, San Diego, CA, USA, 4–8 January 2016; p. 0812. [CrossRef]
12. Chaderjian, N.M. Advances in rotor performance and turbulent wake simulation using DES and adaptive mesh refinement. In Proceedings of the 7th International Conference on Computational Fluid Dynamics, Big Island, HI, USA, 9–13 July 2012.
13. Thibault, S.; Holman, D.; Garcia, S.; Trapani, G. CFD Simulation of a quad-rotor UAV with rotors in motion explicitly modeled using an LBM approach with adaptive refinement. In Proceedings of the 55th AIAA Aerospace Sciences Meeting, Grapevine, TX, USA, 9–13 January 2017; p. 0583. [CrossRef]
14. Guo, Q.; Zhu, Y.; Tang, Y.; Hou, C.; He, Y.; Zhuang, J.; Zheng, Y.; Luo, S. CFD simulation and experimental verification of the spatial and temporal distributions of the downwash airflow of a quad-rotor agricultural UAV in hover. *Comput. Electron. Agric.* **2020**, *172*, 105343. [CrossRef]
15. Zhang, H.; Qi, L.; Wu, Y.; Musiu, E.M.; Cheng, Z.; Wang, P. Numerical simulation of airflow field from a six–rotor plant protection drone using lattice Boltzmann method. *Biosyst. Eng.* **2020**, *197*, 336–351. [CrossRef]
16. Nakai, T.; Van Der Molen, M.K.; Gash, J.H.C.; Kodama, Y. Correction of sonic anemometer angle of attack errors. *Agric. For. Meteorol.* **2006**, *136*, 19–30. [CrossRef]
17. Nakai, T.; Shimoyama, K. Ultrasonic anemometer angle of attack errors under turbulent conditions. *Agric. For. Meteorol.* **2012**, *162*, 14–26. [CrossRef]
18. Liu, C.; Du, L.; Zhao, Z. A directional cylindrical anemometer with four sets of differential pressure sensors. *Rev. Sci. Instrum.* **2016**, *87*, 035105. [CrossRef] [PubMed]
19. Yichen, P.; Zhenyu, L.; Zhen, F.; Lidong, D.; Zhan, Z. An angle-of-attack correction method for a 2-D miniature differential pressure solid-state anemometer. *Measurement* **2021**, *171*, 108833. [CrossRef]
20. Horlock, J.H. *Actuator Disk Theory-Discontinuities in Thermo-Fluid Dynamics*; McGraw-Hill, Inc.: New York, NY, USA, 1978.
21. Kim, Y.H.; Park, S.O. Navier-Stokes simulation of unsteady rotor-airframe interaction with momentum source method. *Int. J. Aeronaut. Space Sci.* **2009**, *10*, 125–133. [CrossRef]
22. Spalart, P.; Allmaras, S. A one-equation turbulence model for aerodynamic flows. In Proceedings of the 30th Aerospace Sciences Meeting and Exhibit, Reno, NV, USA, 6–9 January 1992; p. 439. [CrossRef]
23. Ashford, G.; Powell, K. Adaptive unstructured triangular mesh generation and flow solvers for the Navier-Stokes equations at high Reynolds number. In Proceedings of the 12th Computational Fluid Dynamics Conference, San Diego, CA, USA, 19–22 June 1995; p. 1724. [CrossRef]
24. Pan, Y.; Zhao, Z.; Zhao, R.; Fang, Z.; Wu, H.; Niu, X.; Du, L. High accuracy and miniature 2-D wind sensor for boundary layer meteorological observation. *Sensors* **2019**, *19*, 1194. [CrossRef] [PubMed]
25. Shao, W.Y.; Zhang, Y.P.; Zhu, D.Z.; Zhang, T.Q. Drag force on a free surface-piercing yawed circular cylinder in steady flow. *J. Fluids Struct.* **2013**, *43*, 145–163. [CrossRef]

sensors MDPI

Article

Anonymity Assurance Using Efficient Pseudonym Consumption in Internet of Vehicles

Mehreen Mushtaq [1], Ata Ullah [1], Humaira Ashraf [2], N.Z Jhanjhi [3,*], Mehedi Masud [4], Abdulmajeed Alqhatani [5] and Mrim M. Alnfiai [6]

1 Department of Computer Science, National University of Modern Languages (NUML), Islamabad 44000, Pakistan
2 Department of Computer Science and Software Engineering, International Islamic University Islamabad, Islamabad 44000, Pakistan
3 School of Computer Science SCS, Taylor's University, Subang Jaya 47500, Malaysia
4 Department of Computer Science, College of Computers and Information Technology, Taif University, Taif 21944, Saudi Arabia
5 Department of Information Systems, College of Computer Science & Information Systems, Najran University, Najran 61441, Saudi Arabia
6 Department of Information Technology, College of Computers and Information Technology, Taif University, Taif 21944, Saudi Arabia
* Correspondence: noorzaman.jhanjhi@taylors.edu.my

Abstract: The Internet of vehicles (IoVs) is an innovative paradigm which ensures a safe journey by communicating with other vehicles. It involves a basic safety message (BSM) that contains sensitive information in a plain text that can be subverted by an adversary. To reduce such attacks, a pool of pseudonyms is allotted which are changed regularly in different zones or contexts. In base schemes, the BSM is sent to neighbors just by considering their speed. However, this parameter is not enough because network topology is very dynamic and vehicles can change their route at any time. This problem increases pseudonym consumption which ultimately increases communication overhead, increases traceability and has high BSM loss. This paper presents an efficient pseudonym consumption protocol (EPCP) which considers the vehicles in the same direction, and similar estimated location. The BSM is shared only to these relevant vehicles. The performance of the purposed scheme in contrast to base schemes is validated via extensive simulations. The results prove that the proposed EPCP technique outperformed compared to its counterparts in terms of pseudonym consumption, BSM loss rate and achieved traceability.

Keywords: vehicle anonymization; IoVs; pseudonym consumption; adversary; BSM; traceability

Citation: Mushtaq, M.; Ullah, A.; Ashraf, H.; Jhanjhi, N.Z.; Masud, M.; Alqhatani, A.; Alnfiai, M.M. Anonymity Assurance Using Efficient Pseudonym Consumption in Internet of Vehicles. *Sensors* **2023**, *23*, 5217. https://doi.org/10.3390/s23115217

Academic Editors: Yafei Wang, Chao Huang, Hailong Huang, Zhiqiang Zuo, Bo Leng and Peng Hang

Received: 6 March 2023
Revised: 24 April 2023
Accepted: 23 May 2023
Published: 31 May 2023

1. Introduction

Vehicular ad hoc networks (VANETs) support communication among vehicles to ensure road safety and transportation facilities by using the intelligent transport system (ITS) along with the support of road side units (RSUs) [1]. VANETs are transformed into the Internet of vehicles (IoVs) to provide more flexibility and ease to mankind. The IoVs transportation system is increasing rapidly; it is estimated that 2 billion vehicles will be connected to the IoVs by 2035. The IoVs supports five types of communication including vehicle-to-vehicle (V2V), vehicle-to-RSU (V2R), vehicle-to-infrastructure (V2I), vehicle-to-cloud (V2C) and vehicle-to-pedestrian (V2P). This communication is collectively known as vehicle to everything (V2X) communication [2,3]. The V2X communication is shown in Figure 1. The IoVs provide a set of supporting information for the drivers such as precrash warning, post-crash notification, pedestrian vicinity alert, danger zone alert and amber warning. Because of these timely notifications, the accident ratio is reduced to a large extent [4–6]. Besides these notifications, it provides comfort and entertainment services to both passengers and drivers [4,7].

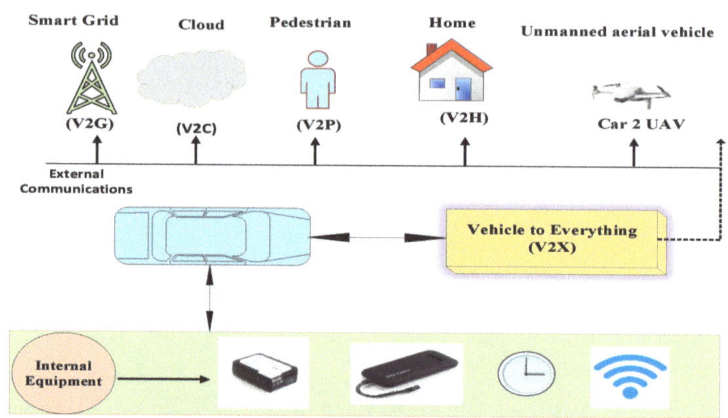

Figure 1. V2X communication.

A basic safety message (BSM) or a beacon is utilized for communication in the network. These BSMs contain all of the important information related to the vehicle (speed, velocity and direction) in plain form [8]. When this BSM is broadcasted, there is a high probability that any adversary can access this BSM. The adversary can be local or global. A local adversary is one that is part of a network, becomes a malicious node and sends network information to any other body. A global adversary is a person who eavesdrops on BSMs from their area of interest by using antennas or other devices [9]. This raises security issues and disturbs the privacy and anonymization of vehicles. An adversary can use this BSM information for bad intentions such as harming users or drivers, blackmailing or threatening them. These issues can cause hesitation in users or drivers and put their lives in danger [10].

Vehicle's anonymity and data privacy are critical factors that cannot be compromised. To avoid these issues, a trusted authority (TA) provides pseudonyms for vehicles. Vehicles use these pseudonyms to communicate with other vehicles or RSUs [11]. These pseudonyms are changed after some time so that if an adversary is tracking a vehicle, they cannot continually trace the target vehicle's whole trajectory. This provides security to some extent, but high pseudonym consumption makes pseudonyms insufficient. In this case, vehicles communicate to the TA directly or indirectly to issue a new set of pseudonyms [12]. This increases pseudonym consumption and computation overhead because only the TA keeps the link between the vehicle's true identity and pseudonym [13]. It also increases the BSM loss rate, and if any safety message is lost, it results in severe consequences. So, it is important to use pseudonyms economically.

This paper presents the efficient pseudonym consumption protocol (EPCP) to use pseudonyms effectively while maintaining vehicle anonymization. In this scheme, neighbor vehicles that exist in a close range, and have the same estimated new location, are considered to be relevant vehicles. A pseudonym-changing alert is broadcasted in an efficient way after calculating the required matrices. The main contributions of our work are as follows:

(1) We explore the literature on pseudonym-based anonymity assurance for messaging in the IoVs.
(2) Next, we propose a solution to estimate the next state of vehicles and their speed and direction before sending the pseudonym-changing alert.
(3) We also deal with the exchange of pseudonyms to reduce costs and ensure anonymity as well.
(4) Finally, simulations are performed to validate the results where the proposed scheme outperforms in contrast to three dominating schemes.

The remaining part of the manuscript is organized as follows: In Section 2, the literature is discussed on pseudonym-based schemes. Section 3 provides a system model and problem statement. Section 4 presents a proposed solution. Section 5 explores the performance of the EPCP. At the end, the conclusion and future work are discussed in Section 6.

2. Literature Review

Many pseudonym-based schemes are presented to enhance vehicle anonymization and provide protection against attacks planned by an adversary. These techniques are majorly divided into two main classes, mix-context-based schemes and mix-zone-based schemes. In this section, schemes of both categories are discussed.

2.1. Mix-Context-Based Schemes

In mix-context-based schemes, vehicles change their pseudonyms together in case specified triggers are satisfied. If such triggers are not fulfilled, vehicles will not change their pseudonyms. These schemes are also known as user-centric schemes.

In [14], Pan et al. proposed a cooperative pseudonym change based on the number of neighbors (CPN) protocol. The idea behind this scheme is that vehicles tend to change pseudonyms after getting triggered. This technique increases anonymity during dense traffic flow; however, it has high pseudonym consumption. Babaghyou et al. proposed a strategy [15] in which the transmission range of vehicles was restricted as per the speed of the neighboring vehicle. The advantage of this scheme is that safety-oriented messages are not neglected. The drawback of this scheme is that pseudonym consumption is high. Vehicles that change lanes also receive BSMs, which lessens the security.

To solve the problem of pseudonym-linking, Xinghua et al. presented a scheme in which vehicles exchange pseudonyms with each other. To exchange its pseudonym, the vehicle broadcasts the request message Req_i and transmits its virtual identity (VID) to the RSU. In case a nearby vehicle receives this, Req_i transmits an assist reply beacon containing all of the information to the RSU [16]. This technique increases the delinking ability among the most recent and former pseudonyms, which reduces the chance of tractability. The shortcomings of the technique include high communication and computation overhead.

To reduce packet loss and reduce adversary linking attacks, Zidani et al. [17] presented a scheme in which vehicles change pseudonyms in case there is a variation in speed and on the basis of surrounding vehicles. The most prominent achievement of this scheme is that it makes use of adaptive beaconing. When the beaconing interval varies, it creates high confusion for the adversary because the adversary cannot identify when vehicles communicate and share information. The benefit of this scheme is that the adversary cannot link correctly to the pseudonyms of target vehicles.

To enhance vehicle confidentiality, cooperative pseudonym exchange and scheme permutation (CPESP) [18] is presented. This technique is a mixture of two separate schemes consisting of cooperative pseudonym exchange (CPE) and scheme permutation (SP). In the first phase, the vehicles which are ready to swap their pseudonym may broadcast a BSM to neighbor vehicles for showing willingness. In scheme permutation, vehicles change their pseudonym using two methods, which are either RSP or the periodical pseudonym-changing procedure. One technique is selected for the time being. The SP technique is considered as being highly valuable in low road traffic. In this scheme, both CPE and SP algorithms work equally. The unutilized set of pseudonyms is used in a hybrid way where one technique is chosen as the RSP, and the periodical pseudonym is considered on behalf of the pseudonym-updating process. This technique has higher protection against linking attacks, and more schemes need to be added for increasing confusion for an adversary. In [19], the technique uses three types of pseudonyms including real, initial and new pseudonyms produced by the TA, RSU and onboard unit (OBU), correspondingly. Each pseudonym is allocated to vehicles before the authentication of the previous one. The advantage of this scheme is that a pseudonym-linking attack is not possible because a pseudonym is updated by three entities, but it increases computation overhead and has very high pseudonym consumption.

To enhance privacy and maintain low traceability, the context-adaptive privacy scheme (CADS) was proposed [20]. Vehicles switch to silence while changing pseudonym; however, this silent mode is smaller to prevent missing important safety-oriented messages. The benefit of this technique is that it much lessens adversary traceability. Another technique, dynamic grouping and virtual pseudonym-changing (DGVP), was recommended to increase anonymization. The idea behind this technique is that vehicles are clustered into groups and any one of them is chosen as the group leader (GL). Each group member is allotted a group identity (GID). When vehicles are higher than a threshold value, vehicles update their pseudonym, or else a virtual pseudonym-updating mechanism is introduced [21]. The benefit of this technique is that external vehicles cannot listen to information from other group members. The problem is that the computation cost rises during the virtual pseudonym exchange due to an extra beacon being created in it.

To reduce the traceability problem, another scheme named crowd-based mix context (CMC) was proposed, in which vehicles with heavy traffic broadcast beacon messages with PU = 1 notify other vehicles to change pseudonyms. When traffic is lower, two pseudonyms are generated and exchanged randomly with each other. The neighbors accept the correct pseudonym and the false one is excluded [22]. The benefit of this technique is that the adversary cannot trace the target vehicle for a long time successfully. The drawback of the technique is that it is applicable only to vehicles moving at low speeds.

In [23], vehicles tend to change pseudonyms in groups, and these groups are monitored by the group head (GH). Pseudonym consumption is lower in this strategy. In [24], the author proposed a mechanism to preserve vehicles' confidentiality throughout the journey to enhance the security of the VANET. When nodes come within the range of an RSU, it broadcasts a BSM. When neighbors receive this beacon, they send a BSM in return, including VID, pseudonym, location and speed. By using this information, the RSU confirms that vehicles are legal. Trip time informs when a vehicle departs from the current RSU. Afterward, trip time T_i is calculated using Equation (1). $Range_{RSU}$ shows the transmission range of the RSU while $Speed_{vehicle}$ represents the vehicle's speed. The vehicle's speed is checked against the threshold speed Vs; if it is less than this, the vehicle enters into the congestion detection phase and transmits a congestion awareness beacon. For the confirmation of congestion, the RSU waits for other vehicles to send congestion messages. The advantage of this scheme is that unauthorized vehicles are reported and quick action is taken so that the adversary cannot listen to the communication of the vehicle. The drawback of this technique is that it is only suitable in heavy traffic.

$$T_i = \frac{Range_{RSU}}{Speed_{vehicle}} \tag{1}$$

Yang et al. [25] presented a technique named the dynamic pseudonym swap zone (DPSZ), in which vehicles exchange their pseudonym by developing a temporary zone. In the case of any malicious activity, that vehicle's credentials are revoked, and its exchanging procedure is also revoked. After it, the target vehicle is notified about it, and then allotted with a novel pseudonym. It will protect nodes from attacks planned by the adversary. The nodes can check their capability to respond according to Equation (2). α shows the likelihood of vehicles to reply to the initiator, $|Þi|$ represents the neighbors of vi, μ is the vehicle's count to create a zone where vehicles can switch their pseudonyms and e is Euler's constant. When $|Þi;| \geq \mu$, in this case, nodes have little chance to response. This scheme is more secure against internal and external attacks. The weakness of this technique is that swapping occurs when vehicles reach a threshold μ. This perfect condition is not possible each time.

$$\alpha = \begin{cases} 1, |Þi| = \mu \\ e^{1 \frac{|Þi|}{\mu}}, |Þi| \geq \mu \end{cases} \tag{2}$$

During the silent mode, there is a great risk that vehicles are unable to receive safety beacons. In order to reduce this issue, vehicles update their pseudonym in the presence of k nodes. Furthermore, road traffic is dynamic and changes frequently; it enhances the anonymity set when more vehicles enter the silent mode. When the anonymity set increases, it ultimately increases adversary confusion. During time t, suppose k neighbors are available to change the pseudonym; then, at $t = t + at$ time, vehicles have a choice to freely decide whether to change their pseudonym or not. If the beacon is transmitted with probability p, it represents vehicles that want to update their pseudonym; this procedure is called flickering. In $t = t + nT$, vehicles set the beacon bit to HT = 1 and inform new neighbors. So, that vehicle updates the pseudonym together at $t = t + (n + 1)$ time. The duration of the silence mode decreases in comparison, to prevent bad effects on safety messages [26]. To prevent linking attacks and to increase privacy, another approach, the synchronized pseudonym-changing protocol (SPCP) [27], was proposed. In this scheme, vehicles change their pseudonym in the group that is monitored by a group head (GH). The advantage of this protocol is that it increases anonymization, and enhances the level of confusion for adversaries. The shortcoming of the scheme is that enormous storage is required for the TA so that the group record information can be handled easily.

2.2. Mix-Zone-Based Schemes

In mix-zone-based schemes, there are some zones (traffic signals, malls, marts, toll plazas) that are predefined. When vehicles enter these zones, they change their pseudonym. K.Emara et al. presented a scheme which allows vehicles to move into silent mode in case they enter the ideal region. When initiator vehicles find any silent node in their surroundings, they switch to silent mode too and then change their pseudonym [28]. This scheme proved to be better in the case of traceability. The drawback of the scheme is that the silent mode reduces safety-oriented applications. Li et al. [29] came up with a strategy to create a mix zone in the red traffic light. When vehicles stop at a red light, they become silent and change pseudonym. During a red light, not many essential beacons are neglected. Vehicles obtain active gain at green traffic lights. The scheme does not make a compromise on safety beacons during silent mode but is effective only with a high density. In [30], vehicles create a virtual cryptographic mix zone for changing pseudonym. In this zone, vehicles broadcast safety messages but in an encrypted format. After changing pseudonyms, vehicles exit from the zone. Safety messages are not neglected in this scheme but the decryption of beacons needs extra time. In [31], vehicles change pseudonyms in parking areas and shopping malls, and these places are considered as zones. Vehicles exit randomly from the zone, which increases the confusion of the adversary. In cases where zones are not available for a long time, vehicles will not change pseudonyms and the attacker can perform linking attacks on target vehicles easily.

In [32], one pseudonym is allotted per vehicle by the pseudonym certificate authority (PCA); after this, more pseudonyms are generated using a Gao algorithm. Pseudonym consumption is very low in this scheme but the randomization process is very challenging. In [33], when vehicles are in traffic, their speed is checked if it is slow (lies within 20 km/h to 40 km/h), and they check their neighbors. After ensuring the existence of neighbors, vehicles update their pseudonym. In order to encourage selfish nodes in the network to take part in the pseudonym-updating mechanism, a motivation procedure is used. Vehicles are given some incentive on changing pseudonym; if they will not change, their incentive value will be detected [34]. The benefit of the scheme is that it increases anonymity. The vehicular location privacy zone (VLPZ) is presented [35] in the network and it is divided into grids. Each grid contains zones where vehicles move and change pseudonyms. The entrance point is known as a router, and from which vehicles move into the zone and exit from the aggregator. The degree of anonymity is calculated using Equation (3), where d shows the degree of anonymity, k represents the capacity of the vehicular zone and $|AS|$ shows the occupancy of the vehicular zone. This scheme needs a separate RSU, which is expensive to deploy.

$$d = \frac{log2(|AS|)}{log2(k)} \tag{3}$$

In [36], vehicles opt for a group as per its velocity and change their pseudonym in cases where $S_{th} > 1$, where S_{th} represents the speed threshold. If a vehicle leaves a group to join another, it is also allowed to change pseudonym. The scheme is appropriate for long journeys but is not suitable for short distances.

3. System Model and Problem Statement

In this section, the system model of the proposed solution is described, which consists of four main entities which are the TA, vehicles, location-based server and RSU.

(1) The TA is used to allocate pseudonyms to vehicles when they enter the network. In case a vehicle is conducting suspicious activities in the network, after receiving the report from the RSU, the TA revokes the pseudonym of that vehicle. So, the main purpose of this entity is to allocate, revoke and keep the link between former and new pseudonyms.

(2) Vehicles are the basic components of the system model, which is equipped with the OBU, GPS and sensors. The vehicles can communicate with each other and the RSU for sharing safety beacons, and share pseudonym information and other information. During traveling on roads, vehicles need to know accurate information about their destination.

(3) The location-based server provides the following facilities: (i) inquiring about vehicle appeal to the RSU, (ii) sends a request to the location-based server (LBS) for providing accurate location information for moving to the desired destination.

(4) The RSU monitors traffic and informs vehicles about it in a timely manner. In this case, the pseudonyms are insufficient, and the RSU requests the TA to provide more numbers. In the case of malicious nodes in the network, the RSU instructs the TA to revoke its pseudonym. The system model of the proposed scheme is shown in Figure 2.

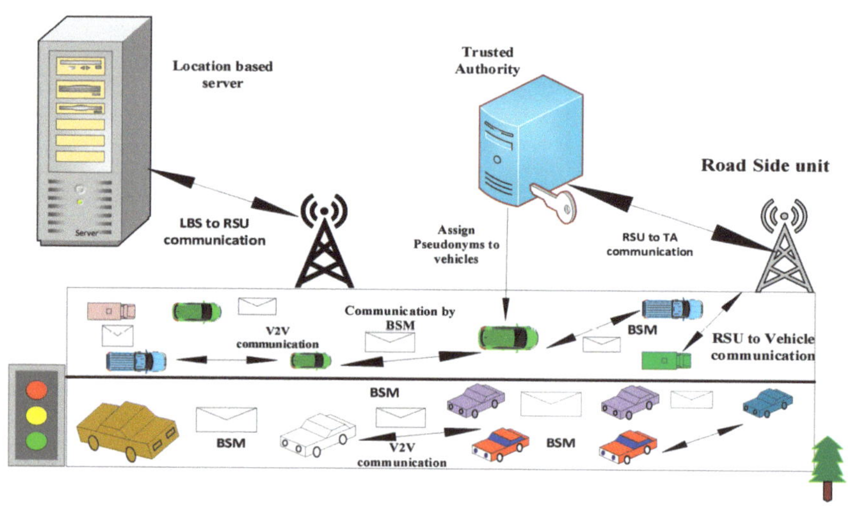

Figure 2. System models.

The core problem before broadcasting is that the vehicle's actual distance is not considered, only the speed of the vehicle is noticed, and the BSM is transmitted. The topology in the IoVs is very dynamic: vehicles move at different speeds and follow different routes and lanes. So, there is a high chance that vehicles that are neighbors at time t will no longer remain neighbors at time Δ + t due to the large distance. However, they still receive a BSM [15]. This problem has a bad impact on pseudonym consumption. High pseudonym consumption increases the chances of an important BSM loss rate. When irrelevant vehicles receive a BSM, it disturbs a vehicle's anonymity.

Adversary Model

An adversary is considered as somebody who spies on vehicles' BSMs to obtain information about a vehicle's location, direction and other sensitive information. The aim behind it is to threaten or trace drivers or passengers and follow the target vehicle's path. After receiving a BSM, an adversary attempts to extract with the vehicle's former pseudonym. With this aim, an adversary installs eavesdropping sensors into the trajectory to gain the BSMs. The adversary passively observes the BSMs from its area of interest but does not change the information available in the adversary model, as shown in Figure 3.

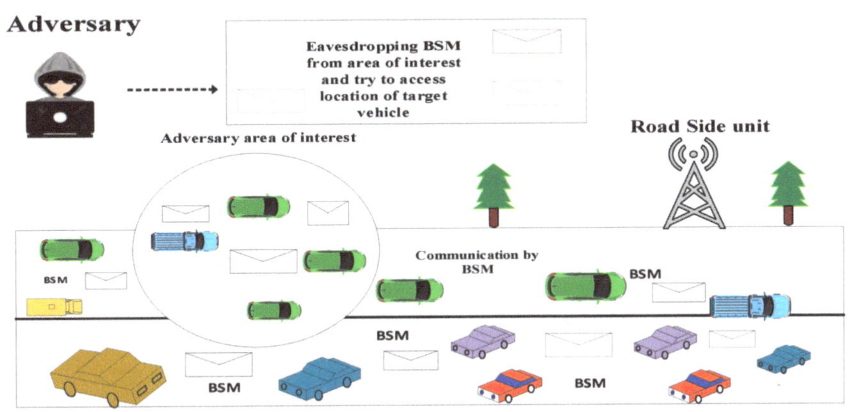

Figure 3. Adversary model.

4. Efficient Pseudonym Consumption Protocol

We present the proposed efficient pseudonym consumption protocol (EPCP) that aims for the efficient utilization of a pseudonym. Vehicles may change their pseudonym when vehicle v has more neighbors. For sparse traffic, vehicles exchange their pseudonyms to avoid pseudonym wastage as well as increase anonymity. Besides the mix-context trend on which the EPCP scheme is based, there are some other methods that pseudonym-changing techniques have used. The silence-based pseudonym-changing trend refers to those cases that become silent for some specified or random time to change pseudonym. During the silent mode, vehicles do not broadcast or receive any safety messages. Fixed-place changing pseudonyms are those that change pseudonym only in front of a traffic red light signal, in parking lots near malls or markets, at road junctions, etc. The group-based changing pseudonym trend refers to those schemes that make groups on the basis of some metric and pseudonym-changing mechanisms that occur within groups. Many cases have used encryption-based pseudonym-changing trends that refer to mechanisms in which vehicles use encrypted beacons to transmit within their transmission range. The receiving vehicles first decrypt the information and then change pseudonym simultaneously, if needed.

The developed solution of the EPCP can be used for smooth and secure long and short journeys. It can be beneficial for military fleets, as the adversary cannot track all of the information all of the time, while such privacy issues exist in traditional transportation. Additionally, the scheme can be implemented for vehicles used for medical emergencies, and for lawyers that have security threats. The EPCP scheme can also be deployed for riding services and public transport, as the proposed scheme is not much more expensive to implement. On the whole, the EPCP is effective to use in all scenarios where anonymity is the main concern of users and passengers.

Before sending a BSM, vehicle v checks some metrics. In the first phase, vehicle v checks its neighbors as per the BSM received in the previous timeframe. After this, the next state is estimated. If the state lies within the premises of a close range then vehicles are considered to be relevant ones that are following the same state.

In the second phase, the speed of vehicles v is checked against two threshold values in contrast to the neighboring vehicles. If the relevant vehicles are moving too slow or too fast, this means that soon they will be far away from the premises of vehicle v. This results in increasing BSM delay. If speed is according to vehicle v, then its direction is checked as the vehicles can change route due to notifications received from the RSU.

In the third phase, if a vehicle's *flagbit* is 1, then the pseudonym will be exchanged or changed as per the density of the road. In the case of sparse traffic when no vehicle lies in the close radius, then the pseudonym time is checked. After the expiry of the lifespan for the current pseudonym, the vehicle is allowed to change the pseudonym. To prevent a pseudonym-linking attack, we reduced the pseudonym lifespan in the proposed scheme. A list of notations used in this scheme is presented in Table 1.

Table 1. List of notations.

Sr.	Notation	Description
1.	k	Number of neighbors
2.	Neigh_dis	Neighbor distance
3.	Neigh_v	Vehicles in locality of vehicle v
4.	$\text{threshold}_{\text{min}}$	Minimum threshold speed
5.	threshold	Neighbor threshold value
6.	Vi	Vehicle v
7.	Vj	Neighboring vehicles
8.	thresholdmax	Maximum threshold speed
9.	Close_R	Close range
10.	N_direction	Direction of neighbor vehicles

The efficient pseudonym consumption algorithm is presented in Algorithm 1. In lines 1–7, when vehicle v obtains the BSM from its neighboring nodes, the position of the sending vehicles is extracted from the received BSM. If it lies within the transmission range, in this case, the BSM is kept; otherwise, it is discarded. The onboard unit of the vehicle helps it in interacting with nearby entities as well as sending and receiving BSMs. In the next time slot, vehicle v intends to send a BSM. After the beacon interval time, the BSM is prepared and important information about vehicle v is included in it. In lines 11–16, the BSM received in the previous time slot is checked, and if at least a single BSM of the vehicle is present, its next state is estimated. For the estimation of the next state, the *Kalman* filter is used. The difference between the present state and the estimated state is checked using Euclidean distance. If it lies in the close range then it is relevant and further parameters are checked.

In lines 20–29, the neighbor vehicle's speed is checked against two speed values. In other schemes, only one threshold value is used, with the reason behind using two values being that vehicles that are too slow or too fast will quickly leave the proximity of vehicle v and will not remain its neighbor. If the road traffic is dense, then the vehicle will change its pseudonym; otherwise, it will be exchanged. In the case of no vehicle existing in proximity, then, after the pseudonym lifetime of the vehicle has expired, the vehicle changes its pseudonym. The pseudonym lifetime is decreased to 50 s to avoid a pseudonym-linking attack.

Algorithm 1: Efficient Pseudonym Consumption Algorithm

```
//When intended vehicle v get BSM
N_position = BSM.pos ();
Neigh_dis = dis(my_position, N_position)
If (Neigh_dis ≤ T) then
    Neigh_v++
    store ← store + Neigh_v;
Else drop BSM.
End if
//intended vehicle v aims to disseminate BSM in upcoming timeslot
while (OBU status is active) do
        wait (beacon interval)
        Ready (BSM);
if (nodes ≥ k) then
        vehicles_trails ← kalman_filter(store);
          for i ← 1 to Neigh_v do
            if (Euclidean (vehicles_trails(i).pos, current_state.pos) ≤ Close_R) then
              adjacent ← adjacent + vehicles_trails(i);
            End if
          End for
          if (!adjacent.empty()) then
            Call Function Neighbor_speed ← BSM.speed()
            if (Neighbor_speed < threshold_min) OR (Neighbor_speed > threshold_max) then
              Call Function BSM (Delay)
Else
              N_direction = Call Function BSM_direction ()
              if (std:: equal(mine_direction, N_direction)) then
                  if (Neigh_v ≥ threshold &&((Neigh_v (Readyflag) && v_readyflag == 1)) then
                    Call Function Update cooperatively pseudonym ()
                    Set Readyflag_bit to 0
                  elseif (Neigh_v < threshold && ((Neigh_v (Readyflag) && v_readyflag == 1))
                        Random exchange of unused pseudonym (Vi, Vj)
                        Set Readyflag_bit to 0
                End if
              End if
          End if
If (adjacent.empty()) then
          Locality ← False //no vehicle is in transmission range of vehicle v
End if
If (v_pseudolife > stable_span) then
        Call Function Update pseudonym ();
        Set Readyflag_bit to 0
    End if
End if
End while
```

5. Results and Discussion

In this section, we present the simulation environment, results and related discussions. To validate the results, we performed extensive simulations using privacy extension (PREXT) [37]. It is built upon the veins framework [38] which includes two main modules, which are Object Modular Network Testbed (OMNet++) version 5.0 [39] for network construction and Simulation of Urban Mobility (SUMO) 0.25.0 [40] for traffic mobility scenarios, as in the real world. The map of Munich city was used by downloading it from Open Street Map (OSM). For creating the vehicles' route, randomTrips was employed. PREXT helped in analyzing crucial factors such as pseudonym consumption, traceability, normalized traceability and confusion rate, which are important factors from an anonymity perspective. For QoS, the BSM loss rate was checked. For simulation, a highway scenario was considered. The minimum and maximum speed thresholds were 5 m/s and 30 m/s, respectively. The base schemes were CPN [14], WHISPER [15] and DGVP [21]. A list of simulation parameters with respected values is shown in Table 2.

Table 2. Simulation parameters and values.

Parameters	Values
Simulation time	300 s
Number of vehicles	50, 100, 150, 200
Transmission range	300 m
Pseudonym stable time	50 s
Minimum speed threshold	5 m/s
Maximum speed threshold	30 m/s
Close range	100 m
Neighbor threshold	40
Operating system	Ubuntu 16.04
Coupling protocol	TraCi

5.1. *Average Percentage of Adversary Attains Traceability*

Traceability is a concept defined as the probability that an adversary will guess the target vehicle's path appropriately using a BSM [28]. If the adversary knows the traces of the target vehicle, this increases its vulnerabilities and security threats. The higher the traceability, the lower the vehicle anonymization. So, it is a crucial parameter from an anonymization perspective; simulation was performed five times, and the average was considered under sparse to dense traffic. Figure 4 shows that the proposed scheme of the EPCP achieved the lowest traceability compared to the base schemes. The reason behind high traceability in CPN is that the techniques do not make use of sufficient triggers for changing pseudonyms. The lack of opting for a suitable context raises the chances of high traceability. WHISPER has relatively low traceability compared to CPN, which limits the transmission range on the basis of the speed of nearby vehicles. In the case of DGVP, initially, the traceability rate surges to 30%, but when the vehicles' densities increase, the traceability factor starts dropping. The reason behind this is that this technique changes the pseudonym in groups. During sparse traffic, the few vehicles remain in the group and do not update the pseudonym until it has expired, whilst high-speed vehicles exit the group, making it easy for adversaries to trace vehicles. A crowd is formed as vehicle density increases, due to crowd vehicles changing to a slow speed and joining groups, changing the pseudonym together, which reduces the traceability factor. As can be observed, when the number of vehicles are 200, the traceability factor reduces to 7%. Our proposed EPCP checks multiple factors (direction, estimated next state of neighbors and direction) to minimize the chances of traceability. Besides this, in the EPCP, the pseudonym lifetime is also reduced to 50 s to lessen the possibility that an adversary creates a connection between a former and a new pseudonym correctly. In the case of sparse traffic on the road with 50 vehicles, CPN achieves 58.4% traceability, whereas WHISPER accomplishes 21.5% traceability, DGVP attains 30% and EPCP accomplishes 14.4% traceability.

5.2. *Average Percentage of Adversary Attains Normalized Traceability*

Some vehicles do not update their pseudonym, and mapping out such vehicles is very easy. Eliminating such vehicles provides a better privacy level. This concept is known as normalized traceability [28]. Under this metric, a simulation can be conducted. Figure 5 depicts that the EPCP has significantly low normalized traceability. Under sparse traffic (when the number of vehicles are 50), after excluding those vehicles, the traceability ratio is reduced in CPN, and it attains normalized traceability of 54.4%. WHISPER lay within 16.5%, DGVP achieved normalized traceability of 23% and the proposed scheme of the

EPCP had 9.5% normalized traceability. The results proved that EPCP and WHISPER had better normalized traceability in comparison to CPN.

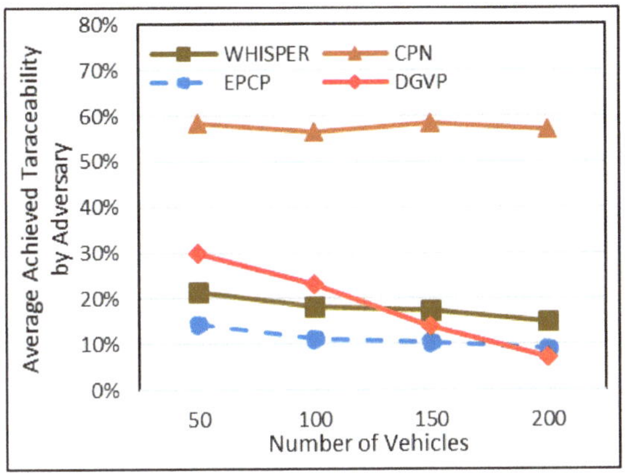

Figure 4. Average percentage of traceability in sparse to dense traffic scenario.

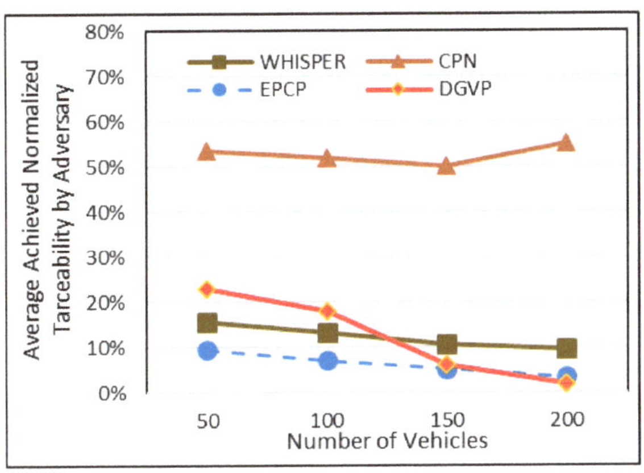

Figure 5. Average percentage of normalized traceability in sparse to dense traffic scenario.

5.3. Pseudonym Consumption

Vehicles interact with other entities using a pseudonym. The TA provides a pair of private and public information to vehicles when they enter into a network for registration. For a pseudonym, the public key is considered. Vehicles have a sufficient set of pseudonyms; so, they must be used wisely. In the case of low pseudonyms, vehicles appeal to the RSU to request the TA to allot them more pseudonyms. In return, the TA provides vehicles more pseudonyms through the RSU. This increases communication and computation overhead and makes the scheme costly to deploy. In CPN, pseudonym utilization is very high; the reason behind this is that when a vehicle wants to update its pseudonym, all neighboring nodes in the network also update their pseudonym even without any need, which ultimately raises pseudonym consumption. Moreover, vehicles also update their pseudonym when they meet a trigger (a trigger is a condition when k number of neighbors are present), and the value of k is kept as 2 within it. Although WHISPER has lower pseudonym consumption than CPN, it should be even less. The WHISPER scheme only uses the metric of speed before sending a BSM, and many neighbor vehicles can change their lanes after some time, but they still change their pseudonyms without any need. In DGVP, vehicles make use of two pseudonyms: one is original, and one is virtual. During the virtual method, two messages are generated with pseudonyms and are transmitted to member vehicles. This mechanism increases pseudonym utilization. The proposed scheme has lower pseudonym consumption, as shown in Figure 6, because only those vehicles that will remain for some time change pseudonyms. If such vehicles do not exist in the network, the BSM is delayed for some time to avoid the wastage of pseudonyms. During a dispersed distribution of vehicles on the road with 50 vehicles, the pseudonym utilization is 440 in CPN. For WHISPER, the pseudonym consumption is 103, in DGVP it lies in the range of 430 and in EPCP it remains at 50.

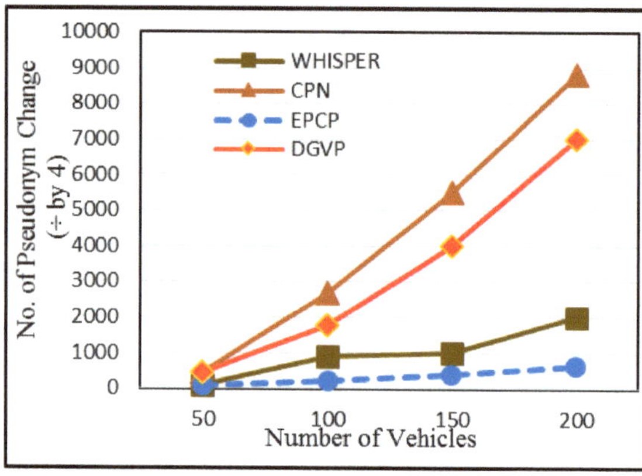

Figure 6. Pseudonym consumption.

5.4. BSM Loss Rate

Vehicles possess a limited buffer to store the beacons received from various entities. The vehicles receive irrelevant BSMs and may keep them for a long time. This results in filling the buffers, which causes emergency messages to be delayed or dropped. The existing techniques retain the BSMs of these vehicles in buffer that takes different paths at time $\Delta + t$ which are not useful now; this rises the BSM loss rate. In the proposed technique of the EPCP, when vehicles receive BSMs outside of the close area, it drops them, which lowers the BSM loss rate. Besides this, the proposed scheme generates and transmits BSMs at a very stable rate, which helps in overcoming the loss rate, while in the CPN and WHISPER strategies, both keep irrelevant BSMs, which increases the chances of important BSMs being lost. In DGVP, vehicles share information about safety only to their group members. So, when few vehicles lie within a group, the BSM loss rate is low, while it increases with increasing increments of group members. The BSM packet loss is stable in the EPCP, compared to WHISPER, DGVP and CPN, as presented in Figure 7. The value on the *X*-axis indicates the total number of vehicles, whereas values on the *Y*-axis represent the BSM loss rate. The loss rate in WHISPER lies within the range of 1500, 3400, 12,000 and 14,000, and the numbers of vehicles are 50, 100, 150 and 200, correspondingly. Similarly, the BSM loss in CPN is up to 4000, 8000, 15,000 and 18,000, with 50, 100, 150 and 200 vehicles. In DGVP, it remains at 2000, 4000, 13,500 and 16,500 under vehicle densities of 50, 100, 150 and 200. In the proposed scheme of the EPCP, the loss rate is up to 200, 1000, 5000 and 7000 under vehicle densities of 50, 100, 150 and 200. The result signifies that the EPCP has a lower loss rate than CPN and WHISPER.

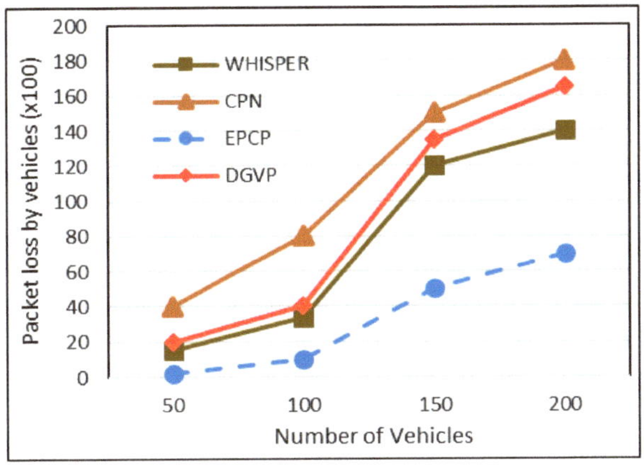

Figure 7. BSM packet loss rate.

5.5. Average Confusion for Attacker Due to Change in Pseudonym

By creating high confusion for an adversary, better anonymity can be achieved that ultimately increases the privacy level. Different vehicle densities (sparse, mediocre and dense) are shown in the *X*-axis, while the *Y*-axis shows the average confusion level for the adversary (the results are shown in Figure 8). The higher confusion rate in the EPCP is because direction and speed threshold factors are considered before sending a pseudonym-changing beacon. It adds the minimum and only relevant vehicles that overcome the possibility of attacks. Apart from this, in sparse situations, pseudonyms are exchanged randomly with each other so that pseudonyms should not be wasted and upsurge the confusion of attackers in tracing the target vehicle. In DGVP, during dense traffic, vehicles slow down their speed which increases the anonymity set, which increases confusion for the adversary in mapping out the target vehicle accurately in the case of the disperse distribution of traffic when the number of vehicles are 50. WHISPER accomplishes a value of 10.2, whereas the proposed scheme of the EPCP maintains an average value of 12.8, DGVP accomplishes an average value of 10.8 and CPN attains an average value of 5.2. During high traffic, the average confusion rate is up to 25.5, 30.5, 33.9 and 20.5 for WHISPER, EPCP, DGVP and CPN, respectively.

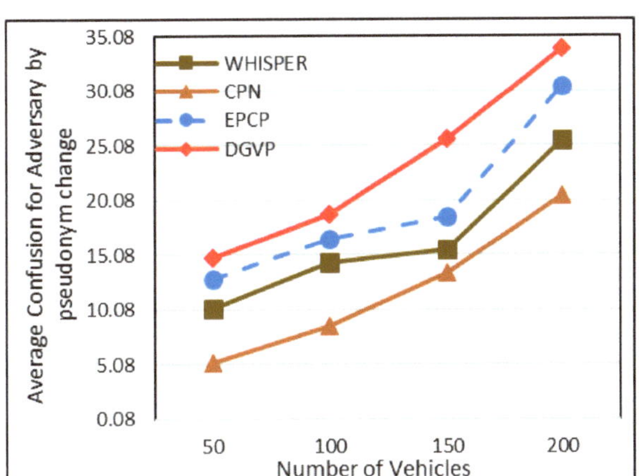

Figure 8. Average confusion for adversary according to pseudonym change.

5.6. Proportion of Vehicles That Changed Pseudonym

When a stable proportion of vehicles updates the pseudonym cooperatively, it surges the efficiency of the technique, while updating the pseudonym very frequently upsurges the communication and computation cost. In the context of the CPN, it had a very high proportion of vehicles that changed pseudonyms because of a trigger (a condition when two vehicles exist in the transmission range), and it changed pseudonyms.

The EPCP had a slightly low proportion of vehicles that changed pseudonyms; because of strict checks, some vehicles showed a lack of interest in changing their pseudonym. WHISPER had a worthy proportion of vehicles that cooperatively updated their pseudonyms. As far as the DGVP is concerned, initially it had a lower vehicle proportion of those changing pseudonyms, but when the density of traffic became heavy, the proportion of vehicles that changed their pseudonym significantly increased. Figure 9 shows that in the EPCP, the proportion of vehicles that changed their pseudonym remained at 70%, 75%, 77% and 80% under traffic of 50, 100, 150 and 200 vehicles, respectively. The WHISPER proportion lay under 77–89% in sparse and dense traffic. The CPN lay within the proportion of 82% in the case of sparse traffic, while this proportion increased up to 90% in dense traffic. In DGVP, the proportion remained at 65%, 71%, 85% and 91% with traffic of 50, 100, 150 and 200 vehicles, correspondingly.

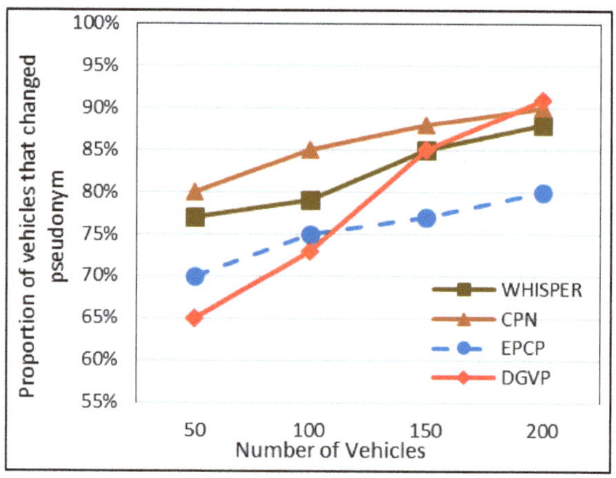

Figure 9. Proportion of vehicles that changed their pseudonym.

Overall, the performance of the proposed scheme remained stable under various metrics from sparse to dense traffic, but the shortcoming of the EPCP scheme is that slightly lower vehicles changed their pseudonym because of selfish nodes in the network. The WHISPER scheme performed fairly for most of the metrics. In the case of the CPN scheme, the pseudonyms were not well utilized, which ultimately increased the computation and communication overheads. As DGVP is a dense-based scheme, it outperforms in dense traffic, while the effectiveness is reduced in distributed traffic. So, DGVP is only acceptable to use in heavy traffic.

6. Conclusions

In this paper, a mix-context technique named the efficient pseudonym consumption protocol was proposed to reduce pseudonym utilization by sending beacons when relevant neighboring vehicles were present on the road. For this purpose, the next state of vehicles, their direction and their speed threshold were checked. In the proposed strategy, vehicles are allowed to exchange pseudonyms in lower traffic and change only when traffic is dense to utilize pseudonyms effectively. Simulation was performed to check the effectiveness of the proposed scheme of the EPCP under the PREXT simulator, along with OMNet++ and SUMO. The results showed that the proposed technique of the EPCP has better pseudonym consumption, a low BSM loss rate and a higher confusion rate for adversaries, and achieved low traceability and normalized traceability compared to the existing schemes of CPN, WHISPER and DGVP when traffic was sparse. The limitation of the scheme is that no motivation mechanism is introduced to encourage selfish nodes to participate in the pseudonym-changing process. For the proposed work, only external adversary was considered, which may not be very efficient for cases of internal adversary. In the near future, an encouragement-based mechanism will be introduced to motivate selfish nodes in the network to participate in the pseudonym-changing process to increase the proportion of vehicles. Besides this, a scenario of an internal adversary should also be checked when some internal entities, i.e., the vehicle or RSU, become semi-honest or malicious. Additionally, the communication cost of the proposed scheme should also be checked, and the EPCP should be compared with other anonymity-based schemes; these are a few of our upcoming plans.

Author Contributions: Conceptualization, M.M. (Mehreen Mushtaq), A.U., H.A. and N.J.; methodology, M.M. (Mehreen Mushtaq), A.U., H.A., N.J., M.M. (Mehedi Masud), A.A. and M.M.A.; software M.M. (Mehreen Mushtaq), A.U., H.A., N.J., M.M. (Mehedi Masud), A.A. and M.M.A.; formal analysis, M.M. (Mehreen Mushtaq), A.U., H.A., N.J., M.M. (Mehedi Masud), A.A. and M.M.A.; investigation, M.M. (Mehreen Mushtaq), A.U., H.A. and N.J.; resources, M.M. (Mehedi Masud), A.U., H.A. and N.J.; data curation M.M. (Mehreen Mushtaq), A.U., H.A. and N.J.; writing—original draft preparation, M.M. (Mehreen Mushtaq) and A.U.; writing—review and editing, M.M. (Mehreen Mush-taq), A.U., H.A., N.J., M.M. (Mehedi Masud), A.A. and M.M.A.; visualization, M.M. (Mehreen Mushtaq), A.U., H.A., N.J., M.M. (Mehedi Masud), A.A. and M.M.A.; supervision, M.M. (Mehedi Masud), A.U., H.A. and N.J.; funding acquisition, M.M. (Mehreen Mushtaq), A.U., H.A., N.J., M.M. (Mehedi Masud), A.A. and M.M.A. All authors have read and agreed to the published version of the manuscript.

Funding: The Research Groups Funding program grant (NU/RG/SERC/12/26), the Deanship of Scientific Research, Najran University, Saudi Arabia.

Institutional Review Board Statement: Not applicable.

Informed Consent Statement: Not applicable.

Data Availability Statement: Will be available on request.

Acknowledgments: The authors are thankful to the Deanship of Scientific Research at Najran University for funding this work under the Research Groups Funding program grant code (NU/RG/SERC/12/26).

Conflicts of Interest: The authors declare no conflict of interest.

References

1. Fadhil, J.A.; Sarhan, Q.I. Internet of Vehicles (IoV): A Survey of Challenges and Solutions. In Proceedings of the 2020 21st International Arab Conference on Information Technology (ACIT), Giza, Egypt, 28–30 November 2020; pp. 1–10. [CrossRef]
2. Sharma, S.; Kaushik, B. A survey on internet of vehicles: Applications, security issues & solutions. *Veh. Commun.* **2019**, *20*, 100182. [CrossRef]
3. Hasan, M.; Mohan, S.; Shimizu, T.; Lu, H. Securing vehicle-to-everything (V2X) communication platforms. *IEEE Trans. Intell. Veh.* **2020**, *4*, 693–713. [CrossRef]
4. Karagiannis, G.; Altintas, O.; Ekici, E.; Heijenk, G.; Jarupan, B.; Lin, K.; Weil, T. Vehicular networking: A survey and tutorial on requirements, architectures, challenges, standards and solutions. *IEEE Commun. Surv. Tutor.* **2011**, *13*, 584–616. [CrossRef]
5. Ji, B.; Zhang, X.; Mumtaz, S.; Han, C.; Li, C.; Wen, H.; Wang, D. Survey on the Internet of Vehicles: Network Architectures and Applications. *IEEE Commun. Stand. Mag.* **2020**, *4*, 34–41. [CrossRef]
6. Khan, M.Z.; Alhazmi, O.H.; Javed, M.A.; Ghandorh, H.; Aloufi, K.S. Reliable internet of things: Challenges and future trends. *Electronics* **2021**, *10*, 2377. [CrossRef]
7. Lin, K.; Li, C.; Li, Y.; Savaglio, C.; Fortino, G. Distributed learning for vehicle routing decision in software defined Internet of vehicles. *IEEE Trans. Intell. Transp. Syst.* **2020**, *22*, 3730–3741. [CrossRef]
8. Talat, H.; Nomani, T.; Mohsin, M.; Sattar, S. A Survey on Location Privacy Techniques Deployed in Vehicular Networks. In Proceedings of the 2019 16th International Bhurban Conference on Applied Sciences and Technology (IBCAST), Islamabad, Pakistan, 8–12 January 2019; Volume 13, pp. 604–613. [CrossRef]

9. Garg, T.; Kagalwalla, N.; Churi, P.; Pawar, A.; Deshmukh, S. A survey on security and privacy issues in IoV. *Int. J. Electr. Comput. Eng.* **2020**, *10*, 5409–5419. [CrossRef]
10. Wang, J.; Shao, Y.; Ge, Y.; Yu, R. A survey of vehicle to everything (V2X) testing. *Sensors* **2019**, *19*, 334. [CrossRef]
11. Talib, M.S.; Hassan, A.; Hussin, B.; Hassan, A.A.H. Vehicular Ad-hoc networks: Current challenges and future direction of research. *J. Adv. Res. Dyn. Control Syst.* **2018**, *10*, 2065–2074.
12. Ferrag, M.A.; Maglaras, L.; Ahmim, A. Privacy-Preserving Schemes for Ad Hoc Social Networks: A Survey. *IEEE Commun. Surv. Tutor.* **2017**, *19*, 3015–3045. [CrossRef]
13. Afzal, Z.; Kumar, M. Security of Vehicular Ad-Hoc Networks (VANET): A survey. *J. Phys. Conf. Ser.* **2020**, *1427*, 012015. [CrossRef]
14. Pan, Y.; Li, J. Cooperative pseudonym change scheme based on the number of neighbors in VANETs. *J. Netw. Comput. Appl.* **2013**, *36*, 1599–1609. [CrossRef]
15. Babaghayou, M.; Labraoui, N.; Ari, A.A.A.; Ferrag, M.A.; Maglaras, L.; Janicke, H. Whisper: A location privacy-preserving scheme using transmission range changing for internet of vehicles. *Sensors* **2021**, *21*, 2443. [CrossRef] [PubMed]
16. Li, X.; Zhang, H.; Ren, Y.; Ma, S.; Luo, B.; Weng, J.; Ma, J.; Huang, X. PAPU: Pseudonym Swap with Provable Unlinkability Based on Differential Privacy in VANETs. *IEEE Internet Things J.* **2020**, *7*, 11789–11802. [CrossRef]
17. Zidani, F.; Semchedine, F.; Ayaida, M. Estimation of Neighbors Position privacy scheme with an Adaptive Beaconing approach for location privacy in VANETs. *Comput. Electr. Eng.* **2018**, *71*, 359–371. [CrossRef]
18. Singh, P.K.; Gowtham, S.N.; Tamilselvan, S.; Nandi, S. CPESP: Cooperative Pseudonym Exchange and Scheme Permutation to preserve location privacy in VANETs. *Veh. Commun.* **2019**, *20*, 100183. [CrossRef]
19. Bouksani, W.; Bensaber, B.A. RIN: A dynamic pseudonym change system for privacy in VANET. *Concurr. Comput. Pract. Exp.* **2019**, *31*, 1–13. [CrossRef]
20. Emara, K.; Woerndl, W.; Schlichter, J. Context-based Pseudonym Changing Scheme for Vehicular Adhoc Networks. *arXiv* **2016**, arXiv:1607.07656.
21. Ullah, I.; Shah, M.A.; Khan, A.; Maple, C.; Waheed, A. Virtual pseudonym-changing and dynamic grouping policy for privacy preservation in vanets. *Sensors* **2021**, *21*, 3077. [CrossRef]
22. Ullah, I.; Shah, M.A.; Khan, A.; Maple, C.; Waheed, A.; Jeon, G. A distributed mix-context-based method for location privacy in road networks. *Sustainability* **2021**, *13*, 12513. [CrossRef]
23. Ullah, I.; Shah, M.A.; Khan, A. Adaptive Grouping and Pseudonym Changing Policy for Protection of Vehicles Location Information in VANETs. In Proceedings of the 2021 IEEE Symposium Series on Computational Intelligence (SSCI), Orlando, FL, USA, 5–7 December 2021; pp. 1–7. [CrossRef]
24. Wahid, A.; Yasmeen, H.; Shah, M.A.; Alam, M.; Shah, S.C. Holistic approach for coupling privacy with safety in VANETs. *Comput. Netw.* **2019**, *148*, 214–230. [CrossRef]
25. Yang, M.; Feng, Y.; Fu, X.; Qian, Q. Location privacy preserving scheme based on dynamic pseudonym swap zone for Internet of Vehicles. *Int. J. Distrib. Sens. Netw.* **2019**, *15*, 1–16. [CrossRef]
26. Zhang, Z.; Feng, T.; Sikdar, B.; Wong, W.C. A Flickering Context-based Mix Strategy for Privacy Protection in VANETs. In Proceedings of the IEEE International Conference on Communications, Montreal, QC, Canada, 14–23 June 2021; pp. 1–6. [CrossRef]
27. Weerasinghe, H.; Fu, H.; Leng, S.; Zhu, Y. Enhancing unlinkability in vehicular ad hoc networks. In Proceedings of the 2011 IEEE International Conference on Intelligence and Security Informatics, ISI, Beijing, China, 10–12 July 2011; pp. 161–166. [CrossRef]
28. Emara, K.; Woerndl, W.; Schlichter, J. CAPS: Context-aware privacy scheme for VANET safety applications. In Proceedings of the 8th ACM Conference on Security & Privacy in Wireless and Mobile Networks, WiSec, New York, NY, USA, 22–26 June 2015; pp. 1–12. [CrossRef]
29. Li, Y.; Yin, Y.; Chen, X.; Wan, J.; Jia, G.; Sha, K. A Secure Dynamic Mix Zone Pseudonym Changing Scheme Based on Traffic Context Prediction. *IEEE Trans. Intell. Transp. Syst.* **2021**, *23*, 9492–9505. [CrossRef]
30. Ying, B.; Makrakis, D.; Mouftah, H.T. Dynamic mix-zone for location privacy in vehicular networks. *IEEE Commun. Lett.* **2013**, *17*, 1524–1527. [CrossRef]
31. Lu, R.; Lin, X.; Luan, T.H.; Liang, X.; Shen, X. Pseudonym changing at social spots: An effective strategy for location privacy in VANETs. *IEEE Trans. Veh. Technol.* **2012**, *61*, 86–96. [CrossRef]
32. Xu, X.; Huang, Q.; Zhu, H.; Sharma, S.; Zhang, X.; Qi, L.; Bhuiyan, M.Z.A. Secure service offloading for internet of vehicles inSDN-enabled mobile edge computing. *IEEE Trans. Intell. Transp. Syst.* **2020**, *22*, 3720–3729. [CrossRef]
33. Shaleesh, I.; Almohammedi, A.; Mohammad, N.; Ahmad, A.; Shepelev, V. Cooperation and radio silence strategy in Mix Zone to Protect Location Privacy of Vehicle in VANET. *Tikrit J. Eng. Sci.* **2021**, *28*, 31–39. [CrossRef]
34. Ying, B.; Makrakis, D. Reputation-based Pseudonym Change for Location Privacy in vehicular networks. *IEEE Int. Conf. Commun.* **2015**, *10*, 7041–7046. [CrossRef]
35. Boualouache, A.; Senouci, S.M.; Moussaoui, S. Towards an efficient pseudonym management and changing scheme for vehicular ad-hoc networks. In Proceedings of the 2016 IEEE Global Communications Conference (GLOBECOM), Washington, DC, USA, 4–8 December 2016; pp. 1–7. [CrossRef]
36. Ullah, I.; Wahid, A.; Shah, M.A.; Waheed, A. VBPC: Velocity Based Pseudonym Changing Strategy to Protect Location Privacy of Vehicles in VANET. In Proceedings of the 2017 International Conference on Communication Technologies (ComTech), Rawalpindi, Pakistan, 19–21 April 2017; pp. 132–137.

37. Emara, K. Poster: PREXT: Privacy extension for Veins VANET simulator. In Proceedings of the IEEE Vehicular Networking Conference (VNC), Columbus, OH, USA, 8–10 December 2016; pp. 3–5. [CrossRef]
38. Sommer, C.; Eckhoff, D.; Brummer, A.; Buse, D.S.; Hagenauer, F.; Joerer, S.; Segata, M. *Veins: The Open Source Vehicular Network Simulation Framework*; Springer: Berlin/Heidelberg, Germany, 2019; ISBN 9783030128425.
39. Varga, A.; Hornig, R. An overview of the OMNeT++ simulation environment. In Proceedings of the SIMUTools 2008—1st International ICST Conference on Simulation Tools and Techniques for Communications, Networks and Systems, Marseille, France, 3–7 March 2008. [CrossRef]
40. Krajzewicz, D.; Erdmann, J.; Behrisch, M.; Bieker, L. Recent Development and Applications of SUMO—Simulation of Urban MObility. *Int. J. Adv. Syst. Meas.* **2012**, *5*, 128–138.

sensors

Article

Trajectory Tracking Coordinated Control of 4WID-4WIS Electric Vehicle Considering Energy Consumption Economy Based on Pose Sensors

Yiran Qiao, Xinbo Chen * and Zhen Liu

School of Automotive Studies, Tongji University, Shanghai 201804, China
* Correspondence: chenxinbo@tongji.edu.cn; Tel.: +86-021-69589120

Abstract: In order to improve the stability and economy of 4WID-4WIS (four-wheel independent drive—four-wheel independent steering) electric vehicles in trajectory tracking, this paper proposes a trajectory tracking coordinated control strategy considering energy consumption economy. First, a hierarchical chassis coordinated control architecture is designed, which includes target planning layer, and coordinated control layer. Then, the trajectory tracking control is decoupled based on the decentralized control structure. Expert PID and Model Predictive Control (MPC) are employed to realize longitudinal velocity tracking and lateral path tracking, respectively, which calculate generalized forces and moments. In addition, with the objective of optimal overall efficiency, the optimal torque distribution for each wheel is achieved using the Mutant Particle Swarm Optimization (MPSO) algorithm. Additionally, the modified Ackermann theory is used to distribute wheel angles. Finally, the control strategy is simulated and verified using Simulink. Comparing the control results of the average distribution strategy and the wheel load distribution strategy, it can be concluded that the proposed coordinated control not only provides good trajectory tracking but also greatly improves the overall efficiency of the motor operating points, which enhances the energy economy and realizes the multi-objective coordinated control of the chassis.

Keywords: 4WID-4WIS EVs; trajectory tracking control; multi-objective coordinated control; mutant particle swarm optimization (MPSO)

Citation: Qiao, Y.; Chen, X.; Liu, Z. Trajectory Tracking Coordinated Control of 4WID-4WIS Electric Vehicle Considering Energy Consumption Economy Based on Pose Sensors. *Sensors* **2023**, *23*, 5496. https://doi.org/10.3390/s23125496

Academic Editor: Ikhlas Abdel-Qader

Received: 10 May 2023
Revised: 29 May 2023
Accepted: 9 June 2023
Published: 11 June 2023

1. Introduction

4WID-4WIS EV is a novel electric vehicle that reduces mechanical components, such as differentials and half shafts. It also integrates four-wheel steering technology based on a distributed four-wheel drive system. This allows four wheels to be driven and steered independently, increasing the controllable degree of freedom. In a word, 4WID-4WIS EV has unique advantages in vehicle dynamics control [1,2].

To fully exploit the control potential of 4WID-4WIS EV and improve the overall performance of the vehicle, coordinated multi-objective control of the chassis has become a current research focus [3–5]. In particular, stability and economy are two important performance characteristics. It is difficult to improve one performance by controlling only one performance, and in extreme cases, the effect deteriorates. Therefore, there is a need for coordinated control of the two performances to make the vehicle stable and economical at the same time, which has been researched and studied by several scholars [6]. In [7], a 4WID EV torque coordination control strategy is developed, which considers both stability and economy and uses MPC to calculate the generalized force. The demand torque of each wheel is determined by an online solution using the control distribution error, tire utilization rate, and power consumption of the drive system as objective functions. In [8], a multi-objective online optimization of energy management strategy for 4WID EV is proposed. It considers the efficiency of the drive system, tire slip energy consumption, wheel torque fluctuation, yaw rate tracking error, etc. The weights of each element are dynamically adjusted using the fuzzy control method, and the effectiveness of the strategy is verified by simulations. In [9], a cooperative game-based actuator fault-tolerant control strategy for a 4WID EV is proposed. This strategy minimizes tire energy dissipation to ensure economic efficiency by designing a two-dimensional game controller that simultaneously satisfies the generalized forces required for vehicle stability.

Due to the great potential of intelligent driving technology to enhance vehicle safety [10,11], improve traffic efficiency, and reduce energy consumption [12], the research on trajectory tracking control of 4WID-4WIS electric vehicles has received increasing attention in the automotive industry, and numerous control methods have been developed [13–15]. In [16], LTV-MPC (Linear Time-Varying Model Predictive Control) based on DYC (Direct Yaw Control) is used to realize velocity tracking and trajectory tracking, which improves stability during trajectory tracking. In [17], a robust path-tracking controller for 4WID-4WIS agricultural robotic vehicles is designed by combining the back-stepping SMC (Sliding Mode Control) theory to improve the robustness of trajectory tracking. In [18], a four-wheel steering controller and a velocity tracking controller for 4WID-4WIS EV are designed based on the SMC algorithm to improve the accuracy of trajectory tracking. In [19], the feedback linear quadratic regulator (LQR) controller is used to realize emergency avoidance under 4WS high-speed conditions.

There are certain shortcomings in the aforementioned research which can be summarized as follows. Most studies on chassis coordinated control is related to traditional 4WID EV, while there is relatively little research on 4WID-4WIS EV. In addition, most existing research establishes multi-objective optimization problems and solves them online to obtain the control variables. This not only makes the objective function too complex and difficult to solve but also has a negative impact on the real-time performance of the controller. Last but not least, many studies on the trajectory tracking control of 4WID-4WIS EV mostly only consider the trajectory tracking performance. Some studies also consider stability, but there is almost no research that simultaneously considers trajectory tracking performance, stability and economy.

To improve the stability and economy of 4WID-4WIS EV in trajectory tracking, the contributions of this paper are summarized as follows:

- A hierarchical architecture for chassis coordinated control is designed including target planning layer, and coordinated control layer.
- The MPC path tracking controller is constructed, which takes into account stability constraints, such as yaw rate and tire slip angle.
- The MPSO algorithm is also used to create a mapping of the distribution coefficients by optimizing the torque distribution between the front and rear wheels offline, thereby reducing the computational cost.

The paper is organized as follows: Section 2 proposes a hierarchical chassis coordination control. Section 3 highlights the simulation validation and compares the simulation results of different torque distribution strategies. Section 4 sums up the conclusions.

2. Trajectory Tracking Coordinated Control

2.1. Vehicle Model

4WID-4WIS EV uses in-wheel motors and steering motors to replace the traditional transmission and steering mechanism, giving it more control freedom and thus enabling more advanced control methods to improve the overall performance of the vehicle. Considering the strong coupling between the vehicle subsystems and the strong nonlinearity of the tires and motor actuators, it is of great significance to establish a reasonable and accurate vehicle model to study the control strategy of 4WID-4WIS EV.

2.1.1. Vehicle Dynamic Model

The vehicle is a complex multi-body dynamic system, and the vehicle dynamic model is the basis for dynamic analysis, active control, function realization, and parameter optimization. According to the research objective and concern, many scholars have proposed various linear and nonlinear vehicle models with different degrees of complexity [20,21]. This work focuses on the longitudinal and lateral coupling control of the 4WID-4WIS EV. The coupling effects between longitudinal motion and lateral motion of the vehicle include three categories: kinematic coupling, tire force coupling, and load transfer coupling. Kinematic coupling refers to the longitudinal motion being influenced by the longitudinal component of the steering wheel lateral deflection force due to the presence of the wheel steering angle. Additionally, the lateral motion is also influenced by the longitudinal velocity. Tire force coupling is the interaction between tire lateral and longitudinal forces, the combined force of which is constrained by the friction ellipse. The vertical load redistribution is caused by longitudinal acceleration or lateral acceleration, which in turn influences the longitudinal and lateral dynamics. Therefore, a vehicle dynamic model with seven degrees of freedom is adopted, i.e., longitudinal motion, lateral motion, yaw motion, and the rotation of four wheels.

The vehicle is assumed to be left-right symmetric about the center plane and driven on a flat horizontal road, ignoring the vertical motion of the body. At the same time, it is assumed that the suspension system is a rigid structure and the body pitch and roll motion is neglected. After conducting the above modifications, the simplified dynamic model is developed, as shown in Figure 1.

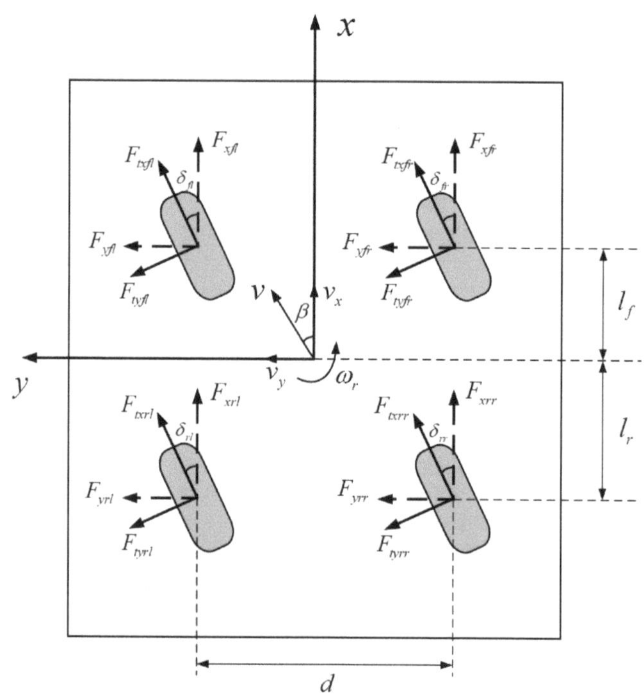

Figure 1. 4WID-4WIS vehicle dynamic model.

The longitudinal motion equation is:

$$m(\dot{v}_x - v_y\omega_r) = F_{xfl} + F_{xfr} + F_{xrl} + F_{xrr} \tag{1}$$

The lateral motion equation is:

$$m(\dot{v}_y + v_x\omega_r) = F_{yfl} + F_{yfr} + F_{yrl} + F_{yrr} \tag{2}$$

The equation of yaw motion is:

$$I_z\dot{\omega}_r = \left(F_{xfr} + F_{xrr} - F_{xfl} - F_{xrl}\right)\frac{d}{2} + \left(F_{yfl} + F_{yfr}\right)l_f - \left(F_{yrl} + F_{yrr}\right)l_r \tag{3}$$

where,

$$F_{xij} = F_{txij}\cos\delta_{ij} - F_{tyij}\sin\delta_{ij} \tag{4}$$

$$F_{yij} = F_{txij}\sin\delta_{ij} + F_{tyij}\cos\delta_{ij} \tag{5}$$

$$i \in \{f, r\},\ j \in \{l, r\} \tag{6}$$

The equation of wheel motion is:

$$I_w\dot{\omega}_{ij} = T_{dij} - F_{txij}R_{eff} - T_{bij} \tag{7}$$

where, m is the vehicle mass, v_x is the longitudinal velocity, v_y is the lateral velocity, ω_r is the yaw rate, F_{xij} and F_{yij} are the longitudinal force and lateral force of each wheel, respectively, I_z is the yaw inertia, l_f is the distance from the center of mass to the front axle, l_r is the distance from the center of mass to the rear axle, d is the wheelbase of the vehicle, F_{txij} and F_{tyij} are the longitudinal and lateral forces of each wheel in the tire coordinate system, respectively, δ_{ij} is the wheel angle, I_w is the wheel moment of inertia, ω_{ij} is the speed of each wheel, T_{dij} is the driving torque of each wheel, T_{bij} is the braking torque of each wheel, and R_{eff} is the effective rolling radius of the wheel.

2.1.2. Tire Model

As the only component that connects the vehicle to the ground, any state of vehicle motion depends on the interaction forces between the tire and the road surface. The tire has a critical impact on the vehicle's performance, so it is necessary to perform accurate dynamic modeling of tires to better describe the effects of tire forces on vehicle dynamics. The tire model describes the relationship between the tire motion parameters and tire forces, which can be mainly divided into three types: theoretical model, empirical model, and semi-empirical model [22,23]. The theoretical model is formed by studying the deformation mechanism of tires based on the physical essence of tire mechanics, but its complex structure is not conducive to simulation research. The empirical model is based on the experimental tire data, but as a result, it also lacks theoretical support and is poorly scalable. The semi-empirical model combines the advantages of both models with theoretical support and experimental data to obtain key parameters that ensure good accuracy and scalability.

In this study, the most commonly used semi-empirical tire model MF (Magic Formula) is selected. The combination formula of trigonometric functions is used to describe the relationship between tire force, slip rate, and slip angle. This model is suitable for the operating condition with combined longitudinal and lateral forces. The general expression of the MF tire model is as follows:

$$\begin{cases} y = D\sin\{C\arctan[Bx - E(Bx - \arctan(Bx))]\} \\ Y(X) = y(x) + S_V \\ x = X + S_H \end{cases} \tag{8}$$

where, B is the stiffness factor, C is the shape factor, D is the peak factor, E is the curvature factor, Y is the output variable, that is, longitudinal force, lateral force or aligning torque, X is the input variable, that is, slip rate or slip angle, S_V is the vertical offset, and S_H is the horizontal offset.

As input variables for the MF tire model, slip rate and side slip angle are important parameters for calculating the tire effect, which must be calculated from the vehicle state as shown in Equations (9) and (10), respectively.

$$s_{ij} = \begin{cases} \frac{\omega_{ij}R_{eff}-u_{ij}}{\omega_{ij}R_{eff}} & \omega_{ij}R_{eff} \geq u_{ij}(Drive) \\ \frac{\omega_{ij}R_{eff}-u_{ij}}{u_{ij}} & \omega_{ij}R_{eff} < u_{ij}(Brake) \end{cases} \tag{9}$$

where, u_{ij} is the speed of the wheel center.

$$\begin{cases} \alpha_{fl} = \tan^{-1}\left(\frac{v_y+\omega_r l_f}{v_x-\omega_r \cdot d/2}\right) - \delta_{fl} \\ \alpha_{fr} = \tan^{-1}\left(\frac{v_y+\omega_r l_f}{v_x+\omega_r \cdot d/2}\right) - \delta_{fr} \\ \alpha_{rl} = \tan^{-1}\left(\frac{v_y-\omega_r l_r}{v_x-\omega_r \cdot d/2}\right) - \delta_{rl} \\ \alpha_{rr} = \tan^{-1}\left(\frac{v_y-\omega_r l_r}{v_x+\omega_r \cdot d/2}\right) - \delta_{rr} \end{cases} \tag{10}$$

2.1.3. Motor Model

The 4WID-4WIS EV drive system consists of four in-wheel motors, and independent four-wheel steering is achieved by four steering motors. Therefore, to investigate the longitudinal and lateral coordinated control strategy, the characteristics of the motors are first analyzed, and suitable motor models are established.

1. In-wheel motor model

The research focuses on the coordinated control of the vehicle with multiple objectives. For the in-wheel motor model, the charging and discharging efficiency of the motor is only for the steady-state characteristics. The transient characteristics of the motor are not considered. Therefore, the charging and discharging efficiency of the motor can be expressed as in Equation (11).

$$\eta_{em} = \eta(n_{em}, T_{em}) \tag{11}$$

where, n_{em} is the motor speed, and T_{em} is the motor torque.

Since the driving state of the vehicle is mainly studied in this paper, the motor efficiency characteristics are shown in Figure 2. The external characteristic curve and the efficiency map are used to characterize the in-wheel motor. The external characteristic curve can be used to determine the maximum torque as a function of wheel speed in real-time. The discharge efficiency of the motor can be determined using a two-dimensional look-up table of motor speed and torque.

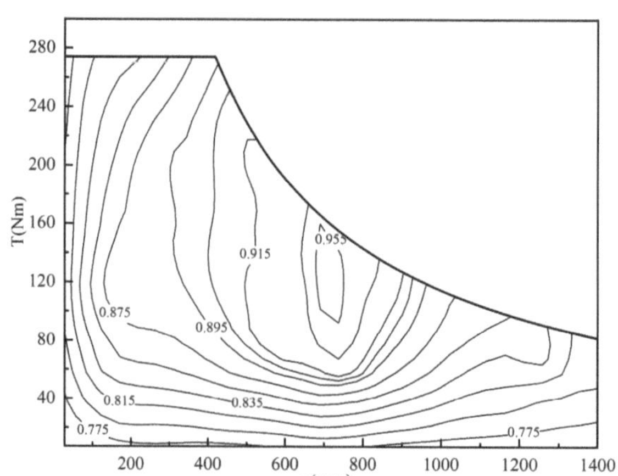

Figure 2. In-wheel motor efficiency map.

2. Steering Motor Model

Since the steering motor in this study is not focused on the efficiency characteristics, it is simplified from the perspective of the motor actuation effect as a wheel angle tracking model. The delay caused by the steering mechanism is considered as shown in Equation (12).

$$\delta_{ij\_out} = \delta_{ij\_req} \cdot \frac{1}{1 + \tau s} \tag{12}$$

where, δ_{ij_out} is the actual output angle, δ_{ij_req} is the required output angle, and τ is the response time constant.

2.2. Vehicle State Acquisition

Before performing chassis coordination control, it is necessary to observe the lateral and longitudinal states of the vehicle, including the vehicle's position information and the body's pose information. The position information of the vehicle consists of the coordinates of the vehicle's lateral and longitudinal axes, velocity and acceleration information, and the pose information of the body includes yaw rate, sides lip angle, pitch angle, roll angle and so on.

The implementation of the control strategy is based on the acquisition of vehicle state information by pose sensors. There have been many studies on how to obtain the pose information of the vehicle. For example, RTK (Real Time Kinematic) and INS (Inertial Navigation System) are combined for positioning. Body combination sensors, including longitudinal acceleration sensors, lateral acceleration sensors and yaw rate sensors are used for pose calculation.

Considering that the signals from pose sensors contain uncertain noise interference, it is necessary to design a reasonable signal filter to obtain accurate pose signals. At the same time, the acquisition of road condition information, such as road adhesion coefficient and road slope, can be realized by designing corresponding state observers based on the information from pose sensors. Since the focus of this paper is on the trajectory tracking coordinated control, this part will not be elaborated in detail.

2.3. Chassis Control Architecture

Based on the vehicle states obtained by pose sensors, the trajectory tracking controller receives the expected path and velocity information sent by the decision planning layer and controls the longitudinal and lateral movement of the vehicle so that the vehicle follows the expected path. In addition, 4WID-4WIS EV has multiple degrees of freedom, and the driving and steering of each wheel are controllable, which provides a basis for realizing multi-objective coordinated control of the chassis. In summary, this research aims to achieve the expected trajectory tracking while improving driving stability and energy economy.

The chassis control architecture designed in this paper is shown in Figure 3, which has a layered structure and is divided into target planning layer and coordinated control layer.

Based on the decentralized control structure, the target planning layer receives the expected path and velocity information and decouples the longitudinal and lateral control targets. In the longitudinal direction, an expert PID control method is used to track the desired velocity and output the generalized longitudinal force. In the lateral direction, the MPC algorithm is used to perform the multi-objective real-time rolling optimization considering the stability constraints, such as yaw rate and tire slip angle. It outputs the generalized yaw moment and the generalized wheel angle.

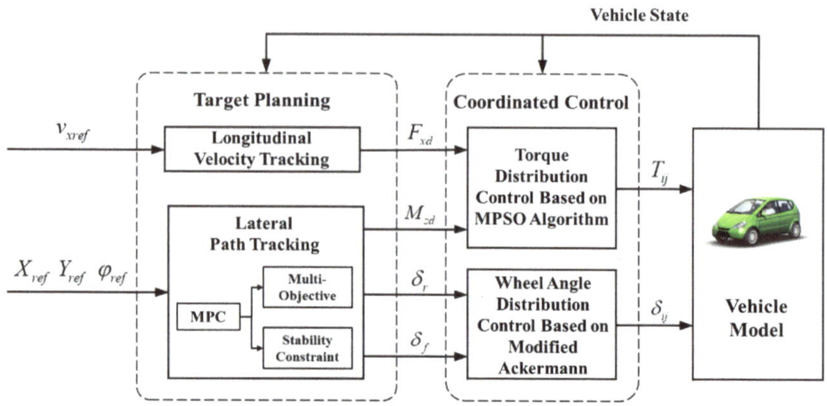

Figure 3. The chassis coordinated control architecture.

The coordinated control layer distributes the torque and wheel angle of each wheel according to the generalized force, generalized yaw moment, and generalized wheel angle, which realizes the calculation of the control objective of the motor actuator. Considering the efficiency characteristics of the in-wheel motor, the torque distribution control uses the MPSO algorithm to optimize the torque of the front and rear wheels offline. Then, the two-dimensional diagram of the optimal torque distribution coefficient is generated and real-time table look-up is performed according to the vehicle state to calculate the torque for each wheel. The wheel angle distribution strategy considers the wheel slip angle and distributes the generalized wheel angle based on the modified Ackerman theory to obtain the wheel angles. Finally, the torque and angle of each wheel are output to in-wheel motors and steering motors to realize the closed-loop control of the strategy.

2.4. Target Planning

The target planning layer completes the resolution of the chassis control targets based on the expected trajectory. The expected vehicle trajectory includes information about the expected path and velocity, i.e., the expected coupling of the vehicle's longitudinal and lateral motion states. Trajectory tracking strategies can be divided into decentralized control and centralized control according to different control structures. Decentralized control [24,25] refers to the decoupling of longitudinal and lateral vehicle motion, i.e., it decomposes the trajectory tracking problem into longitudinal velocity tracking and lateral path tracking problems and designs corresponding control objectives. Centralized control [26–28], on the other hand, means that the longitudinal and lateral motion in the trajectory tracking problem is considered uniformly, with a global controller computing the longitudinal and lateral control targets. The centralized trajectory tracking controller considers the system holistically and can better account for the coupling properties between longitudinal and lateral motion control. However, the higher dimensionality and complexity of the system model make it more challenging to design control laws than decentralized control, and it is also more computationally intensive and costly to implement physically. In contrast, the decentralized trajectory tracking strategy is relatively simple and easy to implement in engineering practice. In this work, the trajectory tracking strategy adopts a decentralized structure, decouples the trajectory tracking problem into a longitudinal velocity tracking problem and a lateral path tracking problem, and designs the tracking strategy to calculate the desired generalized force and moment.

2.4.1. Longitudinal Velocity Tracking

The uncertainty of vehicle system parameters, uncertain external disturbances, and nonlinear coupling between systems lead to difficulties in tuning PID control parameters and low robustness of dynamic and steady-state performance of vehicle velocity tracking [29]. In this work, the expert PID algorithm is used in conjunction with the expert system theory to track the longitudinal velocity.

The expert PID algorithm is based on various knowledge of the controlled object and the control laws. If the precise model of the controlled object is not available, the PID parameters are designed using expert experience. The deviations and increments of deviations are used to determine the current state of the controlled system. Rules are designed to adjust the output of the controller to achieve faster and smoother convergence of the controlled system.

Assume that $e(k) = v(k) - v_d(k)$ is the velocity error, and $\Delta e(k) = e(k) - e(k-1)$ is the velocity error increment. At the same time, based on engineering practice experience, the maximum deviation value is set as M_{max}, the middle deviation value is set as M_{mid} and the minimum deviation value is set as $M_{\min}$. According to the setting of deviation, deviation increment and extreme value of deviation, the setting rules are as follows:

(1) When $|e(k)| > M_{max}$, it means that the velocity error is unacceptably large. At this time, the controller should be directly output at full load, that is, $u(k) = F_{\mathrm{xmax}}$.

(2) When $e(k) * \Delta e(k) > 0,\ \Delta e(k) = 0$, it means that the velocity deviation is changing in the direction of increasing the absolute value of the deviation, or the deviation is a certain fixed value, then

$$u(k) = u(k-1) + k_1\{k_i e(k) + k_p \Delta e(k) + k_d \Delta\Delta e(k)\} \tag{13}$$

If $|e(k)| > M_{mid}$, the velocity deviation is also large, and it may be considered to increase the output gain k_1 of the controller.

(3) When $e(k)\Delta e(k) < 0,\ \Delta e(k)\Delta e(k-1) > 0, e(k) = 0$, it means that the absolute value of the velocity deviation is changing in the direction of decreasing, or has reached the equilibrium state. Then, the controller output remains unchanged, that is, $u(k) = u(k-1)$.

(4) When $e(k)\Delta e(k) < 0,\ \Delta e(k)\Delta e(k-1) < 0$, it means that the velocity deviation is in the limit state, then,

$$u(k) = u(k-1) + k_2 k_i e(k) \tag{14}$$

If $|e(k)| > M_{mid}$, it means that the velocity deviation is also large, and it may be considered to increase the output gain k_2 of the controller.

(5) When $|e(k)| < M_{min}$, it means that the absolute value of the velocity deviation is very small. In order to reduce the static error of the system, PI control is implemented:

$$u(k) = u(k-1) + k_i e(k) + k_p \Delta e(k) \tag{15}$$

To sum up, the generalized longitudinal force is expressed as:

$$F_{xd}(k) = u(k) \tag{16}$$

2.4.2. Lateral Path Tracking

Intelligent driving faces a complex and changing environment where conditions, such as critical safety constraints and actuator constraints, must be met [30–32]. In this paper, MPC is used to realize the lateral path tracking considering the driving stability constraints.

1. Predictive model design

The predictive model is the basis of the MPC used to predict the future output of the controlled system. The vehicle is a complex coupled nonlinear dynamic model. More degrees of freedom can improve the modeling accuracy, but the complexity of the model will also increase, making it difficult to meet the requirements for fast model solutions and real-time control. Therefore, dynamic modeling must strike a balance between improving model accuracy and reducing model complexity. Considering the accuracy and real-time requirements for lateral control of vehicles, a nonlinear three degrees-of-freedom vehicle model with lateral and longitudinal coupling is established in this paper to realize the prediction of system output as shown in Figure 4.

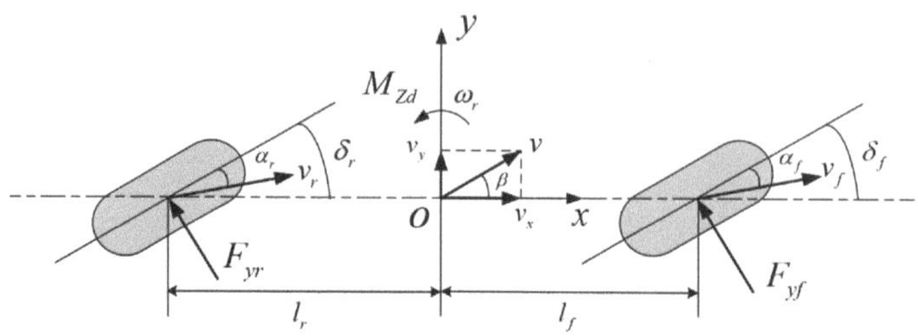

Figure 4. Nonlinear three degrees-of-freedom vehicle model.

Based on the established dynamic model, the force analysis is performed:

$$\begin{cases} m(\dot{v}_y + v_x\omega_r) = C_f\left(\frac{v_y+l_f\omega_r}{v_x} - \delta_f\right) + C_r\left(\frac{v_y-l_r\omega_r}{v_x} - \delta_r\right) - mv_x\omega_r \\ I_z\dot{\omega}_r = l_fC_f\left(\frac{v_y+l_f\omega_r}{v_x} - \delta_f\right) - l_rC_r\left(\frac{v_y-l_r\omega_r}{v_x} - \delta_r\right) + M_{zd} \\ \dot{\varphi} = \omega_r \\ \dot{x} = v_x\cos\varphi - v_y\sin\varphi \\ \dot{y} = v_x\sin\varphi + v_y\cos\varphi \end{cases} \tag{17}$$

where, φ is the heading angle, C_f and C_r are the equivalent cornering stiffnesses of the front axle and the rear axle, respectively, δ_f is the front wheel angle, δ_r is the rear wheel angle, M_{zd} is the generalized yaw moment, x and y are the longitudinal and lateral coordinates of the centroid in the geodetic coordinate system, respectively.

In the path tracking process, the desired heading angle and lateral coordinate can be obtained from the longitudinal coordinate of the current vehicle position. The state variable X, the control variable u and the output variable Y of the controller can be shown as:

$$\begin{aligned} X &= \left[v_y, \varphi, \omega_r, y, x\right]^T \\ u &= \left[\delta_f, \delta_r, M_{zd}\right]^T \\ Y &= [\varphi, y]^T \end{aligned} \tag{18}$$

Assuming that the longitudinal velocity remains constant during path tracking, in order to meet the real-time requirements of high-speed control, this paper uses Taylor's formula to linearize the nonlinear system. The Taylor expansion is performed at the reference point, and the high-order differential components are rounded off. Then, the discretization is carried out using the first-order difference quotient method, as shown in Equation (19).

$$X(k+1) = aX(k) + bu(k-1) + b\Delta u(k) + d(k) \tag{19}$$

where,

$$\begin{aligned} a &= I + T\tfrac{\partial f}{\partial X},\ b = T\tfrac{\partial f}{\partial u} \\ d(k) &= X_r(k+1) - aX_r(k) - bu_r(k) \end{aligned} \tag{20}$$

the augmented state variables are constructed:

$$\xi(k+1) = \begin{bmatrix} X(k+1) \\ u(k) \end{bmatrix} = \begin{bmatrix} a & b \\ 0_{N_u\times N_x} & I_{N_u\times N_u} \end{bmatrix}\xi(k) + \begin{bmatrix} b \\ I_{N_u\times N_u} \end{bmatrix}\Delta u(k) + \begin{bmatrix} d(k) \\ 0_{N_u\times 1} \end{bmatrix} \tag{21}$$

where, N_x is the number of state variables, and N_u is the number of control variables. Equation (21) can be simplified as

$$\xi(k+1) = A\xi(k) + B\Delta u(k) + \bar{d}(k) \tag{22}$$

where,

$$A = \begin{bmatrix} a & b \\ 0_{N_u \times N_x} & I_{N_u \times N_u} \end{bmatrix}, B = \begin{bmatrix} b \\ I_{N_u \times N_u} \end{bmatrix}, \bar{d}(k) = \begin{bmatrix} d(k) \\ 0_{N_u \times 1} \end{bmatrix} \tag{23}$$

Therefore, the discretized output equation can be obtained as:

$$Y(k+1) = C\xi(k+1) \tag{24}$$

where,

$$C = \begin{bmatrix} 0 & 1 & 0 & 0 & 0 & 0 & 0 & 0 \\ 0 & 0 & 0 & 1 & 0 & 0 & 0 & 0 \end{bmatrix} \tag{25}$$

The prediction time-domain length is set as p, and the control time-domain length is set as c ($p \geq c$). Assuming that the control increment outside the control time domain is zero, the predicted output equation of the system can eventually be obtained by iteration, as shown in Equation (26).

$$Y_p(k) = \Phi\xi(k) + \Theta\Delta U(k) + \Gamma D(k) \tag{26}$$

where,

$$Y_p(k) = \begin{bmatrix} Y(k+1|k) \\ Y(k+2|k) \\ Y(k+3|k) \\ \vdots \\ Y(k+c|k) \\ \vdots \\ Y(k+p|k) \end{bmatrix}, \Delta U(k) = \begin{bmatrix} \Delta u(k) \\ \Delta u(k+1) \\ \Delta u(k+2) \\ \vdots \\ \Delta u(k+c-1) \end{bmatrix}, D(k) = \begin{bmatrix} \bar{d}(k) \\ \bar{d}(k+1) \\ \bar{d}(k+2) \\ \vdots \\ \bar{d}(k+c-1) \\ \vdots \\ \bar{d}(k+p-1) \end{bmatrix} \tag{27}$$

$$\Phi = \begin{bmatrix} CA \\ CA^2 \\ CA^3 \\ \vdots \\ CA^c \\ \vdots \\ CA^p \end{bmatrix}, \Theta = \begin{bmatrix} CB & 0_{N_y \times N_u} & 0_{N_y \times N_u} & \cdots & 0_{N_y \times N_u} \\ CAB & CB & 0_{N_y \times N_u} & \cdots & 0_{N_y \times N_u} \\ CA^2B & CAB & CB & \cdots & 0_{N_y \times N_u} \\ \vdots & \vdots & \vdots & \vdots & \vdots \\ CA^{c-1}B & CA^{c-2}B & CA^{c-3}B & \cdots & CB \\ \vdots & \vdots & \vdots & \vdots & \vdots \\ CA^{p-1}B & CA^{p-2}B & CA^{p-3}B & \cdots & CA^{p-c}B \end{bmatrix} \tag{28}$$

$$\Gamma = \begin{bmatrix} C & 0_{N_y \times (N_x+N_u)} & 0_{N_y \times (N_x+N_u)} & \cdots & 0_{N_y \times (N_x+N_u)} & \cdots & 0_{N_y \times (N_x+N_u)} \\ CA & C & 0_{N_y \times (N_x+N_u)} & \cdots & 0_{N_y \times (N_x+N_u)} & \cdots & 0_{N_y \times (N_x+N_u)} \\ CA^2 & CA & C & \cdots & 0_{N_y \times (N_x+N_u)} & \cdots & 0_{N_y \times (N_x+N_u)} \\ \vdots & \vdots & \vdots & \vdots & \vdots & \vdots & \vdots \\ CA^{c-1} & CA^{c-2} & CA^{c-3} & \cdots & C & \cdots & 0_{N_y \times (N_x+N_u)} \\ \vdots & \vdots & \vdots & \vdots & \vdots & \vdots & \vdots \\ CA^{p-1} & CA^{p-2} & CA^{p-3} & \cdots & CA^{p-c} & \cdots & C \end{bmatrix} \tag{29}$$

2. Objective function design

The control objective of lateral path tracking is to ensure that the vehicle tracks the desired path smoothly, accurately, and quickly. Therefore, the objective function in this paper is designed as shown in Equation (30).

$$J = \left[Y_p(k) - Y_{ref}\right]^T Q_Q \left[Y_p(k) - Y_{ref}\right] + \Delta U(k)^T R_R \Delta U(k) \tag{30}$$

where, $Q_Q = I_p \otimes Q$, $R_R = I_c \otimes R$, Q is the weight coefficient matrix of the output of the control system, R is the weight coefficient matrix of the control increment, and $\otimes$ represents the Kronecker product.

The first in the objective function represents the deviation of the heading angle and the deviation of the lateral displacement, which characterizes the accuracy of the vehicle in tracking the desired path. The second limits the control increment, which not only ensures the stability and continuity of the control but also takes into account the response capability of the actuator. Equation (30) can be simplified as,

$$J = [E + \Theta\Delta U(k)]^T Q[E + \Theta\Delta U(k)] + \Delta U(k)^T R\Delta U(k) \tag{31}$$

where,

$$E = \Phi\xi(k) + \Gamma D(k) - Y_{ref} \tag{32}$$

Considering that E is a constant matrix at each sampling moment, it can be omitted. The final objective function in standard quadratic programming form is obtained, as shown in Equation (33).

$$J = \Delta U(k)^T \left(\Theta^T Q\Theta + R\right)\Delta U(k) + 2E^T Q\Theta\Delta U(k) \tag{33}$$

3. Constraints design

A major advantage of MPC over other control methods is that it handles constraints better. MPC is a rolling optimization process that can naturally incorporate constraints into the optimization problem. For the path-following controller in this paper, the constraints are mainly considered in the form of the constraint on the control increment, the constraint on the control set, and the stability constraints.

In order to make the control process more stable and improve the stability and comfort of the vehicle when tracking the desired path, the control increment should be constrained. Due to the constraints of the actuator, there are also constraints on the control variables, such as front wheel angle, rear wheel angle, and generalized yaw moment. So, in summary, the following can be said:

$$U_{\min}(k) \le U_{ct}(k-1) + A_{ct} * \Delta U(k) \le U_{\max}(k) \tag{34}$$

where,

$$U_{ct}(k-1) = \begin{bmatrix} u(k-1) \\ u(k-1) \\ \vdots \\ u(k-1) \end{bmatrix}, \; A_{ct} = \begin{bmatrix} 1 & 0 & \cdots & 0 \\ 1 & 1 & \cdots & 0 \\ \vdots & \vdots & \vdots & \vdots \\ 1 & 1 & 1 & 1 \end{bmatrix} \otimes I_c \tag{35}$$

To ensure that the vehicle has good stability while tracking the desired path, stability constraints must be imposed on the MPC optimization problem. The stability constraints generally start with the two parameters yaw rate and tire slip angle. Therefore, the constraint output variable is defined as $Y_b = [\omega_r, \alpha_r]^T$. The expression for the constraint output can be obtained, as shown in Equation (36).

$$Y_b(k+1) = C_b\xi(k+1) \tag{36}$$

where,

$$C_b = \begin{bmatrix} 0 & 0 & 1 & 0 & 0 & 0 & 0 & 0 \\ \frac{1}{v_x} & 0 & -\frac{l_r}{v_x} & 0 & 0 & 0 & -1 & 0 \end{bmatrix} \tag{37}$$

The predicted output equation for the stability constraint is shown in Equation (38).

$$Y_{p,b}(k) = \Phi_b\xi(k) + \Theta_b\Delta U(k) + \Gamma_b D(k) \tag{38}$$

where,

$$Y_{p,b}(k) = \begin{bmatrix} Y_b(k+1|k) \\ Y_b(k+2|k) \\ Y_b(k+3|k) \\ \vdots \\ Y_b(k+c|k) \\ \vdots \\ Y_b(k+p|k) \end{bmatrix}, \ \Phi_b = \begin{bmatrix} C_bA \\ C_bA^2 \\ C_bA^3 \\ \vdots \\ C_bA^c \\ \vdots \\ C_bA^P \end{bmatrix} \tag{39}$$

$$\Theta_b = \begin{bmatrix} C_bB & 0_{N_{y1}\times N_u} & 0_{N_{y1}\times N_u} & \cdots & 0_{N_{y1}\times N_u} \\ C_bAB & C_bB & 0_{N_{y1}\times N_u} & \cdots & 0_{N_{y1}\times N_u} \\ C_bA^2B & C_bAB & C_bB & \cdots & 0_{N_{y1}\times N_u} \\ \vdots & \vdots & \vdots & \vdots & \vdots \\ C_bA^{c-1}B & C_bA^{c-2}B & C_bA^{c-3}B & \cdots & C_bB \\ \vdots & \vdots & \vdots & \vdots & \vdots \\ C_bA^{p-1}B & C_bA^{p-2}B & C_bA^{p-3}B & \cdots & C_bA^{p-c}B \end{bmatrix} \tag{40}$$

$$\Gamma_b = \begin{bmatrix} C_b & 0_{N_{y1}\times(N_x+N_u)} & 0_{N_{y1}\times(N_x+N_u)} & \cdots & 0_{N_{y1}\times(N_x+N_u)} & \cdots & 0_{N_{y1}\times(N_x+N_u)} \\ C_bA & C_b & 0_{N_{y1}\times(N_x+N_u)} & \cdots & 0_{N_{y1}\times(N_x+N_u)} & \cdots & 0_{N_{y1}\times(N_x+N_u)} \\ C_bA^2 & C_bA & C_b & \cdots & 0_{N_{y1}\times(N_x+N_u)} & \cdots & 0_{N_{y1}\times(N_x+N_u)} \\ \vdots & \vdots & \vdots & \vdots & \vdots & \vdots & \vdots \\ C_bA^{c-1} & C_bA^{c-2} & C_bA^{c-3} & \cdots & C_b & \cdots & 0_{N_{y1}\times(N_x+N_u)} \\ \vdots & \vdots & \vdots & \vdots & \vdots & \vdots & \vdots \\ C_bA^{p-1} & C_bA^{p-2} & C_bA^{p-3} & \cdots & C_bA^{p-c} & \cdots & C_b \end{bmatrix} \tag{41}$$

Based on the road adhesion conditions and the tire slip angle limits, it can be shown in Equation (42).

$$Y_{b,\min} \le \Phi_b\xi(k) + \Theta_b\Delta U(k) + \Gamma_b D(k) \le Y_{b,\max} \tag{42}$$

In summary, the finite time domain optimization problem for MPC in this paper can be transformed into a standard quadratic programming problem:

$$\begin{aligned} &\min_{\Delta U(k)} \Delta U(k)^T(\Theta^T Q\Theta + R)\Delta U(k) + 2E^T Q\Theta\Delta U(k) \\ &s.t. \quad \Delta U_{\min}(k) \le \Delta U(k) \le \Delta U_{\max}(k) \\ &\qquad U_{\min}(k) \le U_{ct}(k-1) + A_{ct} * \Delta U(k) \le U_{\max}(k) \\ &\qquad Y_{b,\min} \le \Phi_b\xi(k) + \Theta_b\Delta U(k) + \Gamma_b D(k) \le Y_{b,\max} \end{aligned} \tag{43}$$

The optimal sequence of control increments is obtained by solving the standard quadratic programming problem, and the first element $\Delta u^*(k)$ is output. The control variable at the current moment can be described as Equation (44).

$$u(k) = u(k-1) + \Delta u^*(k) \tag{44}$$

2.5. Coordinated Control

The coordinated control layer receives the generalized force, generalized yaw moment, and generalized wheel angle from the upper layer. The wheel torque and the wheel angle are calculated and output to in-wheel motors and steering motors. Depending on the actuating subsystem, the coordinated control layer consists of two parts: torque distribution control and wheel angle distribution control.

2.5.1. Torque Distribution Control

The torque distribution control receives the generalized longitudinal force and the generalized yaw moment output from the upper level and distributes and controls the driving torque of the individual wheels. The current research in torque distribution mainly considers the stability index and economic index as the control objective and uses online real-time optimization to complete the torque coordinated distribution [33,34]. However, current research has the following limitations:

- The rule-based torque distribution control does not consider the operating efficiency of motor, resulting in unnecessary power loss.
- The real-time optimization of torque distribution places a large burden on the controller. The solution speed may be slow and this problem may even be unsolvable under certain working conditions.
- The economic evaluation generally uses the size of the control variables as the index, ignoring the efficiency characteristics of the motor. This is mainly due to the nonlinearity of the motor model, which makes it impossible to optimize the system efficiency in real-time.

To address these issues, this paper simplifies the real-time optimization problem of torque distribution in order to balance practical and optimization objectives. A combination of rule-based strategy and offline optimization strategy is used for the distribution strategy. A rule-based strategy is designed to distribute the torque between the left and right wheels. Considering the stability constraints, the optimization of the torque distribution coefficients between the front and rear wheels based on the MPSO algorithm leads to an optimal wheel torque, improving the efficiency while ensuring the response speed of the vehicle. First, the torque distribution between left and right wheels is carried out based on the generalized longitudinal force and generalized yaw moment requirements. Then, the MPSO algorithm is used to optimize the torque distribution offline between the front and rear wheels on the same side to obtain the optimal distribution coefficient of the front axle. Finally, the driving torque of each wheel is calculated.

1. Left-right distribution

A rule-based strategy is used for the generalized longitudinal force to distribute it equally between the left and right sides of the vehicle. The corresponding demand torques are obtained as follows.

$$T_{dlF} = T_{drF} = \frac{F_{xd}}{2} R_{eff} \tag{45}$$

where, T_{dlF} and T_{drF} are the left-hand and right-hand demand torques, respectively, corresponding to the generalized longitudinal forces.

The same uniform distribution is used for the generalized yaw moment. The yaw moment generated by the wheels on both sides is equal and has an opposite direction, which can shorten the response time of the vehicle and speed up the response speed. It can be represented as Equation (46).

$$\begin{aligned} T_{dlM} &= -\tfrac{M_{zd}}{d} R_{eff} \\ T_{drM} &= \tfrac{M_{zd}}{d} R_{eff} \end{aligned} \tag{46}$$

where, T_{dlM} and T_{drM} are the left-hand and right-hand demand torques, respectively corresponding to the generalized yaw moment.

The total demand torques are,

$$\begin{aligned} T_{dl} &= \tfrac{F_{xd}}{2} R_{eff} - \tfrac{M_{zd}}{d} R_{eff} \\ T_{dr} &= \tfrac{F_{xd}}{2} R_{eff} + \tfrac{M_{zd}}{d} R_{eff} \end{aligned} \tag{47}$$

where, T_{dl} and T_{dr} are the left-hand and right-hand demand torques, respectively.

2. Front-rear distribution

For the front and rear wheels on the same side, the total demand torque is fixed. The independent drive function allows the torque of the front and rear wheels to be optimally distributed so that in-wheel motors can operate in the high-efficiency range as far as possible. Considering the high time complexity of online optimization, this paper uses the MPSO algorithm to optimize the distribution coefficient of the front axle offline, and it is put into a tabular form for subsequent table lookup operations which ensures the real-time performance of the system.

(1) The MPSO algorithm

PSO (Particle Swarm Optimization) is a type of swarm intelligent optimization algorithm [35]. In PSO, each solution of the optimization problem is abstracted as a particle, and all particles search for the optimal solution in the solution space. Each particle is assigned a fitness function to determine the fitness of the current location, and a speed property to determine the distance and direction of flight, after which the optimal solution is determined by iteration.

The traditional PSO algorithm has the advantage of fast convergence, but it can easily fall into local optimal solutions in some complex situations. Therefore, the MPSO algorithm is introduced to avoid the problem of premature convergence. The MPSO algorithm combines the traditional PSO algorithm with the idea of mutation in the genetic algorithm. The mutation occurs when the population location is updated, thereby jumping out of the local optimal solution, which is conducive to finding the global optimal solution and reduces the probability of premature convergence. The process of the MPSO is shown in Figure 5.

(2) Distribution coefficient optimization

Since the four in-wheel motors are the same, the offline optimization problems on the left and right sides are essentially the same, so this paper takes the right-side wheel as an example. In order to carry out the optimization, the economy needs to be characterized and this paper uses the overall motor efficiency as the economic index. The expression of the overall motor efficiency is,

$$J = \frac{T_{fr}n_{fr} + T_{rr}n_{rr}}{\frac{T_{fr}n_{fr}}{\eta_{fr}} + \frac{T_{rr}n_{rr}}{\eta_{rr}}} \tag{48}$$

where, T_{fr} and T_{rr} are the output torques of the right front and right rear motors, respectively, n_{fr} and n_{rr} are the output speeds of the right front and right rear motors, respectively, and η_{fr} and η_{rr} are the output efficiencies of the right front and right rear motors, respectively.

To simplify the complexity of the optimization problem, the concept of the front axle distribution coefficient λ is introduced, which is the ratio of the front wheel torque to the total demand torque.

During normal driving, there is little difference in speed between the front and rear wheels on the same side, so the optimization problem can be translated into Equation (49).

$$\min J = \frac{\lambda}{\eta_{fr}} + \frac{1-\lambda}{\eta_{rr}} \tag{49}$$

Considering the constraints on the external characteristics of the in-wheel motor, the following constraints are made:

$$\begin{array}{l} 0 \leq \lambda T_{dr} \leq T_{\max} \\ 0 \leq (1-\lambda)T_{dr} \leq T_{\max} \end{array} \tag{50}$$

where, $T_{\max}$ is the peak torque of the in-wheel motor.

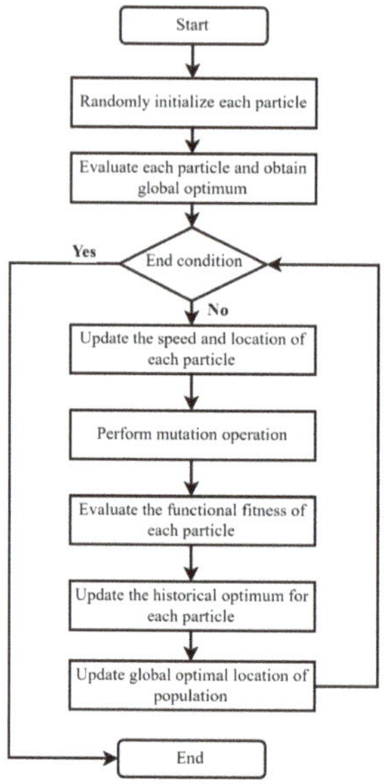

Figure 5. Flowchart of MPSO algorithm.

In addition, depending on the vehicle model, the center of mass is closer to the front axle, which means that the vertical load on the front axle is greater than that on the rear axle. This results in a greater tire adhesion limit for the front wheel, with a greater longitudinal force limit. Therefore, the front wheels should take on a larger portion of the required torque. The following restriction applies to the distribution coefficient of the front axle.

$$0.5 \leq \lambda \leq 1 \tag{51}$$

Factors affecting the front axle distribution coefficient include the velocity and the demand torque, wherein the velocity is represented by wheel speed. Based on the MPSO algorithm, the offline optimization of the front axle distribution coefficient is realized by programming in MATLAB. Finally, the mapping of the optimal distribution coefficient for the front axle is shown in Figure 6. The optimal distribution coefficient for the front axle is determined by the velocity and the demand torque, which realizes the torque distribution between the front and rear wheels on the same side. From the optimization results, when the demand torque is low, the front wheel drive is selected; when the demand torque is high, it tends to be four-wheel drive.

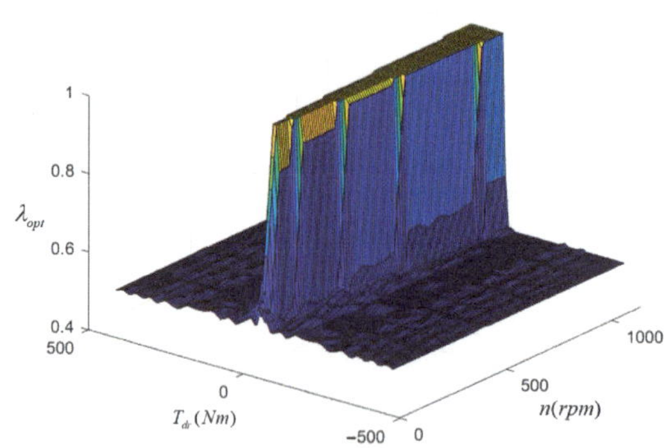

Figure 6. Optimal front axle distribution coefficient.

(3) Torque calculation for each wheel

Based on the optimal coefficient for the distribution of front axle torque determined via offline optimization, the torque for each wheel can be calculated. The offline optimization of the front axle distribution coefficients takes into account the constraints on the peak torque of the in-wheel motors but still requires stability corrections according to the constraints on road adhesion, as shown in Equation (52).

$$\begin{cases} T_{fl} = \min\left(|\lambda_l T_{dl}|, \mu F_{zfl}\right) \cdot \mathrm{sgn}(T_{dl}) \\ T_{fr} = \min\left(|\lambda_r T_{dr}|, \mu F_{zfr}\right) \cdot \mathrm{sgn}(T_{dr}) \\ T_{rl} = \min(|(1-\lambda_l)T_{dl}|, \mu F_{zrl}) \cdot \mathrm{sgn}(T_{dl}) \\ T_{rr} = \min(|(1-\lambda_r)T_{dr}|, \mu F_{zrr}) \cdot \mathrm{sgn}(T_{dr}) \end{cases} \tag{52}$$

where, λ_l and λ_r are the optimal front axle distribution coefficients for the left and right side, respectively. T_{ij} is the torque of each wheel, μ is the coefficient of road adhesion, and F_{zij} is the vertical load of each wheel, which can be obtained using Equation (53).

$$\begin{cases} F_{zfl} = \frac{mgl_r}{2l} - \frac{ma_x h}{2l} - \frac{ma_y hl_r}{dl} \\ F_{zfr} = \frac{mgl_r}{2l} - \frac{ma_x h}{2l} + \frac{ma_y hl_r}{dl} \\ F_{zrl} = \frac{mgl_f}{2l} + \frac{ma_x h}{2l} - \frac{ma_y hl_f}{dl} \\ F_{zrr} = \frac{mgl_f}{2l} + \frac{ma_x h}{2l} + \frac{ma_y hl_f}{dl} \end{cases} \tag{53}$$

2.5.2. Wheel Angle Distribution Control

The generalized wheel angle calculated by the target planning layer is the corresponding angle at the center of the front and rear axles, which is referred to as the equivalent angle of the front and rear wheels. Traditional control methods often assume that the left and right wheels on the same axis have the same angle, which directly corresponds to the equivalent angle. Although this method can simplify the design process, it does not take into account the actual steering geometry relationship, which not only increases tire wear but also causes unnecessary energy consumption. For this reason, the wheel angle distribution method based on Ackermann steering geometry is applied in this paper to convert the equivalent angle into the angle of each wheel.

By distributing the wheel angles based on the Ackermann steering geometry, the tires can be put into a pure rolling condition as much as possible. However, the ideal Ackermann steering principle is limited by the fact that it ignores the cornering characteristics of the tires. To further improve vehicle stability and reduce excessive tire wear, the Ackermann principle must be modified to obtain wheel angles.

When the vehicle is steering, the lateral elastic deformation of the tires results in the slip angles of the individual wheels on the front and rear axles. As shown in Equation (54), the corresponding relationship that each wheel angle should satisfy can be called the modified Ackermann principle.

$$\begin{cases} \tan(\delta_{fl} - \alpha_{fl}) = \frac{\tan\delta_f}{1-\frac{d}{2l}(\tan\delta_f - \tan\delta_r)} \\ \tan(\delta_{fr} - \alpha_{fr}) = \frac{\tan\delta_f}{1+\frac{d}{2l}(\tan\delta_f - \tan\delta_r)} \\ \tan(\delta_{rl} - \alpha_{rl}) = \frac{\tan\delta_r}{1-\frac{d}{2l}(\tan\delta_f - \tan\delta_r)} \\ \tan(\delta_{rr} - \alpha_{rr}) = \frac{\tan\delta_r}{1+\frac{d}{2l}(\tan\delta_f - \tan\delta_r)} \end{cases} \tag{54}$$

3. Simulation and Results

This section is divided into subheadings. It should provide a concise and precise description of the experimental results, their interpretation, as well as the experimental conclusions that can be drawn.

3.1. Environment and Configuration

In this section, the simulation model is built based on the MATLAB R2022a software produced by MathWorks (Natick, MA, USA), as shown in Figure 7. The vehicle model is the seven degrees-of-freedom vehicle model presented in Section 2. The simulations were performed using single-lane change and slalom test. The results verify the effect of trajectory tracking and economic optimization, which fully proves the effectiveness of the proposed control strategy.

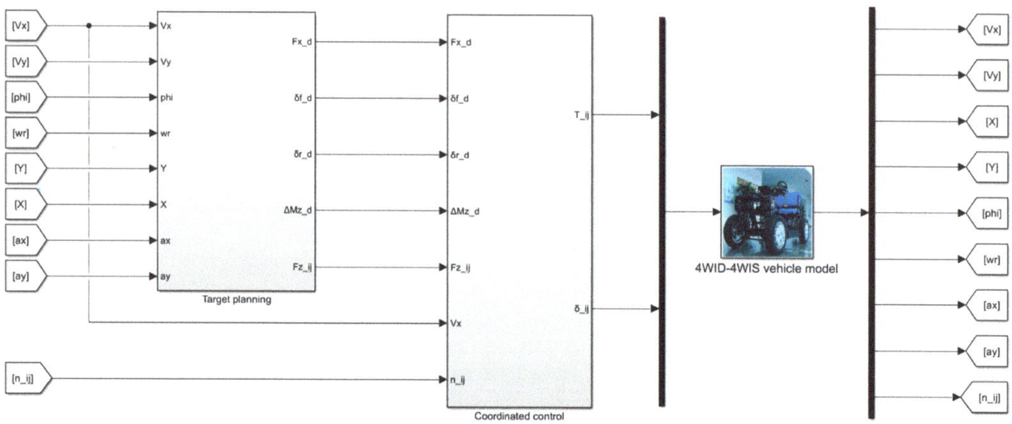

Figure 7. 4WID-4WIS EV simulation model.

To validate the economic performances of the coordinated control, simulations comparing the average distribution strategy and the wheel load distribution strategy are performed. These two strategies are simple and efficient and are widely used in practice. Therefore, the advantages of the proposed distribution control strategy can be further emphasized.

3.1.1. Average Distribution Strategy

The generalized longitudinal force and the generalized yaw moment output from the upper layer are distributed equally to each wheel. The wheel torque in this strategy is as follows:

$$\begin{cases} T_{fl} = T_{rl} = \frac{F_{xd}}{4} R_{eff} - \frac{M_{zd}}{2d} R_{eff} \\ T_{fr} = T_{rr} = \frac{F_{xd}}{4} R_{eff} + \frac{M_{zd}}{2d} R_{eff} \end{cases} \tag{55}$$

3.1.2. Wheel Load Distribution Strategy

The generalized force and yaw moment are distributed equally between the left and right sides of the vehicle. Then, the distribution between the front and rear wheels is carried out on the same side in proportion to the vertical load on the tires. The wheel torque with this strategy is shown in Equation (56).

$$\begin{cases} T_{fl} = \frac{F_{zfl}}{F_{zfl}+F_{zrl}} \left(\frac{F_{xd}}{2} R_{eff} - \frac{M_{zd}}{d} R_{eff} \right) \\ T_{fr} = \frac{F_{zfr}}{F_{zfr}+F_{zrr}} \left(\frac{F_{xd}}{2} R_{eff} + \frac{M_{zd}}{d} R_{eff} \right) \\ T_{rl} = \frac{F_{zrl}}{F_{zfl}+F_{zrl}} \left(\frac{F_{xd}}{2} R_{eff} - \frac{M_{zd}}{d} R_{eff} \right) \\ T_{rr} = \frac{F_{zrr}}{F_{zfr}+F_{zrr}} \left(\frac{F_{xd}}{2} R_{eff} + \frac{M_{zd}}{d} R_{eff} \right) \end{cases} \tag{56}$$

3.2. Results and Analysis

3.2.1. Single-Lane Change

As one of the common conditions, the single-lane change condition is usually used to simulate a vehicle lane change scenario.

1. Trajectory tracking effect

Three speeds of 40 km/h, 80 km/h, and 120 km/h are simulated to verify the effect of the controller. The longitudinal velocity tracking and lateral path tracking results are analyzed to verify the performance of the vehicle at low, medium, and high speeds. The simulation results are shown in Figures 8 and 9.

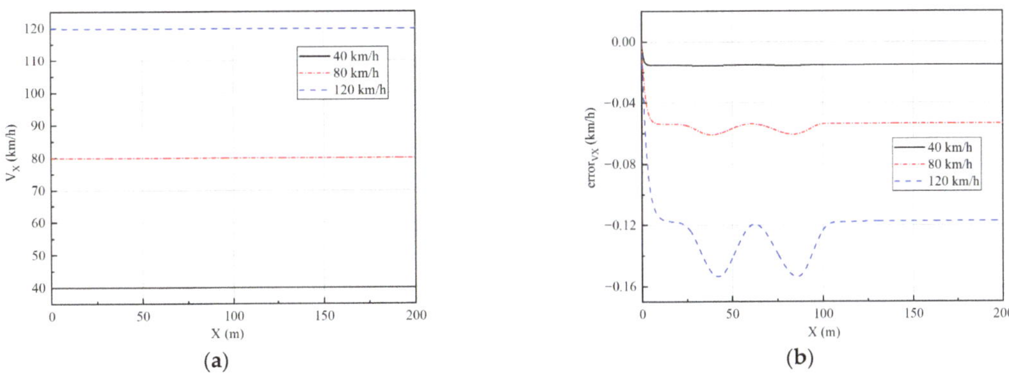

Figure 8. Velocity tracking results under single-lane change: (**a**) The results of velocity; (**b**) The results of longitudinal tracking error.

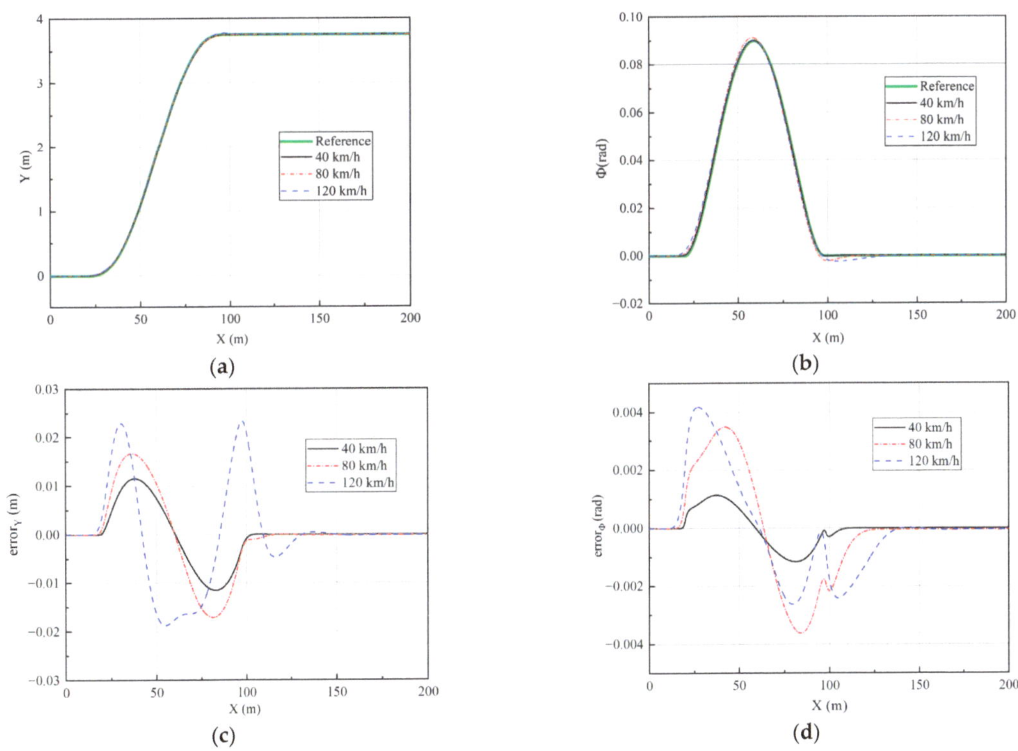

Figure 9. Path tracking results under single-lane change: (**a**) The results of lateral displacement; (**b**) The results of heading angle; (**c**) The results of lateral displacement tracking error; (**d**) The results of heading angle tracking error.

For a single-lane change, the longitudinal velocity tracking algorithm can track the expected velocity very well according to Figure 8. The tracking error is less than 0.2 km/h, which satisfies the longitudinal tracking accuracy requirements. As the velocity increases, the tracking error gradually increases. This is mainly because the nonlinearity of the tire gradually increases and is associated with the strong coupling properties between the subsystems, resulting in a decrease in tracking accuracy.

The data in Figure 9c,d have been processed to obtain the tracking error at different velocities to facilitate quantitative analysis of the error, as shown in Table 1.

Table 1. Comparison of path tracking errors under single-lane change.

	Performance Index	40 km/h	80 km/h	120 km/h
Lateral displacement tracking error (m)	Maximum	0.0115	0.0171	0.0234
	Average	0.0024	0.0036	0.0053
	Standard deviations	0.0040	0.0058	0.0075
Heading angle tracking error (rad)	Maximum	0.0012	0.0036	0.0042
	Average	0.0002	0.0009	0.0009
	Standard deviations	0.0004	0.0012	0.0012

From Figure 9 and Table 1, it can be seen that the lateral path tracking algorithm is very good at tracking the desired path for single-lane change. As the speed increases, the lateral displacement error gradually increases. The maximum lateral displacement error is 0.0115 m at 40 km/h, 0.0171 m at 80 km/h, and 0.0234 m at 120 km/h, which is mainly due to the gradual increase in the nonlinearity of the tire and the enhancement of the coupling effect of the subsystem. At the same time, although the lateral displacement error increases, it is still generally small and within acceptable limits. Similarly, the heading angle error also increases with velocity but remains at a low level overall. The maximum heading angle error at 120 km/h is only 0.0042 rad, which demonstrates the good path-tracking performance of the controller.

The trajectory tracking controller is designed to achieve good stability while providing accurate and reliable path tracking. To check the stability effect, the lateral acceleration of the vehicle, side slip angle, and yaw rate are selected as indicators, and the simulation results are analyzed.

From Figures 10–13, it can be seen that the vehicle is in a stable state due to the stability constraints and no instability occurs even at a high speed of 120 km/h. The lateral acceleration is kept within 0.4 g, which ensures the accuracy of the linear tire model. The side slip angles at different speeds are all within 0.01 rad, and the yaw rate is all within 0.15 rad/s, indicating that the vehicle has good driving stability.

The simulation results shown in Figure 13 illustrate the advantages of 4WIS in terms of dynamic control. At low speeds, the front and rear wheels rotate in the opposite direction, reducing the steering radius and improving the vehicle's maneuverability. At high speeds, the front and rear wheels turn in the same direction, thereby increasing the lateral stability margin and improving the vehicle's driving stability.

2. Economy optimization effect

The above three distribution strategies are simulated at different velocities to validate the economy optimization effect, using the overall efficiency of the motor and the battery SOC as performance indicators. It is assumed that Rule 1 is the average distribution strategy, Rule 2 is the wheel load distribution strategy, and Rule 3 is the torque distribution strategy proposed in this paper. At the same time, the control strategy does not include a braking energy recovery strategy. The initial value of SOC is set to 0.8. The simulation results are shown in Figures 14–16 and Table 2.

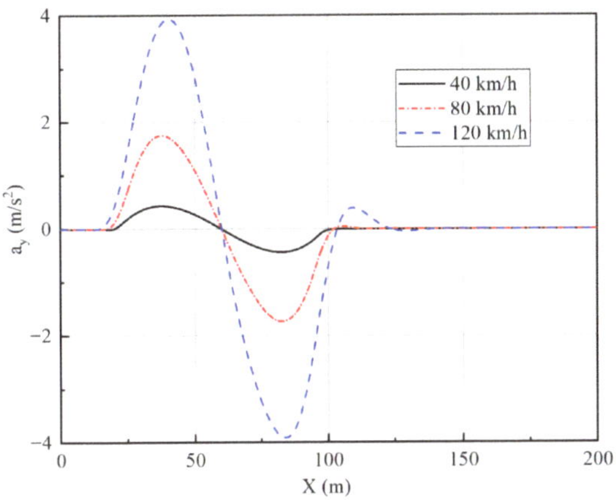

Figure 10. Lateral acceleration under single-lane change.

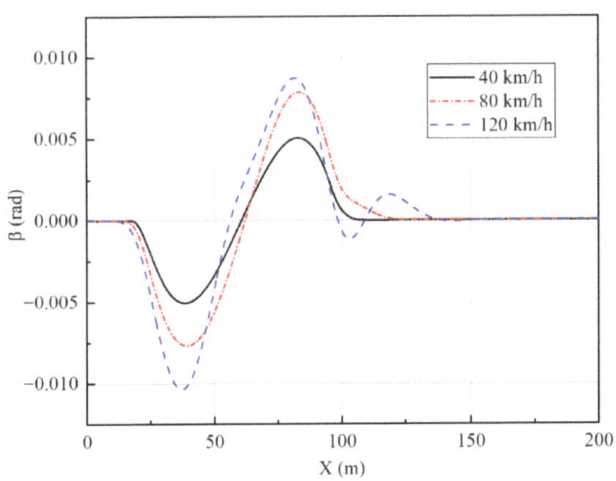

Figure 11. Side slip angle under single-lane change.

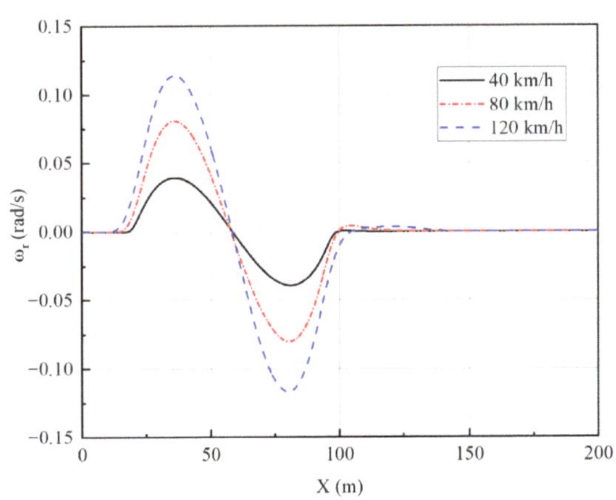

Figure 12. Yaw rate under single-lane change.

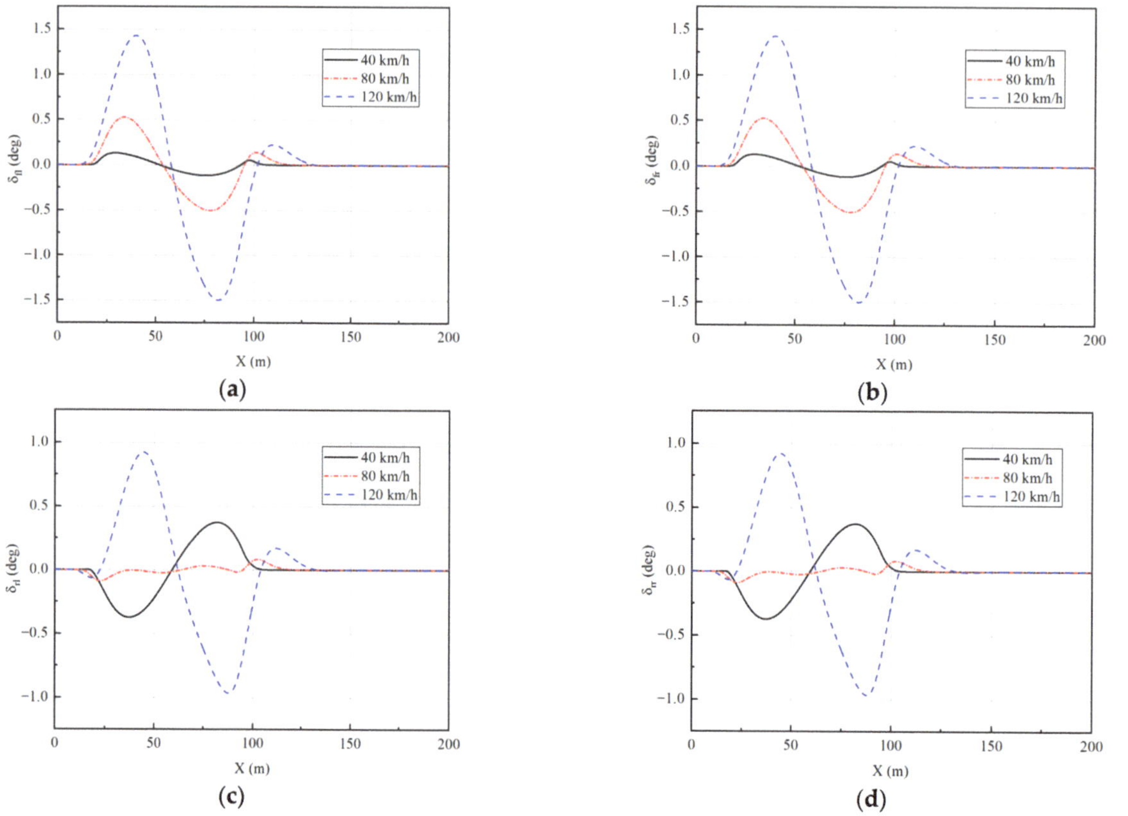

Figure 13. Wheel angles under single-lane change: (**a**) Front-left wheel; (**b**) Front-right wheel; (**c**) Rear-left wheel; (**d**) Rear-right wheel.

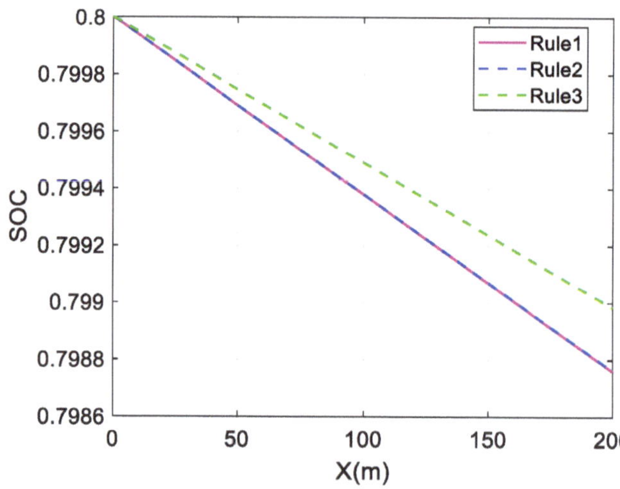

Figure 14. Change in SOC at 40 km/h under single-lane change.

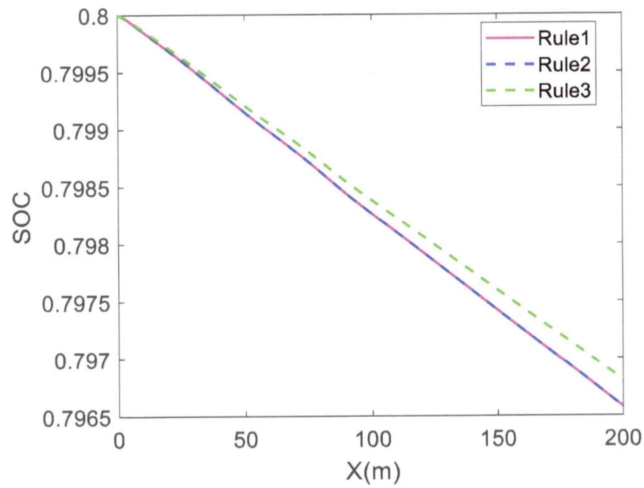

Figure 15. Change in SOC at 80 km/h under single-lane change.

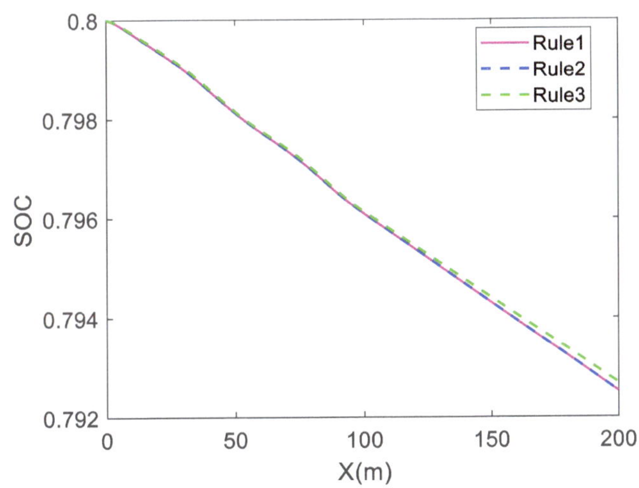

Figure 16. Change in SOC at 80 km/h under single-lane change.

Table 2. Comparison of overall motor efficiency under single-lane change.

Velocity (km/h)	Strategy	Maximum	Average
40	Rule 1	0.7356	0.7303
	Rule 2	0.7347	0.7283
	Rule 3	0.8411	0.8375
80	Rule 1	0.8969	0.8877
	Rule 2	0.8970	0.8882
	Rule 3	0.9276	0.9241
120	Rule 1	0.8973	0.8791
	Rule 2	0.8972	0.8793
	Rule 3	0.9147	0.9107

For a single-lane change, the change in SOC is almost identical for the average and wheel load distribution strategies at different speeds. The maximum and average values of the overall motor efficiency are also almost identical, indicating that the economic effects of these two strategies are almost the same.

The change in battery SOC represents the electrical energy consumed by the in-wheel motors while driving. Comparing the change curves of SOC at different speeds, the SOC change in Rule 3 is smaller than those in Rule 1 and Rule 2, and the SOC in Rule 3 falls relatively slowly. This shows that the in-wheel motors consume less electric energy in the Rule 3 strategy, which proves the energy economy of the proposed strategy.

The average value of the overall motor efficiency of Rule 3 is 14.68% higher than that of Rule 1 at 40 km/h. Compared with the results of Rule 1, the average values at 80 km/h and 120 km/h have increased by 4.10% and 3.5%, respectively. This shows that the coordinated control strategy proposed in this paper can maintain the overall motor efficiency at a relatively high level and effectively improve the vehicle economy.

Comparing the effects of economy optimization at different speeds, it can be found that the economy optimization at low speeds is better, and the difference in SOC decline is significant. This is because the demand torque is relatively low at low speeds. In the average distribution strategy, the torque of each wheel is very low, so in-wheel motors operate in the low-efficiency region. However, the proposed strategy optimizes the torque distribution and tends to select the front wheel drive mode. This allows the front in-wheel motors to operate in a relatively efficient range, and the overall efficiency can be significantly improved. However, at high speeds, the in-wheel motors' speeds are also high. As shown in Figure 2, the motor's efficiency characteristic range gradually narrows as the speed increases when the motor's speed is high. The distribution difference of the motor working points decreases for different strategies, resulting in a less significant improvement effect.

3.2.2. Slalom Test

1. Trajectory tracking effect

Two speeds of 30 km/h and 60 km/h are simulated to verify the effect of the controller. The longitudinal velocity tracking and lateral path tracking results are analyzed to verify the performance of the vehicle at different speeds. The simulation results are shown in Figures 17 and 18.

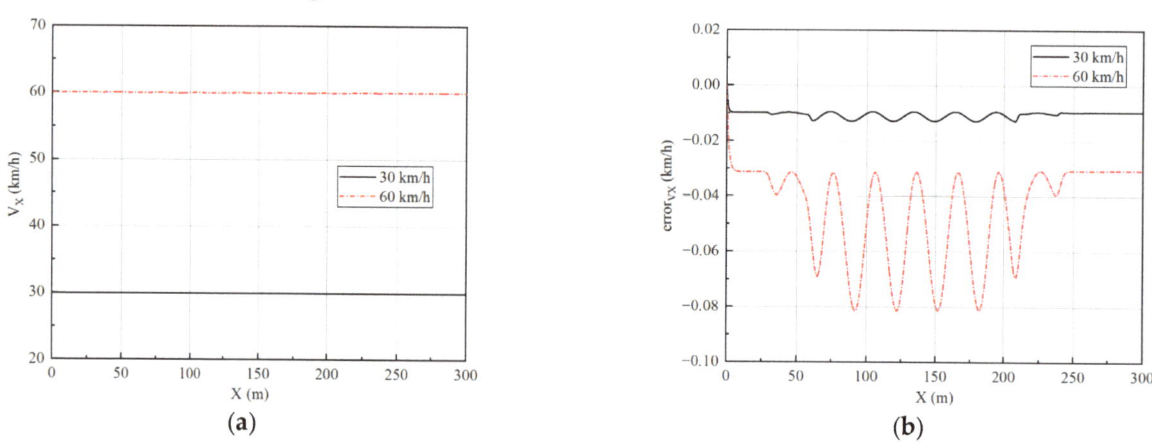

Figure 17. Velocity tracking results under slalom test: (**a**) The results of velocity; (**b**) The results of longitudinal tracking error.

The data in Figure 18c,d have been processed to obtain the tracking error at different velocities to facilitate quantitative analysis of the error, as shown in Table 3.

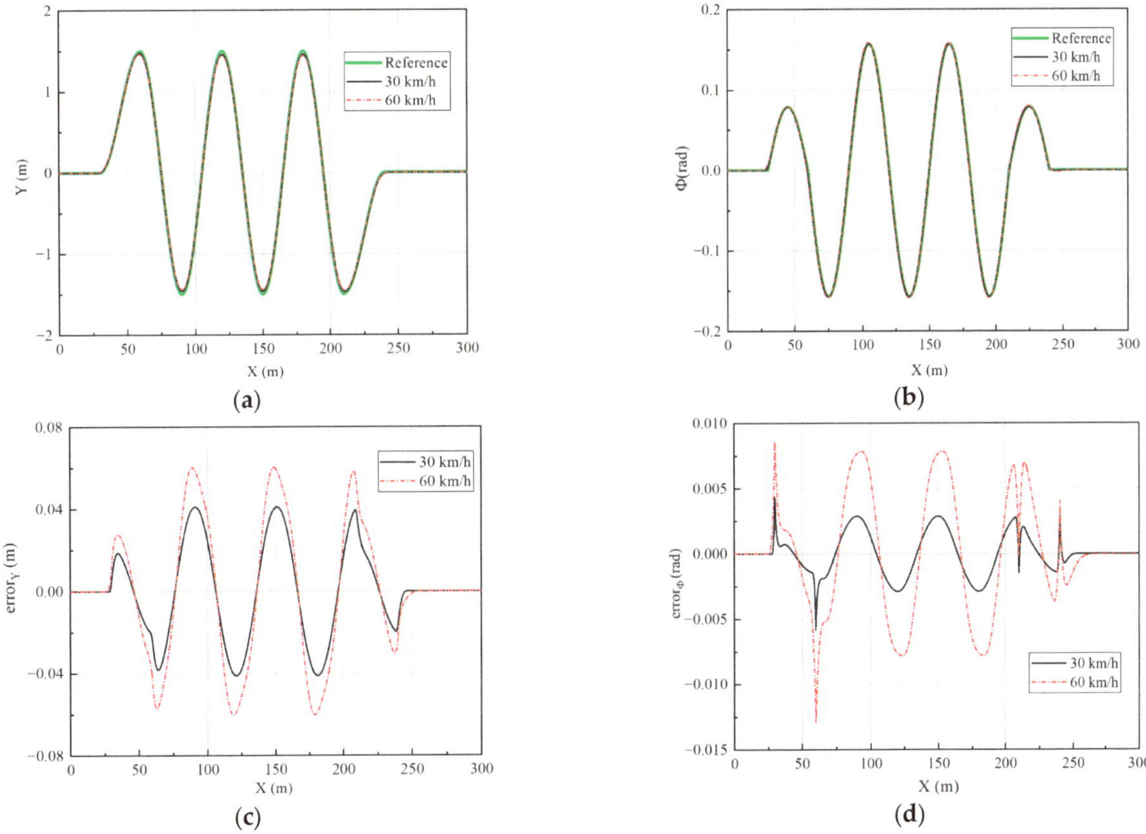

Figure 18. Path tracking results under slalom test: (**a**) The results of lateral displacement; (**b**) The results of heading angle; (**c**) The results of lateral displacement tracking error; (**d**) The results of heading angle tracking error.

Table 3. Comparison of path tracking errors under slalom test.

	Performance Index	30 km/h	60 km/h
Lateral displacement tracking error (m)	Maximum	0.0412	0.0603
	Average	0.0158	0.0241
	Standard deviations	0.0147	0.0214
Heading angle tracking error (rad)	Maximum	0.0058	0.0129
	Average	0.0011	0.0033
	Standard deviations	0.0010	0.0029

From Figure 18 and Table 3, it can be seen that the lateral path tracking algorithm is very good at tracking the desired path in a slalom test. As the speed increases, the path-tracking error gradually increases. The average lateral displacement error is 0.0158 m and the average heading angle error is 0.0011 rad at 30 km/h, while the average lateral displacement error is 0.0241 m and the average heading angle error is 0.0033 rad at 60 km/h. Although the tracking error increases, it remains low overall, which shows the good tracking performance of the designed controller.

Figures 19–22 show that the vehicle is in a stable state due to stability constraints. The side slip angle is maintained at a low level, which provides good tracking capability. The lateral acceleration is kept within 0.4 g, ensuring the accuracy of the linear tire model. The velocities under both simulation conditions are not particularly high, so the front and rear wheels turn in opposite directions, which improves the maneuverability of the vehicle while meeting the stability requirements.

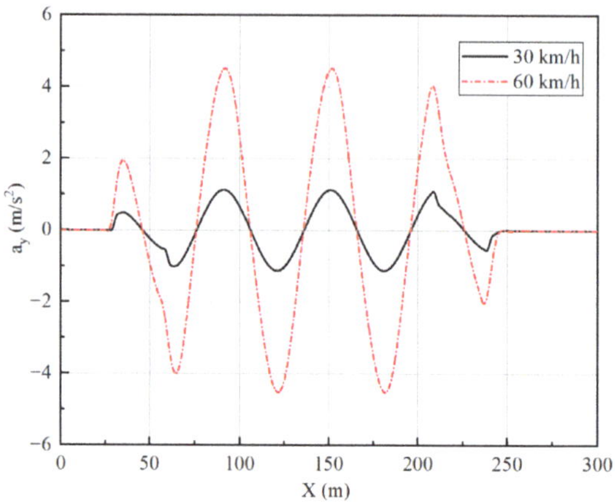

Figure 19. Lateral acceleration under slalom test.

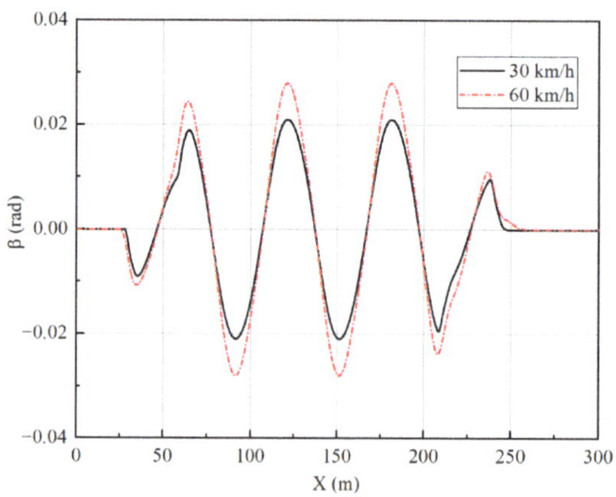

Figure 20. Side slip angle under slalom test.

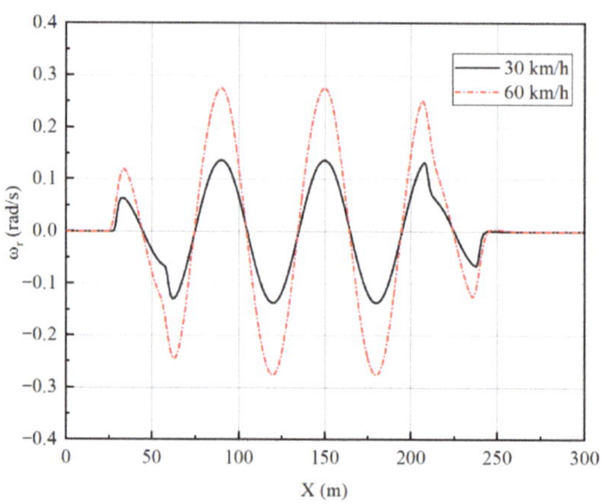

Figure 21. Yaw rate under slalom test.

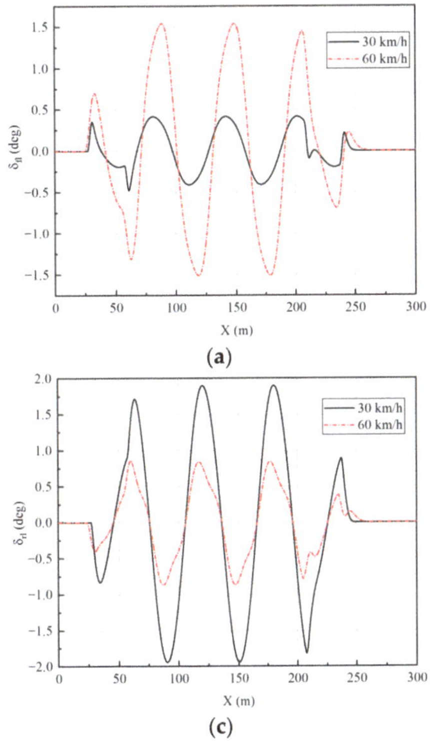

Figure 22. Wheel angles under slalom test: (**a**) Front-left wheel; (**b**) Front-right wheel; (**c**) Rear-left wheel; (**d**) Rear-right wheel.

2. Economy optimization effect

The above three distribution strategies are simulated at different speeds to validate the economy optimization effect, using the overall efficiency of the motor and the battery SOC as performance indicators. The simulation results are shown in Figures 23 and 24 and Table 4.

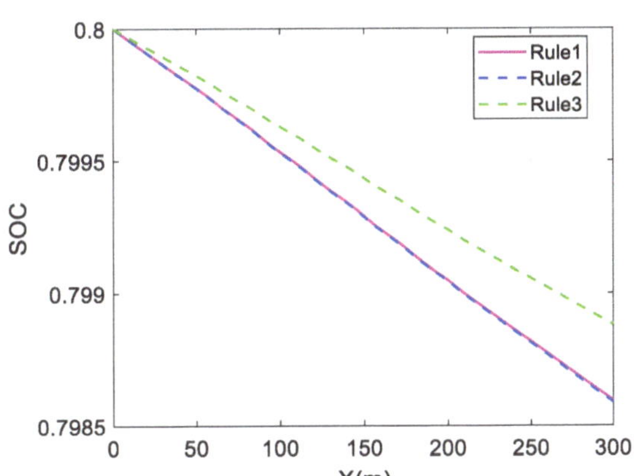

Figure 23. Change in SOC at 30 km/h under slalom test.

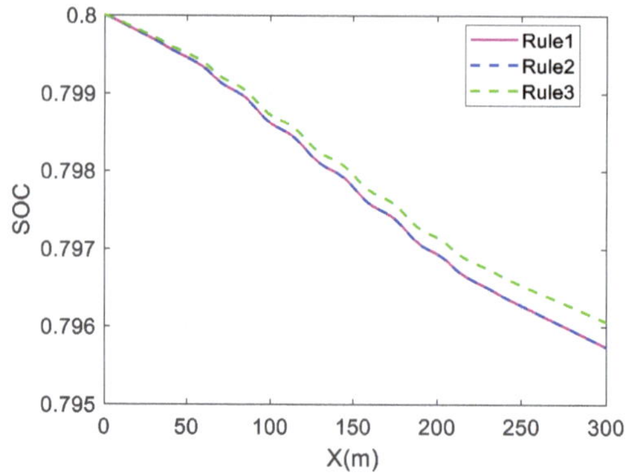

Figure 24. Change in SOC at 80 km/h under slalom test.

Table 4. Comparison of overall motor efficiency under slalom test.

Velocity (km/h)	Strategy	Maximum	Average
30	Rule 1	0.6404	0.6151
	Rule 2	0.6409	0.6121
	Rule 3	0.7857	0.7484
60	Rule 1	0.8973	0.7995
	Rule 2	0.8971	0.7987
	Rule 3	0.9345	0.8829

In the slalom test, the change in SOC is almost identical for the average and wheel load distribution strategies at different speeds. The maximum and average values of the overall motor efficiency are also almost identical, indicating that the economic effects of these two strategies are almost the same.

Comparing the change curves of SOC at different speeds, the change in SOC in Rule 3 is smaller than in Rule 1 and Rule 2, and the SOC in Rule 3 drops relatively slowly. This shows that the in-wheel motors consume electric energy under the Rule 3 strategy, which proves the energy economy of the proposed strategy.

The coordinated control proposed in this paper is significantly better than the other two strategies, as shown in Table 4. Compared with the results of Rule 1, the average value of the overall motor efficiency of Rule 3 has increased by 21.67% at 30 km/h. At 60 km/h, it has increased by 10.43%, which proves that the coordinated control strategy can effectively improve the vehicle economy.

4. Conclusions

In this paper, a 4WID-4WIS EV trajectory tracking coordinated control strategy considering energy consumption economy is proposed to improve vehicle stability and economy during trajectory tracking. A hierarchical chassis coordinated control architecture for 4WID-4WIS EV is designed. In the target planning layer, the longitudinal velocity tracking and lateral path tracking are achieved by using expert PID and MPC, respectively, considering stability constraints, such as yaw rate and tire slip angle. In the coordinated control layer, the optimal torque distribution for each wheel is performed to achieve the optimal overall motor efficiency based on the MPSO algorithm. Then, the angle distribution of each wheel is performed in combination with the modified Ackermann theory. From the simulation results, it can be concluded that the proposed coordinated control strategy not only achieves good trajectory tracking but also ensures driving stability and improves the energy consumption economy.

In future work, the coupling characteristics between multiple subsystems will be considered. The influence of vertical dynamics will be incorporated into the control strategy to achieve unified control of multi-dimensional lateral, longitudinal, and vertical dynamics. At the same time, the motor model will be further improved, and research on redundant control of actuator failures will be conducted. The test with a real vehicle will be conducted to further verify the effectiveness of the control strategy.

Author Contributions: Investigation, Z.L.; Supervision, X.C.; Writing—original draft preparation, Y.Q. All authors have read and agreed to the published version of the manuscript.

Funding: This research was funded by the Jiangsu Provincial Science and Technology Project, grant number BE2021006-3 and National Natural Science Foundation of China, grant number 52275123.

Institutional Review Board Statement: Not applicable.

Informed Consent Statement: Not applicable.

Data Availability Statement: Not applicable.

Acknowledgments: We would like to acknowledge the support of Dongxiao Yin.

Conflicts of Interest: The authors declare no conflict of interest.

References

1. Peng, H.; Chen, X.B. Active Safety Control of X-by-Wire Electric Vehicles: A Survey. *Sae Int. J. Veh. Dyn. Stab. NVH* **2022**, *6*, 115–133. [CrossRef]
2. Tong, Y.W.; Li, C.; Wang, G.; Jing, H. Integrated Path-Following and Fault-Tolerant Control for Four-Wheel Independent-Driving Electric Vehicles. *Automot. Innov.* **2022**, *5*, 311–323. [CrossRef]
3. Tianjun, Z.; Wan, H.; Wang, Z.; Wei, M.; Xu, X.; Zhiliang, Z.; Sanmiao, D. Model Reference Adaptive Control of Semi-active Suspension Model Based on AdaBoost Algorithm for Rollover Prediction. *SAE Int. J. Veh. Dyn. Stab. NVH* **2021**, *6*, 71–86. [CrossRef]
4. Song, P.; Tomizuka, M.; Zong, C.F. A novel integrated chassis controller for full drive-by-wire vehicles. *Veh. Syst. Dyn.* **2015**, *53*, 215–236. [CrossRef]
5. Lai, F.; Huang, C.; Jiang, C. Comparative Study on Bifurcation and Stability Control of Vehicle Lateral Dynamics. *SAE Int. J. Veh. Dyn. Stab. NVH* **2021**, *6*, 35–52. [CrossRef]
6. Zhang, N.; Wang, J.; Li, Z.; Xu, N.; Ding, H.; Zhang, Z.; Guo, K.; Xu, H. Coordinated Optimal Control of AFS and DYC for Four-Wheel Independent Drive Electric Vehicles Based on MAS Model. *Sensors* **2023**, *23*, 505. [CrossRef]
7. Peng, H.N.; Wang, W.D.; Xiang, C.L.; Li, L.; Wang, X.Y. Torque Coordinated Control of Four In-Wheel Motor Independent-Drive Vehicles With Consideration of the Safety and Economy. *IEEE Trans. Veh. Technol.* **2019**, *68*, 9604–9618. [CrossRef]
8. Wang, J.N.; Gao, S.L.; Wang, K.; Wang, Y.; Wang, Q.S. Wheel torque distribution optimization of four-wheel independent-drive electric vehicle for energy efficient driving. *Control Eng. Pract.* **2021**, *110*, 104779. [CrossRef]
9. Zhang, B.H.; Lu, S.B.; Zhao, L.; Xiao, K.X. Fault-tolerant control based on 2D game for independent driving electric vehicle suffering actuator failures. *Proc. Inst. Mech. Eng. Part D-J. Automob. Eng.* **2020**, *234*, 3011–3025. [CrossRef]
10. Xu, X.; Zhang, Y.; Wang, X.; Fang, J.G.; Chen, J.W.; Li, J.X. Searching superior crashworthiness performance by constructing variable thickness honeycombs with biomimetic cells. *Int. J. Mech. Sci.* **2022**, *235*, 107718. [CrossRef]
11. Zhang, Y.; Xu, X.; Fang, J.G.; Huang, W.Z.; Wang, J. Load characteristics of triangular honeycomb structures with self-similar hierarchical features. *Eng. Struct.* **2022**, *257*, 114114. [CrossRef]
12. Li, D.F.; Liu, A.; Pan, H.; Chen, W.T. Safe, Efficient and Socially-Compatible Decision of Automated Vehicles: A Case Study of Unsignalized Intersection Driving. *Automot. Innov.* **2023**, *6*, 1–16. [CrossRef]
13. Xie, F.; Liang, G.; Chien, Y.R. Highly Robust Adaptive Sliding Mode Trajectory Tracking Control of Autonomous Vehicles. *Sensors* **2023**, *23*, 3454. [CrossRef] [PubMed]
14. Oh, K.; Seo, J. Development of a Sliding-Mode-Control-Based Path-Tracking Algorithm with Model-Free Adaptive Feedback Action for Autonomous Vehicles. *Sensors* **2023**, *23*, 405. [CrossRef]
15. Zhang, Y.G.; Chen, G.Y.; Hu, H.Y.; Gao, Z.H. Hierarchical Parking Path Planning Based on Optimal Parking Positions. *Automot. Innov.* **2023**, *6*, 1–11. [CrossRef]
16. Guo, N.Y.; Zhang, X.D.; Zou, Y. Real-Time Predictive Control of Path Following to Stabilize Autonomous Electric Vehicles Under Extreme Drive Conditions. *Automot. Innov.* **2022**, *5*, 453–470. [CrossRef]
17. Mashadi, B.; Ahmadizadeh, P.; Majidi, M. Integrated Controller Design for Path Following in Autonomous Vehicles. In *Autonomous Vehicles for Safer Driving*; SAE: Warrendale, PA, USA, 2011.
18. Li, B.Y.; Du, H.P.; Li, W.H. Trajectory control for autonomous electric vehicles with in-wheel motors based on a dynamics model approach. *Iet Intell. Transp. Syst.* **2016**, *10*, 318–330. [CrossRef]
19. Liu, R.Q.; Wei, M.X.; Zhao, W.Z. Trajectory tracking control of four wheel steering under high speed emergency obstacle avoidance. *Int. J. Veh. Des.* **2018**, *77*, 1–21. [CrossRef]

20. Nah, J.; Yim, S. Vehicle Stability Control with Four-Wheel Independent Braking, Drive and Steering on In-Wheel Motor-Driven Electric Vehicles. *Electronics* **2020**, *9*, 1934. [CrossRef]
21. Seo, Y.; Cho, K.; Nam, K. Integrated Yaw Stability Control of Electric Vehicle Equipped with Front/Rear Steer-by-Wire Systems and Four In-Wheel Motors. *Electronics* **2022**, *11*, 1277. [CrossRef]
22. Xu, N.; Zhou, J.F.; Li, X.Y.; Li, F. Analysis of the Effect of Inflation Pressure on Vehicle Handling and Stability under Combined Slip Conditions Based on the UniTire Model. *Sae Int. J. Veh. Dyn. Stab. NVH* **2021**, *5*, 259–277. [CrossRef]
23. Li, F.; Xu, N.; Wang, Z.; Gruber, P.; Bruzelius, F.; Ran, S. Letter from the Special Issue Editors. *SAE Int. J. Veh. Dyn. Stab. NVH* **2021**, *5*, 229–231. [CrossRef]
24. Yu, Z.T.; Wang, J.M. Automatic Vehicle Trajectory Tracking Control with Self-calibration of Nonlinear Tire Force Function. In Proceedings of the 2017 American Control Conference (ACC), IEEE, Seattle, WA, USA, 24–26 May 2017; pp. 985–990.
25. Attia, R.; Orjuela, R.; Basset, M. Combined longitudinal and lateral control for automated vehicle guidance. *Veh. Syst. Dyn.* **2014**, *52*, 261–279. [CrossRef]
26. Wang, Y.Q.; Ye, Y.H.; Zhang, H. A Stable Tracking Control Method for an Autonomous Welding Mobile Robot. *Chem. Mech. Mater. Eng.* **2011**, *79*, 264–269. [CrossRef]
27. Attia, R.; Orjuela, R.; Basset, M. Coupled longitudinal and lateral control strategy improving lateral stability for autonomous vehicle. In Proceedings of the 2012 American Control Conference (ACC), IEEE, Montreal, QC, Canada, 27–29 June 2012; pp. 6509–6514.
28. Brown, M.; Gerdes, J.C. Coordinating Tire Forces to Avoid Obstacles Using Nonlinear Model Predictive Control. *IEEE Trans. Intell. Veh.* **2020**, *5*, 21–31. [CrossRef]
29. Choi, S.; d'Andrea-Novel, B.; Fliess, M.; Mounier, H.; Villagra, J. Model-free control of automotive engine and brake for Stop-and-Go scenarios. In Proceedings of the European Control Conference, Budapest, Hungary, 23–26 August 2009.
30. Falcone, P.; Borrelli, F.; Asgari, J.; Tseng, H.E.; Hrovat, D. Predictive active steering control for autonomous vehicle systems. *IEEE Trans. Control Syst. Technol.* **2007**, *15*, 566–580. [CrossRef]
31. Falcone, P.; Borrelli, F.; Tseng, H.E.; Asgari, J.; Hrovat, D. Linear time-varying model predictive control and its application to active steering systems: Stability analysis and experimental validation. *Int. J. Robust Nonlinear Control* **2008**, *18*, 862–875. [CrossRef]
32. Falcone, P.; Tseng, H.E.; Borrelli, F.; Asgari, J.; Hrovat, D. MPC-based yaw and lateral stabilisation via active front steering and braking. *Veh. Syst. Dyn.* **2008**, *46*, 611–628. [CrossRef]
33. Wong, A.; Kasinathan, D.; Khajepour, A.; Chen, S.K.; Litkouhi, B. Integrated torque vectoring and power management framework for electric vehicles. *Control Eng. Pract.* **2016**, *48*, 22–36. [CrossRef]
34. Liu, W.; Khajepour, A.; He, H.W.; Wang, H.; Huang, Y.J. Integrated Torque Vectoring Control for a Three-Axle Electric Bus Based on Holistic Cornering Control Method. *IEEE Trans. Veh. Technol.* **2018**, *67*, 2921–2933. [CrossRef]
35. Marini, F.; Walczak, B. Particle swarm optimization (PSO). A tutorial. *Chemom. Intell. Lab. Syst.* **2015**, *149*, 153–165. [CrossRef]

sensors

MDPI

Article

Vehicle Localization Kalman Filtering for Traffic Light Advisor Application in Urban Scenarios

Daniele Vignarca, Stefano Arrigoni and Edoardo Sabbioni *

Department of Mechanical Engineering, Politecnico di Milano, 20156 Milan, Italy; daniele.vignarca@polimi.it (D.V.); stefano.arrigoni@polimi.it (S.A.)
* Correspondence: edoardo.sabbioni@polimi.it

Abstract: The recent advancements in Intelligent Transportation Systems (ITS) have revealed significant potential for enhancing traffic management through Advanced Driver Assist Systems (ADASs), with benefits for both safety and environment. This research paper proposes a vehicle localization technique based on Kalman filtering, as accurate positioning of the ego-vehicle is essential for the proper functioning of the Traffic Light Advisor (TLA) system. The aim of the TLA is to calculate the most suitable speed to safely reach and pass the first traffic light in front of the vehicle and subsequently keep that velocity constant to overcome the following traffic light, thus allowing safer and more efficient driving practices, thereby reducing safety risks, and minimizing energy consumption. To overcome Global Positioning Systems (GPS) limitations encountered in urban scenarios, a multi-rate sensor fusion approach based on the Kalman filter with map matching and a simple kinematic one-dimensional model is proposed. The experimental results demonstrate an estimation error below 0.5 m on urban roads with GPS signal loss areas, making it suitable for TLA application. The experimental validation of the Traffic Light Advisor system confirmed the expected benefits with a 40% decrease in energy consumption compared to unassisted driving.

Keywords: vehicle localization; Kalman filter; ADAS; kinematic model; GPS; TLA; ITS

Citation: Vignarca, D.; Arrigoni, S.; Sabbioni, E. Vehicle Localization Kalman Filtering for Traffic Light Advisor Application in Urban Scenarios. *Sensors* **2023**, *23*, 6888. https://doi.org/10.3390/s23156888

Academic Editors: Yafei Wang, Chao Huang, Hailong Huang, Zhiqiang Zuo, Bo Leng and Peng Hang

Received: 30 June 2023
Revised: 31 July 2023
Accepted: 1 August 2023
Published: 3 August 2023

1. Introduction

Vehicle localization represents a fundamental task in many fields, ranging from Autonomous Vehicles (AV) to Advanced Driving Assistance Systems (ADASs) as well as traffic management [1]. Indeed, the starting point for most of the control logic, on both the vehicle and infrastructure sides, is the knowledge of the vehicle position. This is typically fed to algorithms intended to compute either vehicle optimal trajectory and speed or safety risk indexes to safely overcome dangerous situations, especially in urban scenarios.

Vehicle localization techniques can be distinguished between the onboard sensor-based systems and those relying upon Vehicle-to-Vehicle (V2V) and Vehicle-to-Infrastructure (V2I) communication. The former category can be further split into active sensor based (e.g., LiDAR and RADAR) and passive sensor based (e.g., GPS and IMU). On the one hand, active sensors are generally more costly and computationally expensive than passive ones. On the other hand, Inertial Measurement Units (IMUs) suffer from noisy signals which may lead to integration divergence, while when dealing with Global Positioning Systems (GPS) a typical issue is the signal loss [2]. In vehicle localization, GPS outage is thus a relevant phenomenon to cope with, especially in urban scenarios, where the presence of trees and high buildings limit the sensor capabilities.

The Green Light Optimal Speed Advisory (GLOSA) system is an application that conveys speed references to the driver to achieve lower travel times, fuel/energy consumption, and safer travel conditions [3]. This can be achieved thanks to the knowledge of road data, the vehicle state in terms of position and speed, and traffic light schedules. The speed profile calculation is typically addressed by taking into account different criteria.

One typical approach is to minimize the engine power demand and idling time [4]. Another approach prioritizes driver annoyance reduction by minimizing the difference between the suggested and actual speeds, or by aiming to pass the traffic light in the smallest amount of time [5].

In this framework, the present study proposes a multi-sensor multi-rate vehicle localization technique based on Kalman filtering as the ego-vehicle position is a prerequisite for making ADASs work. The aim is to have an accurate state estimation that provides the vehicle position to the Traffic Light Advisor (TLA) system thought to be used in challenging scenarios such as urban roads, where pure GPS information may be neither present nor reliable. The field tests for the localization algorithm demonstrated good accuracy results in different conditions and GPS signal availability, proving to be consistent for running ADASs, such as the TLA. Furthermore, having implemented the localization algorithm and knowing the working plans of a set of traffic lights on a predefined path of the city of Milan, the paper reports full-scale testing on a trolley-bus in the urban scenario of the TLA developed by [6] at Politecnico di Milano, confirming the previously obtained simulation results.

The remainder of this paper is structured as follows: After the literature review in Section 2, the experimental setup used for the experimental campaigns is presented in Section 3. Section 4 details the proposed localization algorithm, and Section 5 summarizes the main feature of the TLA and presents the scenarios and the metrics adopted for the ADAS validation. The results of both the localization algorithm and the TLA system are reported in Section 6 while Section 7 draws the conclusions of this work, proposing some future development for the implemented systems.

2. State of the Art

In the literature, the criteria employed for the determination of the speed profile in GLOSA applications typically involve the minimization of the total energy consumption and travel time [7–10]. This kind of ADAS can be further distinguished into two categories based on the number of traffic lights they consider in real time to determine the recommended speeds: single-segment GLOSA (S-GLOSA) and multiple-segment GLOSA (M-GLOSA). S-GLOSA systems focus on analyzing the first traffic light encountered by the vehicle, while M-GLOSA systems consider multiple traffic lights along the vehicle's route. In the case of S-GLOSA algorithms, they typically employ modeling approaches incorporating velocity profiles both upstream and downstream of the intersection, as demonstrated in [11]. However, in recent years, data-driven approaches have emerged as a promising alternative, as highlighted in [12]. In that research, a conventional S-GLOSA system is contrasted with reinforcement learning (RL) implementation, which incorporates data from a single traffic light and limited information from the preceding three vehicles. The RL-based approach resulted in an 11% increase in energy savings compared to the standard S-GLOSA system. These developments in data-driven methodologies highlight the potential to enhance the performance and energy efficiency of GLOSA systems.

In the future, Vehicle-to-Infrastructure (V2I) communication will be one of the main drivers making this kind of Advanced Driving Assistance System possible by providing a great amount of data about adjacent vehicles and traffic [13]. In fact, the work in [14] analyzes the effect of the GLOSA system running different simulations varying both infrastructure variables (e.g., cycle times and communication range) and external variables, such as traffic conditions. As far as traffic is concerned, the research conducted in [15] demonstrates the effectiveness of M-GLOSA systems compared to single-segment approaches, especially in free-flow traffic conditions. However, optimizing M-GLOSA systems while considering traffic light phase changes presents challenges, leading to non-convex feasible solution domains. To tackle this issue, the literature proposes various approaches. Studies like [8,15] have implemented Genetic Algorithms (GAs) to address the optimization problem. Additionally, search-based algorithms, employing semi-heuristic or brute-force methods, have also been explored in [11]. A widely adopted alternative is Model Predictive Control (MPC), and the authors in [16] investigated the MPC application for GLOSA implementation in road segments containing multiple traffic lights.

Self-localization has been extensively studied in the literature for a long time, as it serves as a crucial component in the development of Cooperative Active Safety Systems and ADASs in general. In a comprehensive review conducted in [17], various sensor-based and communication-based approaches for localization are thoroughly examined, with a specific focus on accuracy and real-time performance. The findings of the survey show that data fusion techniques, such as the integration of onboard passive sensors and Vehicle-to-Everything (V2X) communication, offer a promising solution due to their robustness, accuracy, and ability to operate in real time.

GPS outages represent a huge limiting point in vehicle localization in urban canyoning; thus, the most adopted strategy to cope with this issue is to use the Kalman filter and its variants to estimate the vehicle state also when GPS is not available, fusing different sensors (e.g., Inertial Measurement Unit and Wheel Speed Sensor) and a dynamic vehicle model [18,19]. In [2], an extended Kalman filter (EKF) fusing a digital map, IMUs, GPS data, and cellular Base Transceiver Stations (BTS) signals is presented. In their work, the authors proposed the use of cellular BTS not only to overcome GPS outages but also to improve localization accuracy.

Recently, LiDAR and vision sensors have been adopted to overcome the challenges of localization in urban scenarios, fusing these types of sensors to cope with the limitation of every single device [20]. In [21], a cascading Kalman filter and dynamic object removal model using multi-GNSS, INS, Precise Point Positioning (PPP), and vision to improve vehicle navigation performances in urban scenarios is presented. In this framework, a novel application in state estimation is Simultaneous Localization and Mapping (SLAM) which can be involved either in AV applications or in Cooperative ITS. On the one hand, Bersani et al. [22] presents an integrated system for vehicle state estimation using unscented Kalman filter fusing data from different passive sensors, such as GPS and IMUs, and from active sensors, like RADAR and LiDAR, which are used to detect and track obstacles as well as improve the localization algorithm. On the other hand, Wang et al. [23] compares different localization systems based on both GNSS and V2X communication for inter-vehicle distance calculation, which is needed for safety ADAS applications.

An interesting challenge in fusing different sensors is dealing with their different sampling frequency. In fact, in asynchronous multi-sensor systems, there is the possibility to miss some data when performing state estimation [24]. The authors in [25] presented different multi-rate multi-sensor models for Kalman filtering with missing measurements. The idea is to run the state estimation algorithm at the fastest sensor frequency and just predict the state vector whenever a sensor is missing or it is considered not reliable.

In addition to the widely studied Kalman filtering techniques, which are extensively covered in the vehicle localization literature, alternative strategies like graph optimization have also been explored. The authors in [26] introduced a novel approach in their work, presenting a multi-sensor fusion method formulated as a graphical model. This model optimizes the integration of onboard sensors to enhance positioning performance, utilizing a kinematic vehicle model as the underlying basis.

Recently, the advent of the 5G network cooperative oriented the research toward new horizons, such as cooperative localization and the Internet of Vehicles (IoV). In fact, thanks to the perception algorithm of the surrounding vehicles, vehicle communication allows gathering localization information that can be used to either improve the ego-vehicle state estimation when GPS is not accurate [27] or to fill the gap in case the GPS signal is missing for long periods [28]. Another approach investigated in the literature in the last decades is the use of Assisted-GPS (A-GPS) systems which exploit the terrestrial communication link to determine the current location [29]. This type of localization is typically employed on cell phones to avoid decoding the GPS messages for each satellite observed, thus using a remote server [30]. More recently, an alternative A-GPS system combining a barometer and accelerometer is proposed in [31] to improve the localization on smartphones.

This research paper introduces a localization technique that is both simple and reliable, offering robustness and accuracy for a TLA application. The approach relies on the use of a Kalman filter and map matching. The filter incorporates a one-dimensional uniformly accelerated kinematic motion model and integrates data from various onboard sensors, including the IMU, GPS, and Electronic Control Unit (ECU), each operating at different frequencies. Moreover, the localization algorithm is integrated into the Traffic Light Advisor (TLA) system, as developed in [6]. This integration optimizes the vehicle's speed, minimizing unnecessary stops and ensuring a smoother driving experience, ultimately reducing energy consumption.

The primary contribution of this study is the development and full-scale experimental testing of a localization algorithm in urban scenarios facing GPS signal loss conditions. In particular, the algorithm utilizes multiple sensors and different sampling rates, making it suitable for implementing Advanced Driver Assistance Systems (ADAS), including Traffic Light Advisor (TLA) systems. The research mainly focuses on the urban road environment, where traditional GPS systems exhibit poor performance due to the challenges posed by urban canyoning. Instead of relying solely on velocity integration, the algorithm leverages the combined information from the Electronic Control Unit (ECU) for speed, Inertial Measurement Unit (IMU) for longitudinal acceleration, and GPS measurements to achieve a smoother output and accurate vehicle localization even in the absence of reliable GPS data. Alongside the presentation of the localization algorithm, the study also showcases the practical application of the proposed method. Specifically, experimental results for the Traffic Light Advisor (TLA) system are presented to further validate the approach introduced in [6] through real-world road tests.

3. Experimental Setup

This section aims at introducing the experimental setup used for the validation of both the Kalman filter and the Traffic Light Advisor system. The localization algorithm relies on a GPS receiver, responsible for locating the vehicle via latitude and longitude measurements, the integrated speed value available from the Electronic Control Unit (ECU) of the trolley-bus, and an Inertial Measurement Unit (IMU) that returns the values of acceleration along its axes. The main navigation system utilized is the GPS which is installed on the front part of the vehicle and provides spatial coordinates in a fixed reference frame. As mentioned, it can be missing for significant portions of the path when the number of satellites is not sufficient or the signal is not reliable, thus not allowing the algorithm to know the measurement of the vehicle's position. The information coming from the GPS receiver must be then fused with other measurements coming from the ECU, providing the longitudinal velocity of the vehicle, and the 5 DoF IMUs measuring the acceleration of a body along the three main axes (x, y, z), as well as the rotational speed around the x and y axes as shown in Figure 1.

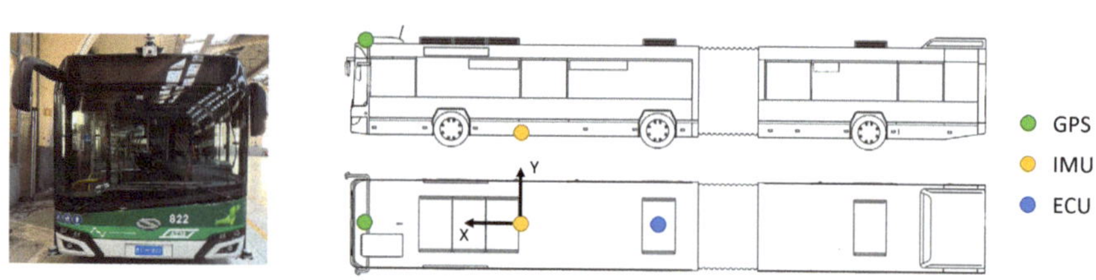

Figure 1. View and schematic representation of the instrumented vehicle used in the testing campaigns.

It is worth mentioning that, as depicted in the architectural diagram in Figure 2, the available sensors have different sampling frequencies. In fact, the GPS rate is 10 Hz, while the ECU returns the speed value at 20 Hz and the acceleration from the IMU comes at 100 Hz. As a consequence, the localization algorithm running at 100 Hz has to deal with these different frequencies.

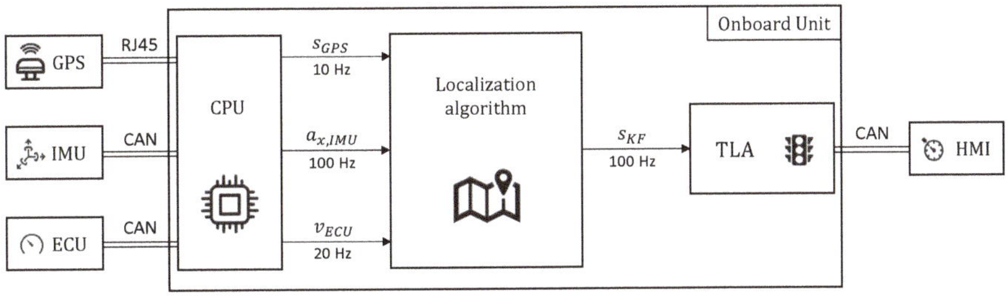

Figure 2. Architectural diagram of the vehicle's sensor acquisition and processing.

On the onboard computational unit, the sensor acquisition, the localization algorithm, as well as the TLA system run on a soft real-time-based Robotic Operating System (ROS) architecture [32], allowing to have a simple framework for managing information coming from different sources using a publisher–subscriber logic. Within this framework, there are different nodes for publishing both the raw sensor data on the vehicle network and the vehicle data that can be read from its Controller Area Network (CAN), such as vehicle speed. This connection to the vehicle CAN-bus is used also to publish the outputs of the Traffic Light Advisor system so that the information can be shown to the driver on the integrated dashboard of the vehicle. Figure 3 reports a sample snapshot of the dedicated Human–Machine Interface (HMI). In particular, in the middle of the dashboard the driver receives a synthetic visual indication (i.e., an arrow indicator) to understand whether to accelerate or decelerate with respect to the current vehicle speed in order to reach the traffic light without stopping. This is done in order to minimize as much as possible the possible distraction source for the driver. However, additional information, such as the current traffic light status (i.e., top left corner), the time-to-change of the upcoming traffic light, and the value of the speed proposed by the TLA algorithm (i.e., bottom left corner), is provided in the periphery of the HMI.

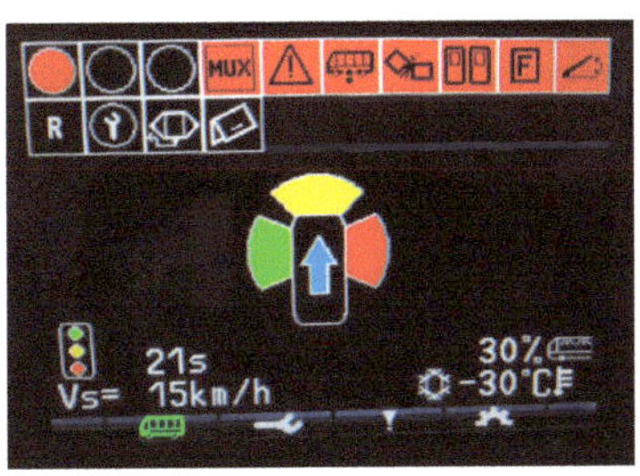

Figure 3. Example view of the Human–Machine Interface used for TLA experimental tests (snapshot with test values for all possible outputs).

The testing area is available on a 4 km long portion of the regular service trolley-bus path in the city of Milan, being mainly covered in a preferential lane for public transportation. The route map (see Figure 4a) includes different scenarios, such as avenues with trees, urban canyoning, mid-narrow turns, and a tunnel 200 m long where the GPS is missing for a relatively long time. Furthermore, in order to have a better assessment of the localization algorithm, additional tests have been performed where a more favorable RTK correction for GPS is available, thus having a ground truth reference to evaluate the algorithm's performances. Moreover, the second testing scenario (see Figure 4b) considered presents more severe testing conditions in terms of curve severity.

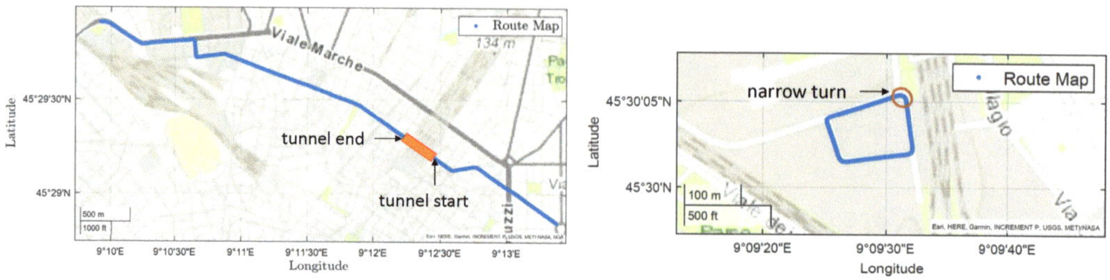

(**a**) (**b**)

Figure 4. Testing areas in the city of Milan. (**a**) Testing Area 1; (**b**) Testing Area 2.

4. Localization Algorithm

In order to minimize the error when estimating the position, a widely spread choice while dealing with linear systems is the use of the Kalman filter, also known as the Linear Quadratic Estimator (LQE). The principle of the Kalman filter (scheme in Figure 5) is to use a dynamic model of a system, with a number of variables constituting the state vector $\boldsymbol{x}$ and describing the system itself and its evolution over time. The system prediction step is taken thanks to known input variables, i.e., control inputs $\boldsymbol{u}$, while the available measurements $\boldsymbol{z}$ are used to properly update the propagated variables in order to minimize the difference between the predicted states and the observed quantities. For the present application, the governing equations are those related to a simple uniformly accelerated 1D model:

$$\begin{cases} s_{t+1} = s_t + v_t \cdot t + \frac{1}{2} a_t \cdot t^2 \\ v_{t+1} = v_t + a_t \cdot t \end{cases} \quad (1)$$

where s, v, and a represent the vehicle's position along the curvilinear coordinate on the path, speed, and longitudinal acceleration, respectively. The choice of such a simple model is justified by the fact that, on the one hand, the speed, and thus the accelerations, are limited. On the other hand, the TLA application needs just longitudinal accuracy along a predefined map of the path which has been obtained with the same algorithm presented in [33] based on the Cubic Hermite Spline (CHS).

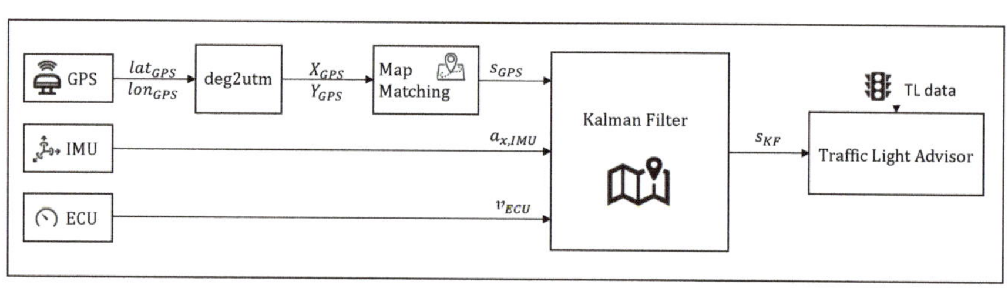

Figure 5. Scheme of the localization algorithm.

Figure 5 summarizes the scheme of the implemented Kalman filter for localizing the trolley-bus along a known map of the route followed by the vehicle. In the following, the general mathematical description of the state estimator is reported, dividing the algorithm into four stages for the sake of clarity.

4.1. Process Equation

The first step reports the process equation, defining a discrete-time linear time-varying system as

$$\boldsymbol{x}_k = \boldsymbol{F}_{k-1}\boldsymbol{x}_{k-1} + \boldsymbol{G}_{k-1}\boldsymbol{u}_{k-1} + \boldsymbol{w}_{k-1} \quad (2)$$

with it being possible to assume that the system matrices $\boldsymbol{F}_{k-1}$ (i.e., state transition matrix), $\boldsymbol{G}_{k-1}$ (i.e., control-input matrix), and control action $\boldsymbol{u}_{k-1}$ are known without errors. In fact, according to Equation (1), the state transition and control-input matrices can be defined as

$$\boldsymbol{F}_{k-1} = \begin{bmatrix} 1 & dt \\ 0 & 1 \end{bmatrix} \qquad \boldsymbol{G}_{k-1} = \begin{bmatrix} \frac{1}{2}dt^2 \\ dt \end{bmatrix} \tag{3}$$

where dt stands for the integration time step equal to 0.01 s.

The process noise $\boldsymbol{w}_{k-1} \sim (0, \boldsymbol{Q}_{k-1})$ assumes a random Gaussian zero-mean covariance $\boldsymbol{Q}_{k-1}$, and it accounts for the noise related both to the modeling and to the input variables. The covariance $\boldsymbol{Q}_{k-1}$ indicates how much the system model can be trusted for the prediction of the estimate. In fact, higher values of $\boldsymbol{Q}_{k-1}$ indicate lower accuracy of the model, so less weight on the estimate.

In this work, the noise due to the model is assumed to be negligible and, as an additional assumption, the covariance matrix $\boldsymbol{Q}_{k-1}$ is considered constant and diagonal:

$$\boldsymbol{Q}_{k-1} = \begin{bmatrix} Q_{pos} & 0 \\ 0 & Q_{vel} \end{bmatrix} \tag{4}$$

in which the diagonal elements of the matrix are calculated as

$$Q_{pos} = dt^2 \cdot \sigma_{a_x}^2 \cdot dt^2 = 10^{-10} \tag{5}$$

$$Q_{vel} = dt \cdot \sigma_{a_x}^2 \cdot dt = 10^{-6} \tag{6}$$

where $\sigma_{a_x}^2$ represents the longitudinal acceleration variance, obtained from measurements.

4.2. Measurement Equation

The measurement equation of the system is written as

$$\boldsymbol{z}_k = \boldsymbol{H}_k \boldsymbol{x}_k + \boldsymbol{v}_k \tag{7}$$

where $\boldsymbol{H}_k$ represents the measurement matrix structured as

$$\boldsymbol{H}_k = \begin{bmatrix} H_{gps} & 0 \\ 0 & H_{ecu} \end{bmatrix} \tag{8}$$

with H_{gps} and H_{ecu} being Boolean values depending on GPS and ECU data availability, as depicted in Figure 6.

As far as the $\boldsymbol{H}_k$ definition is concerned, in order to deal with the GPS outages and the multi-rate multi-sensor setup illustrated above, H_{gps} and H_{ecu} are Boolean variables defined depending on each sensor's availability at time k. Both H_{gps} and H_{ecu} are initially set to zero; if a GPS measure is available and it is considered reliable as the number of satellites received is greater than 7, then H_{gps} is set to 1. As for the speed, if a new speed measurement arrives, H_{ecu} is set to 1. It is worth noting that the speed value read from the ECU is considered always reliable, as the vehicle is thought to run in a standard adherence condition with a limited average speed. In case no measurement is available for the update, $\mathbf{H}_k$ remains null; thus, the state keeps on just being predicted. Performing this check every 0.01 s, the consequence, in the best-case scenario, is that the GPS update occurs just once every 10 time steps, while the ECU update happens once in 5 time steps.

Gaussian measurement noise $\boldsymbol{n}_k \sim (0, \boldsymbol{R}_k)$ is associated with the sensors used for the measurement update. It is characterized by zero mean and covariance $\boldsymbol{R}_k$ defined as

$$\boldsymbol{R}_k = \begin{bmatrix} R_{gps} & 0 \\ 0 & R_{ecu} \end{bmatrix} \tag{9}$$

where $R_{gps} = 10^{-1}$ and $R_{ecu} = 10^{-5}$ are tuning parameters for the filter related to the reliability of the sensors, as they are obtained by computing the variance of the signals of the two sensors. In fact, more reliable sensors lead to lower $\mathbf{R}_k$ values, while sensors introducing more noise will be responsible for higher $\mathbf{R}_k$ values, thus leading to a lower impact of the measurement update on the state estimation. These values are then adjusted to obtain additional stability for the estimate, especially when the vehicle stands still. The idea behind the tuning is based on the following principles:

- $R_{gps} > R_{ecu}$ in the driving condition: the availability and the accuracy of the GPS depend on many different factors, while the Wheel Speed Sensor is much more reliable and accurate, as the values of the variance confirm;
- $R_{gps} >> R_{ecu}$ in the standing-still condition: the value of R_{gps} has been increased to 1 when the vehicle speed is lower than 1 km/h; otherwise, the covariance values remain the default ones. In this way, when the vehicle stands still, the GPS data are less considered as they are less reliable and accurate for the update, while the ECU speed value becomes much more important.

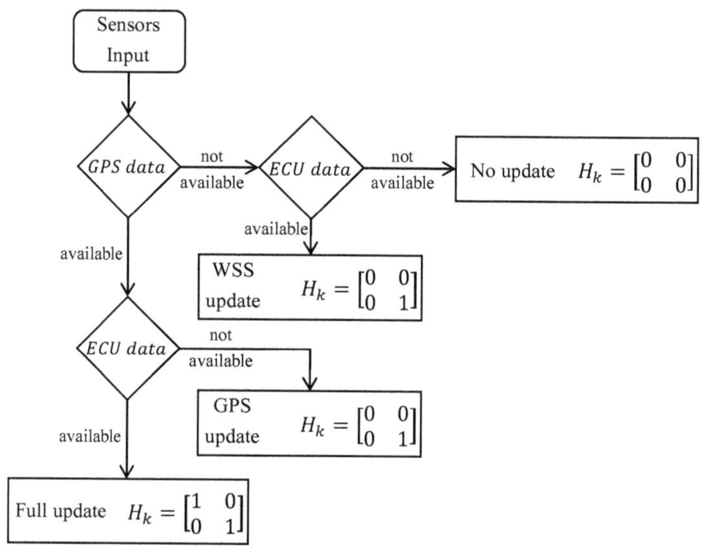

Figure 6. Measurement matrix $\boldsymbol{H}_k$ workflow definition.

4.3. Time Update Equations

In the prediction step defined in (2) at time k, the predicted state $\hat{\boldsymbol{x}}_k^-$ and corresponding covariance $\boldsymbol{P}_k^-$ are calculated according to the model:

$$\begin{cases} \hat{\boldsymbol{x}}_k^- = \boldsymbol{F}_{k-1}\hat{\boldsymbol{x}}_{k-1}^+ + \boldsymbol{G}_{k-1}\boldsymbol{u}_{i-1} \\ \boldsymbol{P}_k^- = \boldsymbol{F}_{k-1}\boldsymbol{P}_{k-1}^+\boldsymbol{F}_{k-1}^T + \boldsymbol{Q}_{k-1} \end{cases} \tag{10}$$

4.4. Measurement Update Equations

When the measurements are available at instant k, then the measurement step consists of the following equations:

$$\begin{cases} \boldsymbol{K}_k = \boldsymbol{P}_k^- \boldsymbol{H}_k^T \left[\boldsymbol{H}_k\boldsymbol{P}_k^-\boldsymbol{H}_k^T + \boldsymbol{R}_k\right]^{-1} \\ \hat{\boldsymbol{x}}_k^+ = \hat{\boldsymbol{x}}_k^- + \boldsymbol{K}_k\left[\boldsymbol{z}_k - \boldsymbol{H}_k\hat{\boldsymbol{x}}_k^-\right] \\ \boldsymbol{P}_k^+ = \left[\boldsymbol{I} - \boldsymbol{K}_k\boldsymbol{H}_k\right]\boldsymbol{P}_k^- \end{cases} \tag{11}$$

These equations represent the updated estimations for the state $\hat{x}_k^+$ and the covariance P_k^+, respectively, and these two quantities will be required for the following step $k+1$ in the prediction step equations shown above.

5. Traffic Light Advisor Experimental Validation

The Traffic Lights Advisor (TLA) system is an auxiliary tool for the driver, which is able to provide real-time information about the following traffic light's phase while suggesting the speed to cruise through the intersections on the path during the green light phase. Entering into the details of the TLA as an ADAS, it is expected to deal with the typical situations faced approaching a traffic light:

- Stop&Go: the algorithm is intended to properly modulate the vehicle speed in order to avoid a complete stop (when possible) in front of the traffic light.
- Last-second braking: the algorithm should inform the driver about the need to slow down, as an acceleration maneuver is not feasible.
- Unnecessary stop: the algorithm aims at suggesting to the driver the recommended speed (compliant with road limits and vehicle safety) in order to pass the upcoming intersections during the green phase of the traffic light.

As a result, the main goal of this algorithm is to save both travel time and the energy used by the vehicle. This is done by considering the 4 traffic lights ahead on the path closest to the vehicle. In particular, the algorithm is thought to consider a uniformly accelerated motion model for the vehicle to reach the suggested speed within the first traffic light of the series and then keep that velocity to safely pass the upcoming intersections without stopping. In the following, the most relevant features of the functioning of the TLA system are reported; for further details, the reader can refer to the work in [6] where the full design of the algorithm is presented.

The TLA system is based on an iterative algorithm for selecting the recommended velocity which has to fulfill the following constraints:

- The velocity has to be below the maximum allowed speed limit for the road.
- The vehicle should be able to reach the target velocity following a uniformly accelerated motion model within the end of the first available green phase of the first traffic light ahead, with the the maximum vehicle acceleration limited to 1 m/s^2 for both safety and comfort concerns.
- Considering the generic ith traffic light, the vehicle's admissible speed range is obtained from the intersection between the required velocity to reach the upcoming intersection and the admissible speed range computed for the $(i-1)$th traffic light.

Figure 7 represents the scheme of the Traffic Light Advisor system: the inputs, coming from onboard sensors for the localization algorithm presented in Section 4; the traffic lights; and the map enter the algorithm that produces an output that is shown in the HMI.

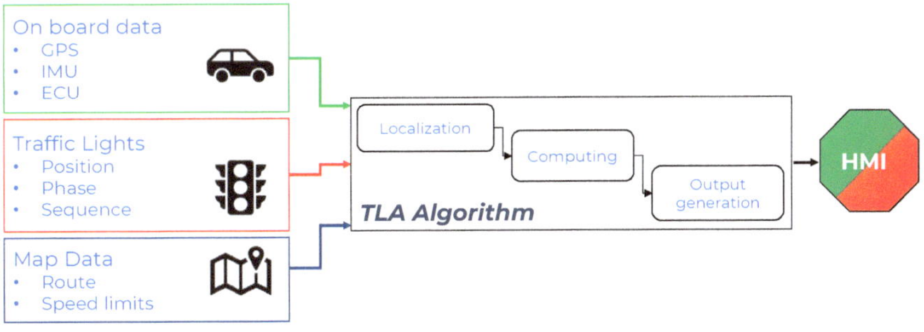

Figure 7. Schematics of the Traffic Light Advisor algorithm.

As far as the traffic light data are concerned, through the municipality of Milan there has been the possibility to know the traffic light plans of the intersections along the route.

These data are fed to a "Traffic Light Generator" which is in charge of emulating the real traffic lights on the road. This emulator, implemented in the Matlab-Simulink environment, communicates the traffic lights' phases, sequences, and positions to the TLA algorithm.

In this work, the aim is to add an experimental validation of the non-optimal Traffic Light Advisor algorithm presented in [6]. To validate the functioning of the algorithm, similarly to what is performed in [34], three different cases are considered:

- **Base Case**: test run without the TLA used to establish a benchmark. This is obtained by letting the driver behave as usual, with no additional information available to the driver with respect to the typical driving case.
- **TLA Case**: test run with the TLA running and showing information to the driver who tries to follow the instructions. The algorithm, running in real time, conveys the information to the driver through a specifically designed HMI (see Figure 3).
- **Ideal Case**: simulation run in post-process on the basis of the TLA Case data experimentally acquired, aiming to assess the behavior of the algorithm, assuming an ideal driver able to perfectly follow the algorithm's instructions. This is proposed to check whether the algorithm is working correctly, observing how it would operate if no constraints set by external factors such as traffic, the driver's reflexes, and other interference were to impact its ideal behavior.

In Section 6, the results of the experimental campaign are reported, proposing a comparison between the Base Case and TLA Case. Subsequently, the real test performed on the road is compared with the aforementioned Ideal Case. The comparison is carried out not only by looking at the kinematic performances in terms of the covered distance, average speed, and acceleration but also from an energetic point of view. In fact, knowing the vehicle mass (i.e., $m = 19{,}800$ kg) and the vehicle acceleration a and velocity v from the onboard sensors, it is possible to obtain a rough evaluation of the power consumption expressed in (kW) starting from the inertial force as

$$P = \frac{m \cdot a \cdot v}{1000} \tag{12}$$

From the power consumption, it is possible to derive then the following index, i.e., Instantaneous Energy Consumption (IEC) expressed in (kWh/100 km), which allows quantifying the energy being drained by the vehicle:

$$IEC = P\,\frac{\Delta t}{\Delta s} \tag{13}$$

where Δt is the time step of the algorithm and Δs is the distance covered by the vehicle in the time span.

6. Results

This section is devoted to the presentation of the validation results of both the localization state estimator and the Traffic Light Advisor algorithm. The tests have been performed on two different paths, as shown in Figure 4, because of the availability of the traffic light information during the test runs.

6.1. Localization

The state estimator presented in Section 4 runs at 100 Hz as a standalone C++ ROS node generated in the Matlab-Simulink environment [35], subscribing to the sensors' topics present in the ROS network of the vehicle. In the following, the results of the state estimator experimental data on the two Testing Areas are reported. It is worth noticing that the measurement used by the Kalman filter is the raw data of the GPS device, as it represents a more general and applicable condition in urban scenarios. As far as the localization accuracy is concerned, the estimation error is evaluated using as ground truth the data coming from the real-time kinematic (RTK)-corrected GPS data which reaches cm level precision.

The analysis is focused on three main conditions:

- Vehicle standing still: this condition is quite frequent in urban scenarios because of the high number of intersections and the stops along the path of a local public transportation vehicle. The typical GPS behavior in this situation is to fluctuate around the actual position, causing the vehicle localization to change, both forward and backward.
- Prolonged GPS outage: this condition is usually faced because of urban canyoning, high trees, or tunnels.
- Regular driving with curved path: this is the general scenario to be considered for the localization accuracy assessment.

The first two items are analyzed in Testing Area 1 (i.e., Figure 4a) which covers the regular service path of the vehicle. Figure 8a depicts the trend of the covered distance (i.e., the curvilinear coordinate along the path) in correspondence to a stop. As can be seen, on the one hand, the GPS trend is floating; on the other hand, the state estimator is able to provide a constant s value thanks to the increase in the R_{gps}. This allows having a constant value for the curvilinear coordinate when the vehicle is stopped, which is fundamental for the TLA application as the change in the distance of the traffic light ahead affects the TLA speed calculation. The plot reported in Figure 8b shows the state estimator trend during a prolonged GPS outage because of a 200 m long tunnel. The distance between the state estimator value s_{KF} and the GPS coordinate s_{GPS} as soon as it becomes available is equal to 1.86 m. Although this value is relatively high, the GPS benchmark after a long outage has to be considered as not reliable, as it requires some time (i.e., 20–30 m) to obtain acceptable GPS accuracy.

The testing campaign on the second Testing Area (i.e., Figure 4b) aims to evaluate the actual localization accuracy that can be obtained with the proposed localization algorithm. In fact, this area is characterized by better GPS coverage as there are no urban canyons that typically affect the GPS signal. Furthermore, the route path has four narrow turns, allowing to assess the behavior of the state estimator in the turn condition as well. In this case study, the raw data of the GPS, i.e., before the RTK correction, are fed to the localization algorithm. The state estimator output is then compared with the RTK correction GPS data which provides the ground truth value for the accuracy assessment.

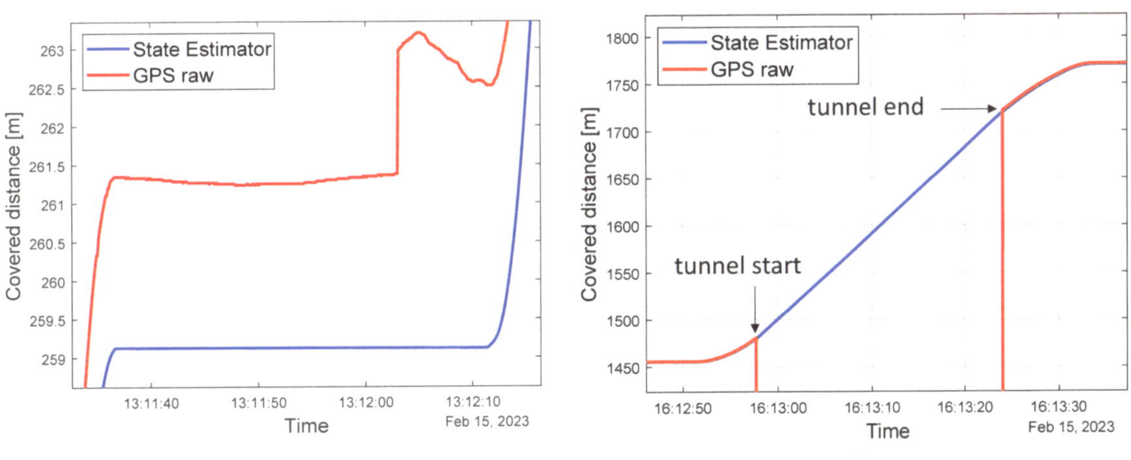

Figure 8. Testing Area 1: state estimator results analysis. (**a**) Vehicle standing still; (**b**) long GPS outage.

In Figure 9, a portion of the trend of the covered distance on the closed path of the Testing Area 2 is shown. In this scenario, it is possible to appreciate a much smoother trend of the state estimator with respect to the step-wise GPS signal because of the higher frequency of the localization algorithm, i.e., 100 Hz. When considering the entire path, the estimation error can be calculated as

$$\epsilon_s = s_{RTK} - s_{KF} \tag{14}$$

where s_{RTK} and s_{KF} are the curvilinear coordinate along the map of the GPS RTK corrected and the proposed Kalman filter, respectively. Taking the Root Mean Square value of the error trend over the whole path, it turns out to be equal to 0.28 m. Figure 10 reports the trajectory of the vehicle estimated by the proposed Kalman filter in the narrowest turn on the path. The color indicates the estimation error ϵ_s, with the estimation error limited to 0.4 m also in the turn condition.

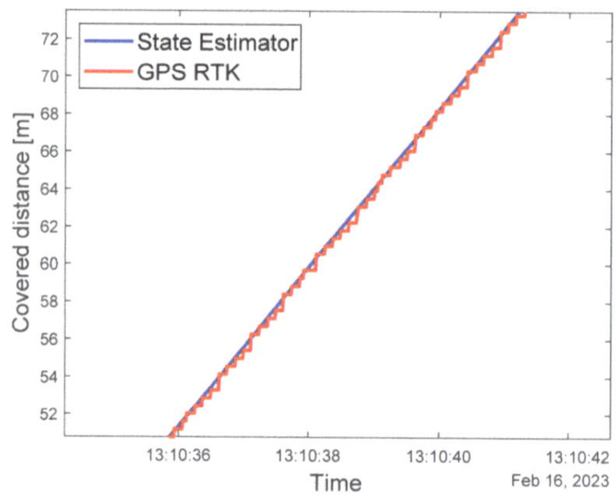

Figure 9. Testing Area 2: state estimator result analysis.

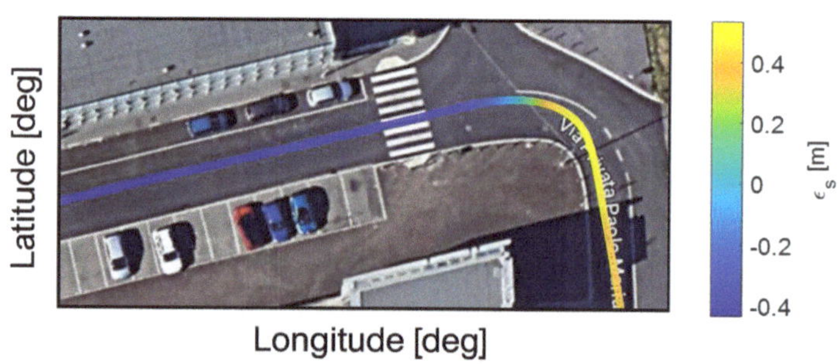

Figure 10. Testing Area 2: state estimator accuracy in curve condition.

The obtained experimental results show good accuracy for the TLA application, which just needs the vehicle location along the path to compute the distance from the upcoming traffic lights. Furthermore, as the vehicle considered is a large local public transportation vehicle, changes in speed and direction are typically not so harsh, so the choice of a one-dimensional model, on the one hand, allows to minimize the implementation effort. As far as the behavior in the curve condition is concerned, the results are promising and allow running the TLA algorithm in a general urban scenario, although the model appears to not be accurate enough for control logic dealing with vehicle lateral dynamics.

6.2. *Traffic Light Advisor*

As mentioned in Section 5, the assessment of the TLA performances is performed both from a kinematic and an energetic point of view. In particular, in Figure 11, the plot of the covered distance of one test run, comparing the Base Case of the driver with the TLA

Case, is shown. The horizontal lines represent the traffic lights status on the path, with the sequence between the red and green phases. It is worth mentioning that the yellow phase has been included in the red one for adding an extra safety margin to the algorithm.

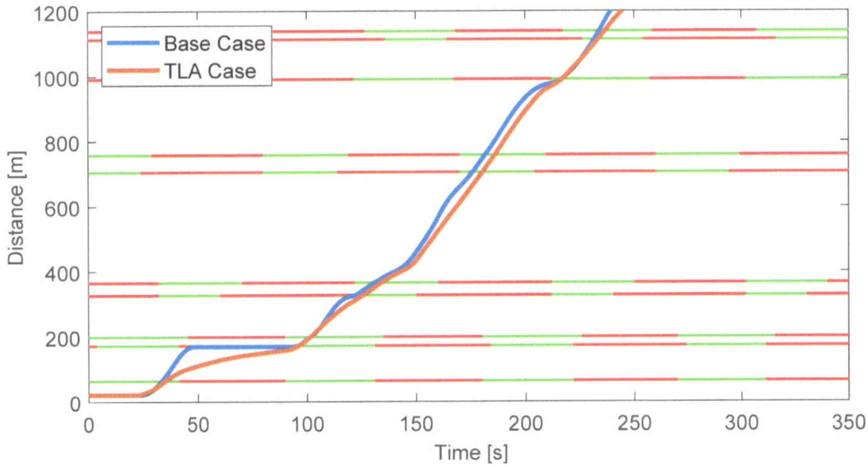

Figure 11. TLA Case comparison with respect to Base Case.

From the graph, it is possible to observe how the TLA system, besides avoiding the stop at the second traffic light encountered by the vehicle, allows having a smoother trend for the entire path, meaning that the velocity has a lower fluctuation, and thus lower acceleration.

Table 1 proposes a similar analysis from a quantitative point of view, showing the Root Mean Square values of velocity, acceleration, and IEC, previously defined. The results indicate that, when the TLA is active, the vehicle proceeds with a lower average speed with respect to the Base Case. This can sound unexpected, but it is consistent with the lower acceleration value, as the driver tends to accelerate more than required to cruise through all the intersections without stops. Besides more comfort for passengers, having lower accelerations guarantees lower energy consumption, with a 40% reduction in the energetic indicator IEC when the TLA is active. Furthermore, these values confirm the results obtained from the simulations conducted in [6].

Table 1. RMS values for speed, acceleration, and IEC: comparison between Base Case and TLA Case.

	Base Case	TLA Case	Reduction
v_{RMS} [m/s]	6.50	4.95	−23.7%
a_{RMS} [m/s^2]	0.53	0.31	−44.3%
IEC_{RMS} $\left[\frac{\text{kWh}}{100\text{ km}}\right]$	291.39	175.92	−39.6%

The plot in Figure 12 is intended to investigate in which situation the algorithm has an improvement margin and how it could be adjusted to further enhance its impact.

In fact, the spotlight is on the algorithm's reaction to the inability of the driver to perfectly follow the speed reference. The plot reports the previously shown trend of the TLA Case and it is compared with a set of ideal behaviors of the vehicle obtained by running different simulations 8 s long, each one having the vehicle position and velocity in correspondence to the start of the simulation (i.e., current actual vehicle position and velocity) as the initial conditions. This allows to see how the vehicle would have proceeded in an ideal case, thus highlighting, on the one hand, the effect of external factors as well as driver behaviors, and, on the other hand, the capability of the TLA system to adapt to the conditions the vehicle is facing. Two main points of interest are indicated in the plot, those in which the ideal response to given real conditions (i.e., blue solid line) is suggesting a different behavior. In both cases, the vehicle should accelerate right before crossing the facing traffic light, but the driver is more confident in waiting for some seconds because the traffic in front of the vehicle is still showing a red light. This is a natural tendency of human drivers to not fully trust an indication of the HMI. In fact, although based on actual real-time data about the traffic light time-to-change, in some cases the TLA is perceived unsafe as it may also suggest accelerating when a common driver, without knowing when the light is going to change, would not.

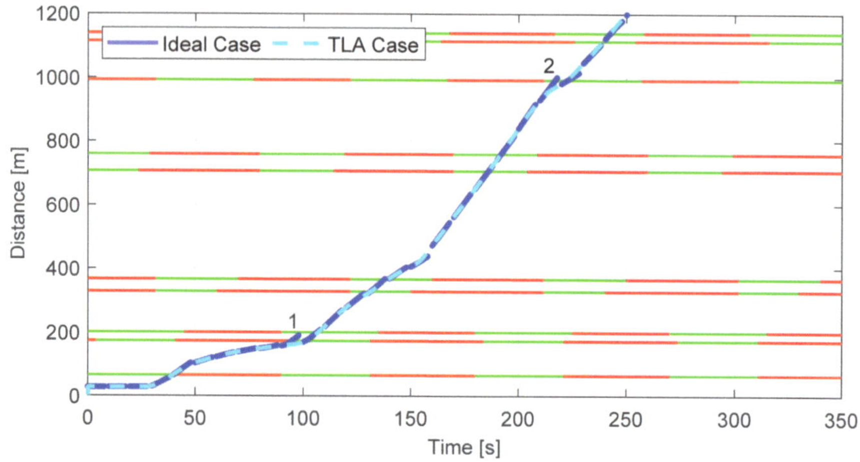

Figure 12. TLA Case comparison with respect to Ideal Case.

7. Conclusions

In this work, a localization algorithm based on Kalman filtering for a Traffic Light Advisor application and the TLA experimental validation are proposed. The implemented Kalman filter is designed to run at 100 Hz with a simple 1D kinematic model and measurements from sensors having different sampling frequencies. Real-world tests provided results accurate enough to be integrated into the TLA algorithm with an average error lower than 0.5 m, having robust behavior both in long GPS outage situations and standing-still and curve conditions thanks to the filter weight tuning and map matching. Regarding the TLA validation, the experimental campaign confirmed the positive impact on both comfort, service regularity, and energy consumption with respect to unassisted driving. The comparison with a simulated ideal case highlighted the areas of improvement for the actual implementation of the system, such as the presence of traffic in front of the vehicle, and external factors, like road unevenness, that make the driver slow down, as well as driver difficulties in following and trusting the HMI indications.

As future developments, on the one hand, it would be interesting to extend the state estimator to a 2D vehicle model, aiming at using the implemented localization algorithm for other ADAS applications also considering later dynamics of the vehicle. On the other hand, the TLA experimental validation would need additional testing campaigns to perform a larger statistical assessment of the algorithm. Furthermore, it would be valuable to introduce information about the traffic ahead in the TLA algorithm to cope with the challenges that emerged from the test.

Author Contributions: Conceptualization, E.S.; Methodology, S.A.; Software, D.V.; Validation, S.A.; Formal analysis, S.A.; Data curation, D.V.; Writing—original draft, D.V.; Writing—review & editing, D.V., S.A. and E.S.; Supervision, E.S.; Funding acquisition, E.S. All authors have read and agreed to the published version of the manuscript.

Funding: This research received no external funding.

Institutional Review Board Statement: Not applicable.

Informed Consent Statement: Informed consent was obtained from all subjects involved in the study.

Data Availability Statement: Not applicable.

Acknowledgments: This work is part of the project "Joint Research Lab per la Mobilità urbana". A particular acknowledgment goes to all the partners involved in the task, ATM S.p.A., Solaris Bus & Coach sp. z o.o., AMAT, Vodafone Italia S.p.A., and Comune di Milano for supporting the project initiative.

Conflicts of Interest: The authors declare no conflict of interest.

References

1. Montanaro, U.; Dixit, S.; Fallah, S.; Dianati, M.; Stevens, A.; Oxtoby, D.; Mouzakitis, A. Towards connected autonomous driving: Review of use-cases. *Veh. Syst. Dyn.* **2019**, *57*, 779–814. [CrossRef]
2. Kassas, Z.Z.M.; Maaref, M.; Morales, J.J.; Khalife, J.J.; Shamei, K. Robust Vehicular Localization and Map Matching in Urban Environments through IMU, GNSS, and Cellular Signals. *IEEE Intell. Transp. Syst. Mag.* **2020**, *12*, 36–52. [CrossRef]
3. Stahlmann, R.; Möller, M.; Brauer, A.; German, R.; Eckhoff, D. Exploring GLOSA systems in the field: Technical evaluation and results. *Comput. Commun.* **2018**, *120*, 112–124. [CrossRef]
4. Barth, M.; Mandava, S.; Boriboonsomsin, K.; Xia, H. Dynamic ECO-driving for arterial corridors. In Proceedings of the 2011 IEEE Forum on Integrated and Sustainable Transportation Systems, Vienna, Austria, 29 June–1 July 2011; pp. 182–188. [CrossRef]
5. Cai, L.S.; Ning, G.B. Adaptive Driving Speed Guiding to Avoid Red Traffic Lights. *Appl. Mech. Mater.* **2013**, *347–350*, 3832–3836. [CrossRef]
6. Khayyat, M.; Gabriele, A.; Mancini, F.; Arrigoni, S.; Braghin, F. Design and Simulation-Based Testing of Connected Traffic Light Guidance Systems. *arXiv* **2022**, arXiv:2209.05783.
7. De Nunzio, G.; Canudas de Wit, C.; Moulin, P.; Di Domenico, D. Eco-driving in urban traffic networks using traffic signal information. In Proceedings of the 52nd IEEE Conference on Decision and Control, Firenze, Italy, 10–13 December 2013; pp. 892–898. [CrossRef]
8. Seredynski, M.; Dorronsoro, B.; Khadraoui, D. Comparison of Green Light Optimal Speed Advisory approaches. In Proceedings of the 16th International IEEE Conference on Intelligent Transportation Systems (ITSC 2013), The Hague, The Netherlands, 6–9 October 2013; pp. 2187–2192. [CrossRef]
9. Guan, T.; Frey, C.W. Predictive fuel efficiency optimization using traffic light timings and fuel consumption model. In Proceedings of the 16th International IEEE Conference on Intelligent Transportation Systems (ITSC 2013), The Hague, The Netherlands, 6–9 October 2013; pp. 1553–1558. [CrossRef]
10. Suramardhana, T.A.; Jeong, H.Y. A driver-centric green light optimal speed advisory (DC-GLOSA) for improving road traffic congestion at urban intersections. In Proceedings of the 2014 IEEE Asia Pacific Conference on Wireless and Mobile, Bali, Indonesia, 28–30 August 2014; pp. 304–309. [CrossRef]
11. Zhang, R.; Yao, E. Eco-driving at signalised intersections for electric vehicles. *IET Intell. Transp. Syst.* **2015**, *9*, 488–497. [CrossRef]
12. Wegener, M.; Koch, L.; Eisenbarth, M.; Andert, J. Automated eco-driving in urban scenarios using deep reinforcement learning. *Transp. Res. Part C Emerg. Technol.* **2021**, *126*, 102967. [CrossRef]
13. Xu, B.; Chen, X.; Li, K.; Hu, M.; Bian, Y.; Yu, Q.; Wang, J. Double-layer speed optimization for reducing fuel consumption with vehicle-to-infrastructure communication. *J. Intell. Transp. Syst.* **2019**, *23*, 513–524. [CrossRef]
14. Coppola, A.; Costanzo, L.D.; Pariota, L.; Santini, S.; Bifulco, G.N. An Integrated Simulation Environment to test the effectiveness of GLOSA services under different working conditions. *Transp. Res. Part C Emerg. Technol.* **2022**, *134*, 103455. [CrossRef]
15. Seredynski, M.; Mazurczyk, W.; Khadraoui, D. Multi-segment Green Light Optimal Speed Advisory. In Proceedings of the 2013 IEEE International Symposium on Parallel & Distributed Processing, Workshops and Phd Forum, Cambridge, MA, USA, 20–24 May 2013; pp. 459–465. [CrossRef]
16. Kural, E.; Jones, S.; Parrilla, A.F.; Grauers, A. Traffic light assistant system for optimized energy consumption in an electric vehicle. In Proceedings of the 2014 International Conference on Connected Vehicles and Expo (ICCVE), Vienna, Austria, 3–7 November 2014; pp. 604–611. [CrossRef]
17. Lu, Y.; Ma, H.; Smart, E.; Yu, H. Real-Time Performance-Focused Localization Techniques for Autonomous Vehicle: A Review. *IEEE Trans. Intell. Transp. Syst.* **2022**, *23*, 6082–6100. [CrossRef]
18. Gao, L.; Xiong, L.; Xia, X.; Lu, Y.; Yu, Z. Vehicle Localization with Vehicle Dynamics During GNSS Outages. In Proceedings of the Advances in Dynamics of Vehicles on Roads and Tracks, Gothenburg, Sweden, 12–16 August 2019; Klomp, M., Bruzelius, F., Nielsen, J., Hillemyr, A., Eds.; Springer International Publishing: Cham, Switzerland, 2020; pp. 1071–1079.
19. Gao, L.; Xiong, L.; Xia, X.; Lu, Y.; Yu, Z.; Khajepour, A. Improved Vehicle Localization Using On-Board Sensors and Vehicle Lateral Velocity. *IEEE Sens. J.* **2022**, *22*, 6818–6831. [CrossRef]

20. Alrousan, Q.; Alzu'bi, H.; Pfeil, A.; Tasky, T. *Autonomous Vehicle Multi-Sensors Localization in Unstructured Environment*; SAE Technical Paper; SAE International: Warrendale PA, USA, 2020; Volume 2020. [CrossRef]
21. Gu, S.; Dai, C.; Mao, F.; Fang, W. Integration of Multi-GNSS PPP-RTK/INS/Vision with a Cascading Kalman Filter for Vehicle Navigation in Urban Areas. *Remote Sens.* **2022**, *14*, 4337. [CrossRef]
22. Bersani, M.; Mentasti, S.; Dahal, P.; Arrigoni, S.; Vignati, M.; Cheli, F.; Matteucci, M. An integrated algorithm for ego-vehicle and obstacles state estimation for autonomous driving. *Robot. Auton. Syst.* **2021**, *139*, 103662. [CrossRef]
23. Wang, F.; Zhuang, W.; Yin, G.; Liu, S.; Liu, Y.; Dong, H. Robust inter-vehicle distance measurement using cooperative vehicle localization. *Sensors* **2021**, *21*, 2048. [CrossRef]
24. Mahmoud, M.S.; Emzir, M.F. State estimation with asynchronous multi-rate multi-smart sensors. *Inf. Sci.* **2012**, *196*, 15–27. [CrossRef]
25. Kordestani, M.; Dehghani, M.; Moshiri, B.; Saif, M. A New Fusion Estimation Method for Multi-Rate Multi-Sensor Systems with Missing Measurements. *IEEE Access* **2020**, *8*, 47522–47532. [CrossRef]
26. Bai, S.; Lai, J.; Lyu, P.; Cen, Y.; Ji, B. Improved preintegration method for gnss/imu/in-vehicle sensors navigation using graph optimization. *IEEE Trans. Veh. Technol.* **2021**, *70*, 11446–11457. [CrossRef]
27. Héry, E.; Xu, P.; Bonnifait, P. Distributed asynchronous cooperative localization with inaccurate GNSS positions. In Proceedings of the 2019 IEEE Intelligent Transportation Systems Conference (ITSC), Auckland, New Zealand, 27–30 October 2019; pp. 1857–1863. [CrossRef]
28. Wang, F.; Xu, L.; Zhuang, W.; Yin, G.; Pi, D.; Liang, J.; Liu, Y.; Lu, Y. Geometry-Based Cooperative Localization for Connected Vehicle Subject to Temporary Loss of GNSS Signals. *IEEE Sens. J.* **2021**, *21*, 23527–23536. [CrossRef]
29. Korb, M.; Huang, Q.; Stockel, P.; Kappen, G.C.; Weber, B.; Garcia, M. A Cellular-Modem-Hosted Low-Cost Single-Shot Dual-Mode Assisted-GNSS Receiver for the Internet of Things. In Proceedings of the 2020 IEEE/ION Position, Location and Navigation Symposium (PLANS), Portland, OR, USA, 20–23 April 2020; pp. 1273–1279. [CrossRef]
30. Zandbergen, P.A. Accuracy of iPhone Locations: A Comparison of Assisted GPS, WiFi and Cellular Positioning. *Trans. GIS* **2009**, *13*, 5–25. [CrossRef]
31. Sheng, L.; Tian, R.; Ye, H. Assist GPS to Improve Accuracy Under Complex Road Conditions Using Sensors on Smart Phone. In Proceedings of the MLICOM 2020: International Conference on Machine Learning and Intelligent Communications, Shenzhen, China, 26–27 September 2020; Springer International Publishing: Cham, Switzerland, 2021; pp. 287–303.
32. Quigley, M.; Gerkey, B.; Conley, K.; Faust, J.; Foote, T.; Leibs, J.; Berger, E.; Wheeler, R.; Ng, A. ROS: An open-source Robot Operating System. In Proceedings of the IEEE International Conference on Robotics and Automation (ICRA) Workshop on Open Source Robotics, Kobe, Japan, 12–17 May 2009.
33. Arrigoni, S.; Braghin, F.; Cheli, F. MPC trajectory planner for autonomous driving solved by genetic algorithm technique. *Veh. Syst. Dyn.* **2022**, *60*, 4118–4143. [CrossRef]
34. Chen, H.; Rakha, H.A. Developing and Field Testing a Green Light Optimal Speed Advisory System for Buses. *Energies* **2022**, *15*, 1491. [CrossRef]
35. The MathWorks, Inc. Generate a Standalone ROS Node from Simulink. Available online: https://it.mathworks.com/help/ros/ug/generate-a-standalone-ros-node-from-simulink.html (accessed on 31 July 2023).

sensors

MDPI

Article

Using System Dynamics Approach to Explore the Mode Shift between Automated Vehicles, Conventional Vehicles, and Public Transport in Melbourne, Australia

Yilun Chen, Peter Stasinopoulos, Nirajan Shiwakoti * and Shah Khalid Khan

School of Engineering, RMIT University, Melbourne, VIC 3000, Australia; s3598361@student.rmit.edu.au (Y.C.); peter.stasinopoulos@rmit.edu.au (P.S.); s3680269@student.rmit.edu.au (S.K.K.)
* Correspondence: nirajan.shiwakoti@rmit.edu.au

Abstract: With the increasing use of automated vehicles (AVs) in the coming decades, government authorities and private companies must leverage their potential disruption to benefit society. Few studies have considered the impact of AVs towards mode shift by considering a range of factors at the city level, especially in Australia. To address this knowledge gap, we developed a system dynamic (SD)-based model to explore the mode shift between conventional vehicles (CVs), AVs, and public transport (PT) by systematically considering a range of factors, such as road network, vehicle cost, public transport supply, and congestion level. By using Melbourne's Transport Network as a case study, the model simulates the mode shift among AVs, CVs, and PT modes in the transportation system over 50 years, starting from 2018, with the adoption of AVs beginning in 2025. Inputs such as current traffic, road capacity, public perception, and technological advancement of AVs are used to assess the effects of different policy options on the transport systems. The data source used is from the Victorian Integrated Transport Model (VITM), provided by the Department of Transport and Planning, Melbourne, Australia, data from the existing literature, and authors' assumptions. To our best knowledge, this is the first time using an SD model to investigate the impacts of AVs on mode shift in the Australian context. The findings suggest that AVs will gradually replace CVs as another primary mode of transportation. However, PT will still play a significant role in the transportation system, accounting for 50% of total trips by person after 2058. Cost is the most critical factor affecting AV adoption rates, followed by road network capacity and awareness programs. This study also identifies the need for future research to investigate the induced demand for travel due to the adoption of AVs and the application of equilibrium constraints to the traffic assignment model to increase model accuracy. These findings can be helpful for policymakers and stakeholders to make informed decisions regarding AV adoption policies and strategies.

Keywords: system dynamics; driverless vehicles; future transportation; transport policy; smart mobility

Citation: Chen, Y.; Stasinopoulos, P.; Shiwakoti, N.; Khan, S.K. Using System Dynamics Approach to Explore the Mode Shift between Automated Vehicles, Conventional Vehicles, and Public Transport in Melbourne, Australia. *Sensors* **2023**, *23*, 7388. https://doi.org/10.3390/s23177388

Academic Editors: Yafei Wang, Chao Huang, Hailong Huang, Zhiqiang Zuo, Bo Leng and Peng Hang

Received: 20 July 2023
Revised: 21 August 2023
Accepted: 23 August 2023
Published: 24 August 2023

1. Introduction

Automated driving technologies (e.g., artificial intelligence and remote sensing) have received much attention for their research and developments [1]. Automated driving technologies can transfer vehicle driving functions from human drivers to computers, and the automation level is divided into six levels [2]. Simply defined, level 0 means no driving automation, while level 5 demonstrates full driving automation without any human intervention. Moreover, AVs could improve road safety by eradicating traffic accidents, as most accidents are due to human errors, such as driving too fast and driver fatigue. In short, the upcoming automated vehicles will benefit the broader society by decreasing traffic congestion, offering new mobility choices, and reducing road accidents [3].

The rise in AVs is expected to significantly affect the transportation sector by changing the way people travel. AVs have the potential to revolutionise mobility by reducing traffic congestion, improving road safety, and increasing energy efficiency [4]. However, there are concerns that the widespread adoption of AVs could lead to an increase in vehicle kilometres travelled and a decrease in the use of public transport (PT) and active transportation modes, ultimately increasing energy consumption, emissions, and congestion [4]. Therefore, it is important to investigate the potential mode shift between AVs, CVs, and PT to evaluate the effect of AVs on the transportation system and plan accordingly. Most studies employed a static approach to investigate the effect of AVs on the transportation system without considering the dynamic interactions between different travel modes and the feedback loops that could affect the mode shift behaviour [5,6].

If we want to manage future road networks to meet the demands of automated vehicle trips due to the shift from public transport and conventional vehicle trips, we need to understand how AV trips change over time due to a range of reasons, such as policy implementation, AV cost, and psychological factors. As such, the main contributions of this study are as follows:

1. We developed a system dynamic (SD)-based model to explore the mode shift between conventional vehicles (CVs), AVs, and public transport (PT) by systematically considering a range of factors, such as road network, vehicle cost, public transport supply, and congestion level. This model addresses the knowledge gaps on the impact of AVs towards mode shift by considering a range of factors at the city level.
2. Inputs such as current traffic, road capacity, public perception, and technological advancement of AVs are used to assess the effects of different policy options on the transport systems. An SD approach has been adopted for the present study because it can incorporate the dynamic interactions [7] between different travel modes and the feedback loops that could affect the mode shift behaviour. To our best knowledge, this is the first time using an SD model to investigate the impacts of AVs on mode shift in the Australian context.
3. The SD model provides a valuable contribution to the methodological understanding of the effects of AVs on transportation by considering various system-level factors. The model can be used to explore the effects of AV adoption on mode shift, changes in traffic congestion, and other transportation-related factors, supporting policy decision making to achieve a sustainable, equitable, and accessible transport system, especially for the long term. This model also presents significant advantages. The SD model not only comprehensively considers various factors and their quantitative relationships, but it also allows for sensitivity analysis of individual variables. This capability enables us to thoroughly investigate the influences of each variable, enhancing the model's comprehensiveness and utility. Additionally, the SD model is a powerful tool for analysing the complex interactions between different components of the transportation system and identifying potential solutions to the challenges posed by AV adoption. By providing a detailed analysis of the effects of AV adoption on modal shift behaviour, the proposed model can help policymakers develop policies that promote the adoption of AVs while also minimising the negative effects on PT and congestion.

The paper is organised as follows. Section 2 presents the literature review for the transport modelling and system dynamics approach. It is followed by Section 3 that describes the SD model developed for this study, while Section 4 discusses the results for different scenarios. A list of the abbreviations used in this study is shown in Table 1.

Table 1. A list of abbreviations used in this study.

Abbreviation	Explanation
AVs	Automated Vehicles
CVs	Conventional Vehicles
PT	Public Transport
LoS	Level of Service
CAVs	Connected and Autonomous Vehicles
VITM	Victorian Integrated Transport Model
EVs	Electric Vehicles
SD	System Dynamic
VISTA	Victorian Integrated Survey of Travel and Activity
EVs	Electric Vehicles
DTP	Department of Transport and Planning
CBD	Central Business District
VKT	Vehicle Kilometre Travelled

2. Literature Review

In this section, we review the relevant works under two subsections: transport modelling and system dynamics modelling.

2.1. Transport Modelling

Some past studies investigated the effects of AVs on traffic flow and traffic safety using microscopic traffic simulations of individual vehicles [8,9]. Other interesting studies also investigated the effects of AVs using microscopic traffic modelling. For example, [10] researched lane assignment strategies of AVs and their effects on overall traffic efficiency and safety in a highway scenario. Reference [11] used the traffic simulation package VISSIM to investigate the congestion effects of shared AVs on urban traffic by modelling the peak morning period in 2040. Several shared AV market penetrations were modelled: 0 per cent, 3 per cent, 25 per cent, 50 per cent, and 100 per cent. Similarly, [12] developed an efficient stochastic optimisation framework to find optimal shares between CVs and AVs by considering factors of CAVs (e.g., VKT, the value of time, and automation cost). This framework was successfully applied to the Chicago network, and the system costs were optimised. Reference [13] studied a mixed traffic system to control the density and ratio of CVs and AVs to avoid large-scale traffic congestion using a cellular automation model. It was suggested that the findings would have practical implications for traffic management control. From a traffic safety perspective, mixed traffic flow was simulated to identify the frequency of dangerous situations and the value of time to collision under different penetration levels [14]. The results revealed that smooth driving increases with the CAV penetration rate. Another study conducted a detailed assessment of the effects of CAVs on a freeway using a microsimulation [15]. The findings showed that CAVs could reduce delay and emissions by 38 per cent and 52 per cent in shared lanes. Shared lanes performed better at low traffic volumes, while dedicated lanes performed at high volumes. A recent study by [16] used SDs to optimise mobility by understanding mode choice between rail, car, bus, and air. In addition, the SD method was implemented in EV adoption by incorporating cost, infrastructure supply, vehicle technology, and social utility [17]. In policy developments, various factors (e.g., GDP, capital investment, and solid waste emission) were modelled in SDs to evaluate policy effects on the urban economy [18].

Further, a study by [19] proposed a multi-stage modelling approach to enhance network performance to cater to the growing demand for AVs. While the AV-related subnetwork could improve network performance, it also increased the total travel distance. In a recent study, [20] proposed a new business model for AVs called 'AV crowdsourcing', which involved renting out privately owned AVs to gain profits. The study tested the feasibility of this business scenario using an equilibrium model. However, the optimal price for AV crowdsourcing needs to be investigated further by considering individuals' sensitivity to the utility function and additional costs associated with AVs. This business model has the potential to provide a high return for private AV users. However, more research is needed to fully understand its effect on the adoption rate of AVs and the overall transportation system. Reference [21] predicted that urban households would see a 2.8 per cent increase in commuting trips using private AVs by coupling North Carolina's demand and choice models to capture household preference. However, the result varied by different penetration rates and fuel types.

In summary, most past studies explored the effects of AVs on traffic flow, efficiency, and safety using microsimulations. Given that Australia will establish an AV safety law in 2026 to facilitate the deployment of AVs [22], this suggests a need for further research into the external factors affecting future AV trips. However, few studies have considered the effect of AVs on mode shift by considering a range of factors at the system level.

2.2. System Dynamics Modelling

This literature review primarily focuses on the application of SD modelling to transport planning, especially the effects of AVs and EVs. Reference [23] examined the possible implications of implementing AVs by employing an SD approach to three scenarios: (1) no change in behaviour and ownership, (2) change in behaviour and no change in ownership, and (3) complete change in ownership in which all vehicles are shared AVs. However, the investigation did not consider the adoption process, including factors such as penetration level and level of service, which may change over time. Additionally, the data were gathered through a workshop setting. In addition, while a study conducted in the Netherlands used an SD modelling approach to examine the adoption process and policy tests across four scenarios (i.e., AV in bloom, demand, doubt, and standby), it failed to account for the potential traffic congestion resulting from AV usage and the associated policy effects, such as congestion charging policies [24]. A comparable study by [25] employed an SD modelling approach to assess the effects of AVs on mode choice, focusing on levels 1 to 3. The study analysed two scenarios: AVs and cooperative/connected vehicles, which can communicate with infrastructure and other vehicles. However, the base year data used in the study were from 2013, which may not accurately reflect the current state of AV technology, as it has been rapidly advancing in recent years. Some past research leveraging SD modelling to solve complex interactive transport problems is shown in Table 2.

Table 2. System Dynamic Approaches Review.

Purpose	Variable	Strength	Conclusion	Future Study Suggestion/Limitation
To evaluate the construction scale of urban rail for traffic, economy, and society [26]	GDP, population, accident, gas emission, congestion degree; construction scale was a policy variable	Presented the effect of the urban rail system on urban traffic, economy, society, and environment; guided transportation infrastructure planning	As the mileage of urban rail increased, the number of cars increased; appropriate construction of urban rail would help	Some variables need more research, such as sociology, economics, and demography
To evaluate the effects of AVs on mode choice and broader transportation system [23]	Travel time, public transit fare, traffic volume, adequacy of PT, etc.	Three different scenarios to investigate the effect on mode choice and mobility	Better to obtain public acceptance of AVs as shared-use vehicles or PT tools before establishing the mindset of private vehicles	Public discussion should be initiated to fully understand views on AVs when AVs are in the market
To evaluate the innovation diffusion of AVs in the long term [24]	Technology maturity, research and development funds, attractiveness, purchase price, and fleet	Complex and dynamic innovation systems of AVs and six levels of AVs were represented	System was highly uncertain due to different market penetration levels and policies adopted	Further research could focus on gaining more knowledge of factors affecting the diffusion of AVs by leveraging this model
To evaluate the mobility effects of AVs [25]	Mode choice, travel time, and time of day choice	Uncertainties were incorporated into penetration rates, capacity, and value of time	AVs could cause increased car trips and level of congestion	Extend the model by considering travel time reliability, road pricing policy, and ride-sharing
A useful approach for optimising individuals' mobility and guiding city planners [16]	Rail, car, bus, and air customers (mode choice)	What factors influence people's choices and can model their behaviours; several scenarios were included (sensitive to price, trip duration, and need to stay overnight)	Customers were not sensitive to price, trip duration, need to stay overnight, or need to use additional means of transport	Future research should be parametrised to identify more details for individual platforms

Table 2. *Cont.*

Purpose	Variable	Strength	Conclusion	Future Study Suggestion/Limitation
Adoption of EVs [17]	Economic utility (cost, infrastructure convenience, and vehicle technology) and social utility	Complex interaction and how feedback can affect EV adoption	Consumers' vague perceptions and pilot of EV projects led to delays in EV adoption; however, social commerce helped	Future research should focus on EV adoption through combinations of incentive plans
To evaluate the effects of AV adoption on greenhouse gas emissions [27]	Emissions, fleet, and adoption	Life cycle assessment to assess the various scenarios in the medium to long term	To decrease greenhouse gas emissions, the government should manage vehicle travel speeds, provide subsidies, and increase the renewable electricity supply	Further research needs to focus on developing the model in conjunction with other methods to support the investigation of greenhouse emission process

As shown in Table 2, in previous studies, SD modelling has been used to explore various facets of AV adoption and its influence on transport planning. However, few studies considered factors, such as network capacity and current transport characteristics, to evaluate the adoption of AVs at the system level (i.e., city level). As AV technology is continuously evolving and maturing, it is crucial to conduct further research into adoption rates of AVs compared with trips from other modes, such as PT and CVs, in a city-level context. Currently, there is a lack of a comprehensive framework to systematically consider mode shift change due to the upcoming AVs, especially in an Australian context.

3. Methods

The SD model in this study considers the interaction of AV and CV adoption in a mixed-vehicle fleet along with PT. It is developed using VENSIM PLE (version 8.2.0) and simulated from 2018 to 2068, with AV adoption starting in 2025 [28]. The model dynamically computes parameters through feedback loops to determine their impact. System dynamics modelling involves the creation of stock-and-flow models, where flows are divided into inflows and outflows, representing the rates at which quantities are added or subtracted from a specific stock. Consequently, the integral of the net flow, combined with the initial stock value at time "*a*0", yields the total stock at time "*a*". The net flow, calculated by subtracting outflows from inflows, represents the derivative of the total stock concerning time, as shown in Equation (1).

$$Stock\ (a) = \int_{a0}^{a} [inflow(a) - outflow(a)]da + Stock(a0) \quad (1)$$

3.1. Description of the System in This Study

The system under consideration is described in Figure 1. External factors are important in determining future AV demand, such as technological advancement and infrastructure capacity of AVs. In this study, AVs represent level 2 and above. Different policies can affect an individual's mode of choice between PT, AVs, and CVs. In addition, if more individuals use CVs or AVs instead of PT, the results could prevent them from using private vehicles because of increased travel time. Thus, the government will adjust the policy (dotted line) once more network traffic is causing congestion. Therefore, policy decision making is important to achieve a sustainable, equitable, and accessible transport system by satisfying equilibrium in the system.

3.2. Model Explanation

3.2.1. Data Input

We obtained the transport data from VITM developed by the Victorian Department of Transport and Planning (DTP), Melbourne, Australia. VITM is a strategic four-step transport demand model developed to predict future travel demand and travel patterns as a result of land use changes, population changes, travel behaviour changes, and major infrastructure projects [29]. VITM is based on the Victorian Integrated Survey of Travel and Activity, including individual trips within family households. The model incorporates the complicated interactions within the transport system (e.g., private vehicle trips, PT, and other modes) and land use changes. Specifically, the VITM is a comprehensive transport demand model that operates on multiple time periods, trip purposes, and modes of travel. This model encompasses car, public transport, and active transport modes and is designed to estimate transportation demand over a typical school day. Employing population, employment, and enrolment projections, VITM assesses the forthcoming impacts of changes in Victoria's road and public transport infrastructure. Further, we considered data from the existing literature [30,31] and the national survey conducted in Australia [32], and where the relevant data were unavailable, we made realistic assumptions, which are shown in the following tables.

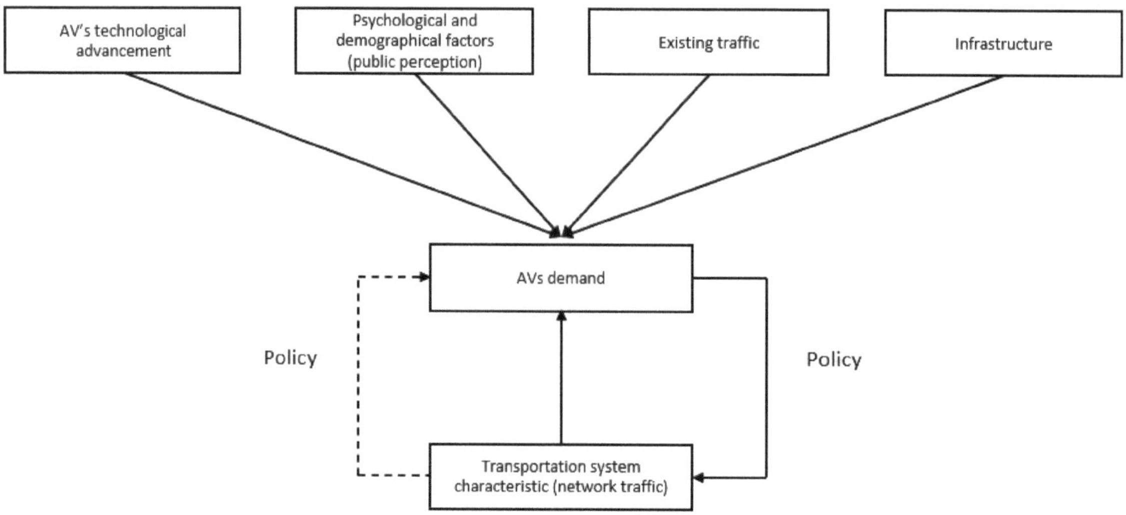

Figure 1. System effects in this study.

3.2.2. Calculations

The study proposed a stock-and-flow-based SD model to simulate the distribution of trips among CVs, AVs, and PT. The model incorporates four sub-models: network capacity, CV trips, AV trips, and PT trips, as shown in Figure 2. As shown in Figures 3–5, the stocks (e.g., CV adopters) are represented by boxes, while the double-lined arrows represent flows (e.g., the total delay in the network). The 'tap' symbol denotes flow rates (e.g., from CV adopters to AV adopters), and the single-lined arrows represent influence links (e.g., local collectors influencing road infrastructure). An encircled R represents the reinforcing feedback loop, while an encircled B represents a balancing feedback loop. The model takes inputs and time series data to generate outputs. The simulation model was used to explore the effects of AV adoption on the transportation system at a city level, including the shifts in mode share and changes in traffic congestion. The model outputs include CV adopters, AV adopters, and PT adopters, representing any 15 min period during a typical weekday AM (7 am to 9 am) peak, signifying that the system dynamics model simulates 15 min segments within the AM peak of a standard weekday. Consequently, we have performed a straightforward calculation to derive 15 min boardings, achieved by dividing the 2 h duration by 8. The simulation period is 50 years, starting from the base year.

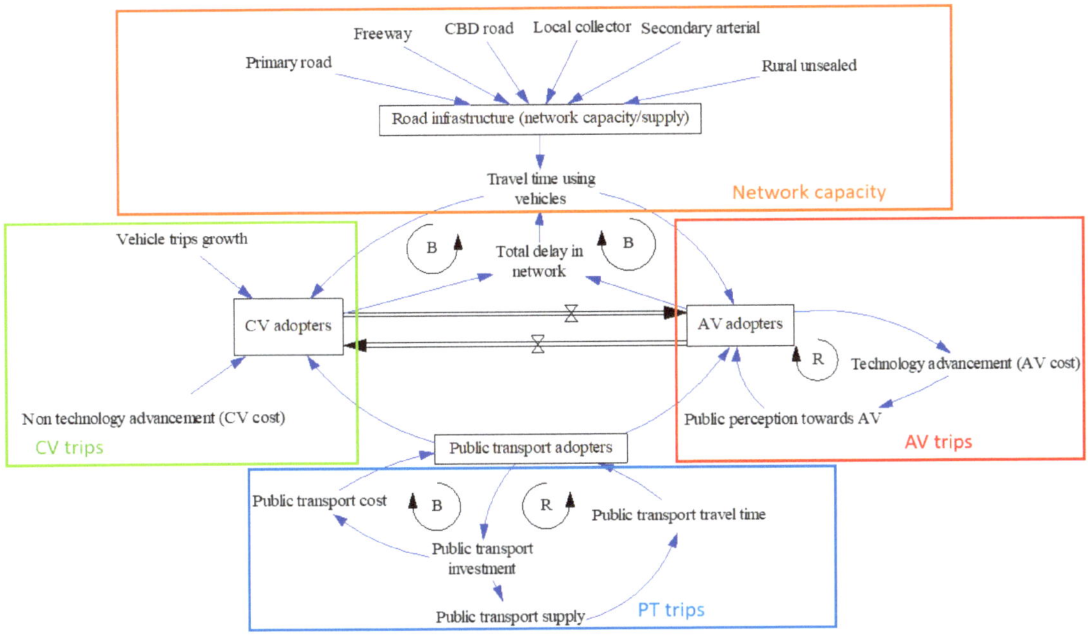

Figure 2. Simplified architecture of the stock-and-flow model.

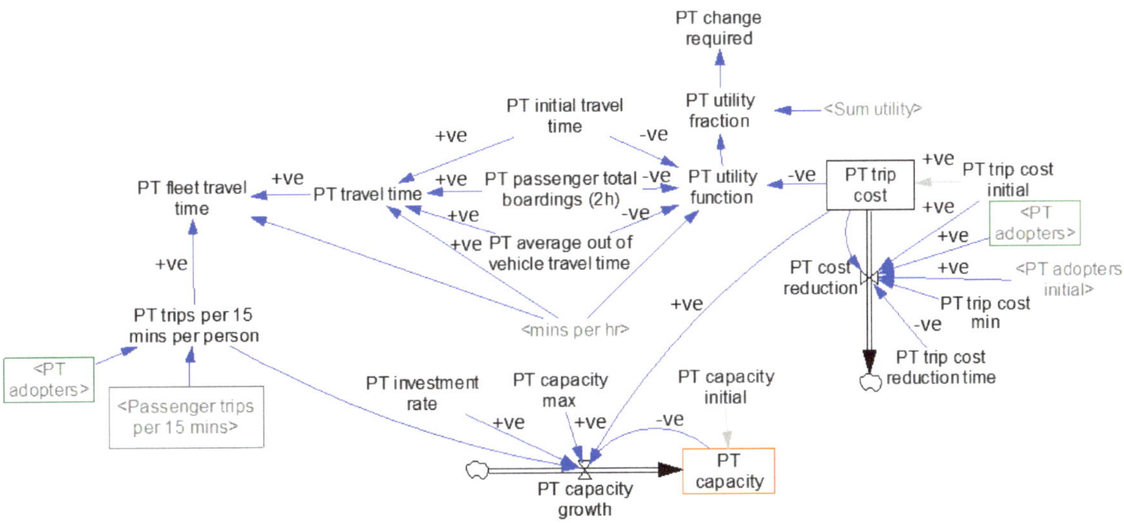

Figure 3. Public transport adoption sub-model. (Notes: 1. <> symbols signify their repeated occurrence within the system dynamics model, whereas variables lacking <> symbols appear only one time in the system. 2. <min per h> denotes a conversion factor of 60 min per h, facilitating unit conversion within this model).

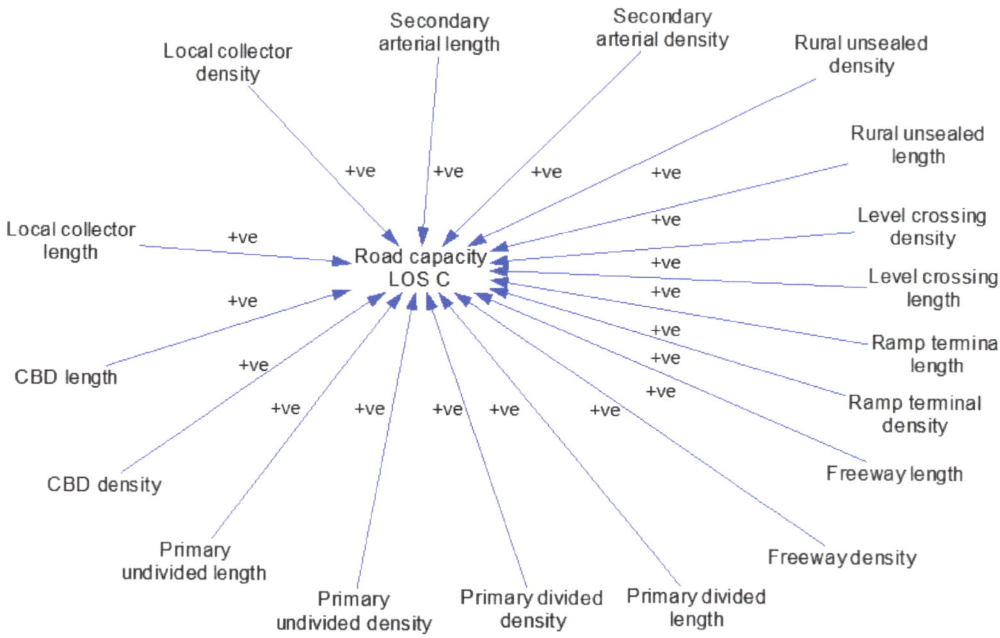

Figure 4. Network capacity sub-model.

3.3. Sub-Model Explanation

The sub-model of the study includes public transport, network capacity, and CV transitions to AV, which are used to evaluate the various factors influencing AV adoption in a closed transportation system.

3.3.1. Public Transport Sub-Model

Figure 3 shows the public transport sub-model. The trips generated by public transport depend on the adoption by AV and CV users. Table 3 shows the sub-model's equations, values, and units for the key variables and stocks.

The utility function determines the number of people who choose PT as their primary mode of transportation [33]. The choice of PT as a primary mode of transportation is influenced by various factors, including travel time, travel cost, and standard deviation of travel time. In this study, VITM was used to determine the total PT travel time, which was used to obtain the average PT travel time by dividing the total PT boarding. The average PT travel time also included the average out-of-vehicle travel time, such as the time spent walking to the tram/train station. For example, the "PT utility function" hinges on two pivotal factors: the "PT trip cost" and "PT travel time". The "PT trip cost" element can be influenced by the uptake of PT by individuals, denoted by "PT adopters" and "PT adopters initial", thereby influencing what we term as "PT cost reduction". Furthermore, the "PT travel time" is sourced from the VITM model's 2018 dataset. As more individuals embrace PT, it might drive the "PT investment rate", thereby impacting "PT capacity growth" and, subsequently, the number of "PT adopters". Additionally, 'PT utility function' refers to a mathematical construct employed for assessing passenger modal preferences, while 'PT utility fraction' utilises this function to ascertain the likelihood of selecting the PT mode over AV and CV modes.

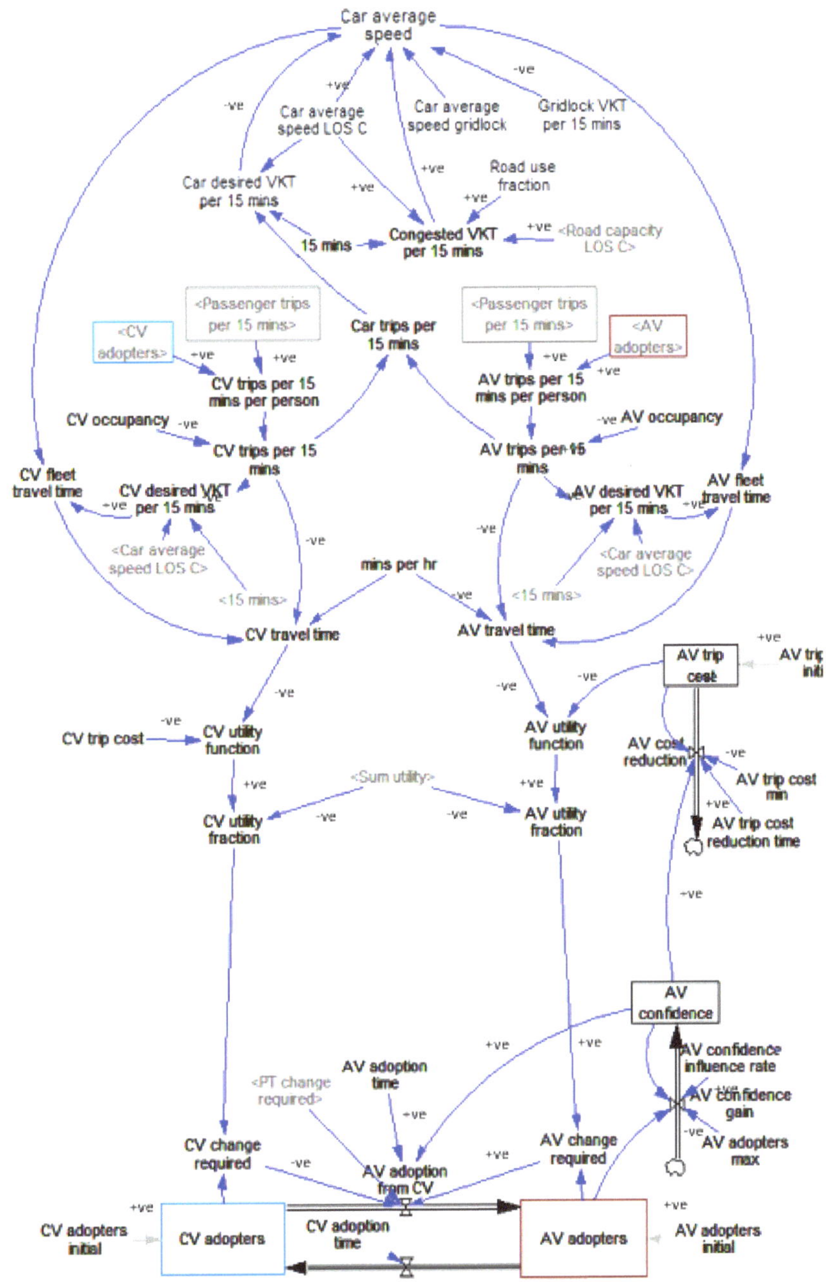

Figure 5. Transitions between CV and AV sub-model. (Notes: 1. <> symbols signify their repeated occurrence within the system dynamics model, whereas variables lacking <> symbols appear only one time in the system. 2. 'AV trip cost' is represented as a stock due to its dependency on other variables like 'AV trip cost initial' and 'AV cost reduction'. In contrast, 'CV trip cost' maintains a more consistent cost due to its mature technology. Consequently, 'AV trip cost' is categorised as a stock, while 'CV trip cost' is a variable unaffected by other factors).

Table 3. The public transportation sub-model's parameters, equations, and values.

Parameter Name	Unit	Value (Equation)	Source/Explanation
Sum utility	N/A	EXP (CV utility) + EXP (AV utility) + EXP (PT utility)	Utility function [34].
PT change required	N/A		Percentage changes in individuals opting for public transport as their primary mode during each simulation interval.
PT utility function	N/A	−0.049 × (PT initial travel time × min per h/"PT passenger total boardings (2 h)") − 0.05 × PT average out-of-vehicle travel time − 0.0038 × PT trip cost	Probability of choosing PT as commuting mode based on travel time and cost during any 15 min at AM peak.
PT initial travel time	Person × hour	165,795	Collective travel duration via various modes such as trains, trams, and buses, as supplied by the VITM model from DTP for input into this system dynamics model.
PT passenger total boarding (2 h)	Person	508,420	Cumulative count of person boardings on public transport encompassing train, bus, and tram trips. This information is furnished by the VITM model from DTP during the AM peak period spanning 2 h.
PT average out-of-vehicle travel time	Minute	11	VISTA provided by DTP.
PT travel time	Minute	PT initial travel time × min per h/"PT passenger total boardings (2 h)" + PT average out-of-vehicle travel time	PT travel time includes in-vehicle travel time and out-of-vehicle travel time.
PT fleet travel time	Person × hour	PT trips per 15 min per person × PT travel time/min per h	Total public transport fleet travel time including trains, trams, and buses.
PT trips per 15 min per person	Person	Passenger trips per 15 min × PT adopters	It is to determine the number of people who choose PT modes across total people.
PT investment rate	Dmnl/Year/Person/dollar	1×10^{-9}	Amount by which 'PT capacity' grows each year for each dollar spent on PT.
PT capacity max	Dmnl	0.5	Fraction of passenger travel that PT can ultimately service.
PT capacity growth	Dmnl/Year	PT trip cost × PT trips per 15 min per person × PT investment rate × (PT capacity max − PT capacity)/PT capacity max	

The adoption of PT by users can contribute to PT-related revenue, which can then be used to increase PT capacity by providing more services. The PT investment rate is the amount by which PT capacity grows each year for each dollar spent on PT. However, the PT capacity max sets the maximum proportion of individuals who will adopt PT as their primary mode of transportation for commuting purposes. For instance, if the PT capacity max is set at 0.6, it means that a maximum of 60 per cent of all adopters (i.e., AV, CV, and PT) are PT adopters.

This study highlights the importance of considering various factors when modelling PT demand. The results can be used to inform policy decisions and transportation planning.

3.3.2. Network Capacity Sub-Model

The road network capacity sub-model is illustrated in Figure 4. In this model, the variable 'road capacity' represents the total number of cars in the road network that can travel without congestion, and its unit is cars. Level of service is a qualitative measure used to evaluate traffic flow based on factors such as speed, congestion, and density. As a result, the number of vehicles in a given time period in level C condition, "road capacity", is calculated by multiplying the road length by the level of service C. This sub-model plays a significant role in determining the capacity of the road network and its ability to accommodate the increased use of AVs and CVs. It also assists in identifying potential road congestion and areas where road infrastructure may require upgrades to handle the influx of AVs and CVs. The network capacity sub-model evaluates the combined length of distinct road types within Melbourne. It establishes the overall road capacity, a critical factor influencing the adoption of CVs and AVs, and subsequently impacts the volume of individuals choosing for these vehicle types as the number of trips grows. Table 4 shows the sub-model equations, values, and units for the key variables and stocks.

Table 4. Network capacity sub-model's parameters, equations, and values.

Parameter Name	Unit	Value (Equation)	Source/Explanation
Local collector density	Car/km	11.2	LoS C standard (HCM 2016)
Local collector length	km	5572.5	VITM provided by DTP
Secondary arterial density/Rural unsealed density/Ramp terminal density/Primary divided density/Primary undivided density/CBD density	Car/km	13.7	LoS C standard
Secondary arterial length	km	3626.84	VITM provided by DTP
Rural unsealed length	km	741.2	VITM provided by DTP
Level crossing length	km	84.83	VITM provided by DTP
Ramp terminal length	km	29.58	VITM provided by DTP
Freeway density	Car/km	16.2	LoS C standard (HCM 2016)
Freeway length	km	2707.51	VITM provided by DTP
Primary divided length	km	4113.7	VITM provided by DTP
Primary undivided length	km	4010.57	VITM provided by DTP
CBD length	km	64.04	Sourced from the VITM model to provide input for this analysis, signifying the road length within Melbourne's central business district (CBD) in kilometres
CBD density	Car/km	13.7	Acquired from the traffic engineering standard, specifically the level of service C standard, to ascertain the optimal traffic density for vehicle movement to travel smoothly

Notes: DTP (Department of Transport and Planning); VISTA (Victorian Integrated Survey of Travel and Activity).

3.3.3. CV Transitions to AV Sub-Model

Figure 5 shows the CV transitions to the AV sub-model, which presents the transition model between CVs and AVs, where the number of trips generated by each mode depends on the trip cost and time spent. Therefore, if AVs can travel faster and become cheaper and AVs are more attractive than current CVs, people change from public transit (e.g., PT) to auto-based modes [35].

A previous study used two-stage stochastic programming that assumed AV cost and commuting travel time would simultaneously affect AV ownership and adoption rates [36]. Similar to the study conducted by [28], this research assumed that AVs would be available in the market after 2025.

The utility function determines the proportion of people who choose AVs or CVs as their primary mode. 'AV trip cost' is expected to decrease over time, represented by the 'AV trip cost min' variable and 'AV trip cost reduction time' variable. Similarly, 'AV confidence' is expected to increase over time with more people adopting AVs and matured technology. 'AV trip cost' and 'AV confidence' are the two main factors affecting the adoption of AVs compared to the adoption of CVs.

Additionally, the actual VKT will decrease as the 'car average speed' decreases due to congestion. The threshold of VKT is determined by the level of service C, called 'congested VKT per 15 min', which is calculated based on the total road network capacity ('Road capacity LOS C') and the fraction of frequently used road networks during the AM peak (please refer to Section 3.3.2). In contrast, the 'car desired VKT per 15 min' variable represents the actual VKT, including both CVs and AVs, which affects the 'car average speed'. The 'car average speed' and 'Car desired VKT per 15 min' then influence the 'CV travel time' and 'AV travel time', ultimately affecting the proportion of individuals choosing these modes ('CV/AV utility function'). Further, the 'AV confidence influence rate' refers to the proportion of individuals positively influenced to choose AVs. This is because those interested in owning AVs tend to rely on their friends for information and recommendations [37].

Therefore, the variables in the transition model are interconnected, and the changes in one variable will affect the other variables, affecting users' mode choices. Table 5 shows the sub-model equations, values, and units for the key variables and stocks.

3.4. *Testing*

Different tests build confidence for stock-and-flow models [38]. To ensure the reliability and validity of the model, we conducted a series of tests, as recommended by [38], including a model structure test, behavioural test, and boundary test. The model structure test assessed the parameters, boundaries, and overall structural adequacy of the model. For instance, in structure assessment, all the parameters align with the actual system, such as increasing the AV trip cost could make less people choose the AV mode. Similarly, for the boundary assessment, the stock-and-flow model behaviour is sensitive to the removal of existing endogenous elements but insensitive to adding new endogenous elements.

The behavioural test examined the model's ability to capture and simulate the behaviour of the transportation system realistically. For example, we have changed the value of a single parameter (e.g., PT investment rate) in extreme conditions. Then, the model performs realistically, as the impacted variable (e.g., PT adopters) is within range.

Finally, the boundary test evaluated the sensitivity of the model to changes in the input parameters and boundaries. By passing these tests in the present study, the researchers were confident in the model's ability to represent the transportation system and evaluate different policy scenarios realistically.

Figures 6 and 7 illustrate the sensitivity analysis conducted on AV occupancy and AV initial trip cost. In Figure 6, different scenarios for AV occupancy—low (average 1.1 person per AV car), medium (average 1.3 person per AV car), and high (average 1.5 person per AV car)—were evaluated. Interestingly, there was minimal variation in AV adoption rates across these scenarios, suggesting that AV occupancy has a minor influence compared to factors like cost, travel time, and social influence.

Table 5. CV transitions to AV sub-model's parameters, equations, and values.

Parameter Name	Unit	Value (Equation)	Source/Explanation
AV/CV desired VKT per 15 min	Car × km	AV/CV trips per 15 min × Car average speed LoS C × "15 min"	Maximum car capacity in the network that does not lead to congestion
AV/CV occupancy	Person/Car	1.1	Average number of persons per car
AV/CV trips per 15 min	Car	AV/CV trips per 15 min per person/AV/CV occupancy	Number of AV/CV trips for any 15 min during AM peak
AV/CV trips per 15 min per person	Person	Passenger trips per 15 min × AV/CV adopters	Number of AV/CV trips among total trips generated by private vehicle trips and PT trips
AV/CV fleet travel time	Car × hour	AV/CV desired VKT per 15 min/Car average speed	Vehicle × km/km/h equals vehicle × h
AV/CV travel time	Minute	AV fleet travel time × min per h/AV trips per 15 min	Average AV/CV travel time per vehicle
AV/CV utility function	N/A	−1.55–0.066 × AV/CV travel time − 0.004 × AV/CV trip cost	It is an AV/CV utility function to determine the probability of choosing AV/CV mode
Car average speed	km/hour	Car average speed LoS C − (Car desired VKT per 15 min − Congested VKT per 15 min)× (Car average speed LoS C − Car average speed gridlock)/(Gridlock VKT per 15 min − Congested VKT per 15 min)	Vehicle speed decreases as VKT exceeds the congestion threshold
Car average speed LoS C	km/hour	48.1	VITM provided by DTP
Congested VKT per 15 min	Car × km	Road capacity LoS C × Road use fraction × Car average speed LoS C × "15 min"	Threshold for congestion in a network level depends on average vehicle speed (travel in a smooth way) and road capacity
Road use fraction	N/A	0.62	This is the assumed value as there are some roads that are seldomly used in Victorian network
AV adopters initial	Dmnl	0.01	This must be greater than zero to avoid a 'floating point error' due to division by zero in 'AV travel time' at t = 0

Notes: AV (automated vehicle); CV (conventional vehicle); DTP (Department of Transport and Planning); VITM (Victorian Integrated Transport Model); and VKT (vehicle kilometres travelled).

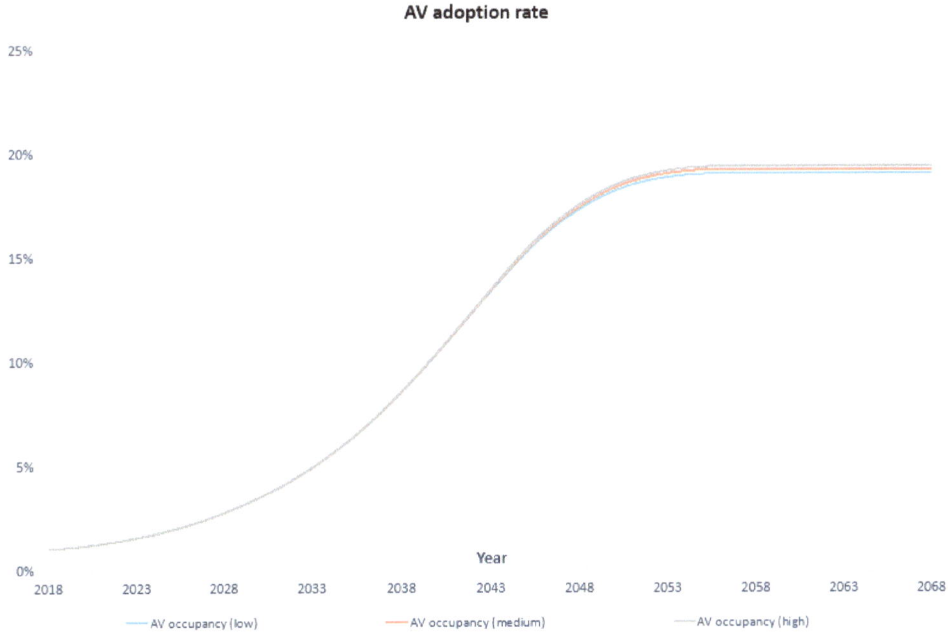

Figure 6. Sensitivity test of AV occupancy.

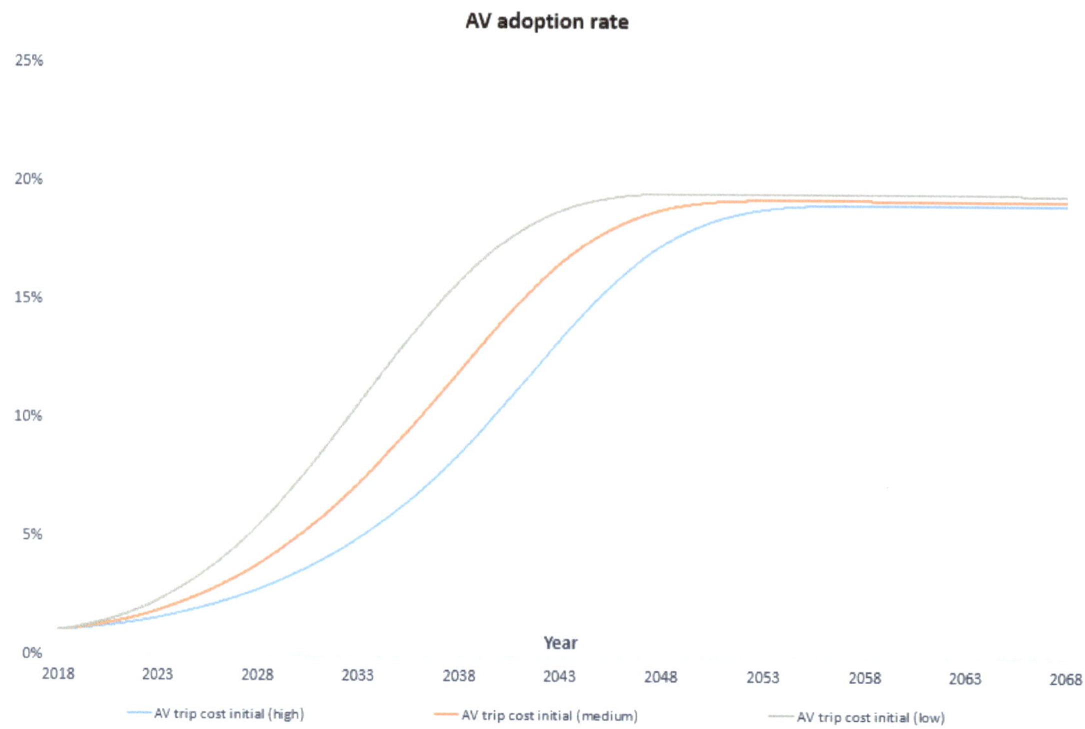

Figure 7. Sensitivity test of AV initial trip cost.

In Figure 7, the sensitivity test explored AV initial cost through high (800), medium (700), and low (600) scenarios in comparison to the CV trip cost (400). The high-cost scenario exhibited a slow growth in AV adoption rates from 2018 to 2048 due to fewer individuals embracing AVs at a higher cost. However, after 2048, all scenarios converged to the same AV adoption rate, aligning with the decreasing cost trend. Consequently, these sensitivity tests regarding "AV occupancy" and "AV trip cost initial" enhance the model's credibility and reinforce its validity.

3.5. Scenarios

Table 6 outlines the scenarios tested in the model and their respective assumptions. In the base scenario, the maximum fraction of AV adopters remained at 90 per cent, while PT capacity was assumed to be at a 50 per cent fraction level, and the minimum AV trip cost was set at 400. Setting the maximum fraction of AV adopters at 90% is a practical choice, considering that not everyone may fully switch to AVs due to personal preferences or concerns about new technology. As technology improves and people become more confident, a significant portion of the population is expected to embrace AVs. So, in practice, we assume 90% AV adoption instead of 100%. Choosing 50% for PT capacity makes sense because many big cities with well-used public transportation, like New York and London, hover around this utilisation level. Thus, we assume Melbourne's PT capacity to be 50%, given the city's current PT usage being below 30%. Additionally, we set AV trip cost min equal to CV trip cost (400) to ensure AV trips remain affordable and competitive, aligning with the cost of conventional trips.

Table 6. Various scenarios implemented in the model.

Scenario	Parameter Name	Unit	Value		
			Low	Neutral	High
Baseline	AV adopters max	N/A			90%
	AV trip cost min	N/A		400	
	PT capacity max	N/A		50%	
1a	AV adopters max	Fraction	40%		
1b				60%	
1c					100%
2a	AV trip cost min	N/A	360		
2b		N/A			430
3a	PT capacity	Fraction	30%		
3b					60%
Lower	AV adopters max	Fraction	40%		
	AV trip cost min	N/A			430
	PT capacity max	Fraction			60%
Upper	AV adopters max	Fraction			100%
	AV trip cost min	N/A	360		
	PT capacity max	Fraction	30%		

Notes: AV, automated vehicle; na, not applicable; and PT, public transport.

Over time, the model predicted that more CV adopters would transition to AV adopters as trust in AV technology increases and the cost of AVs decreases due to technological advancement. The model also predicted that as network congestion increases, the percentage of trips taken via PT would increase. Other scenarios tested included varying the AV trip cost, PT investment rate, and PT capacity max assumptions to evaluate their effects on mode choice and travel behaviour. These scenarios provide insight into potential future outcomes and the effects of different policy and technological interventions on the adoption of AVs and travel behaviour in Melbourne.

Scenario 1 in the model included the base case with a neutral assumption of 60 per cent for 'AV adopters max' because approximately 60 per cent of participants surveyed who had heard of AVs held a positive view of them [32]. The high scenario assumed that 'AV adopters max' would be 100 per cent, as it was believed that 100 per cent of individuals could adopt AVs in the next 50 years. In contrast, the low scenario assumed that only 40 per cent of individuals would adopt AVs [32].

In scenario 2, the base case for 'AV trip cost min' for calculating the utility function was assumed to be 400 (baseline scenario), the same as the 'CV trip cost'. However, the Australia-wide survey results showed that around 20 per cent of respondents preferred shared AVs for daily work, reducing the average cost of an AV trip [32]. Thus, the 'AV trip cost min' for the low scenario was set to 360. Conversely, for the high scenario, the Australia-wide survey results revealed that the respondents thought AVs were worth more than CVs [32]. As a result, the 'AV trip cost min' for the high scenario was assumed to be 430. It was essential to consider preferences and perceptions towards AVs in determining the AV trip cost, as these play a significant role in influencing the adoption rate of AVs. Cost factor was one of the critical factors affecting the adoption rate of AVs. Thus, it was necessary to examine various scenarios to identify the potential effects of AV trip costs on the adoption rate of AVs.

Scenario 3 examined the effect of PT capacity on the transportation system. The assumption for the base case (same as the baseline scenario) was that 50 per cent of the adopters would choose PT as their primary mode, and the PT capacity max was set at

50 per cent. For the low scenario, the PT capacity max was decreased to 30 per cent because currently, PT trips only account for 30 per cent of total trips during the morning peak in Melbourne. This is due to research finding that individuals who already rely on PT as their primary mode of transportation are more inclined to continue using it in the future [39].

The high scenario assumed that 60 per cent of trips were generated by PT in Melbourne for the 24 h period from VITM 2018, and the PT capacity max was set at 60 per cent during the AM morning peak. These assumptions reflect the potential for increased PT usage and the need to ensure that PT capacity can meet growing demand.

The lower and upper scenarios in the study represent the minimum and maximum possible scenarios for AV adopters based on three variables: AV adopters max, AV trip cost min, and PT capacity max. These scenarios help to explore these variables' potential effects (boundary) on the adoption of AVs and the use of PT.

Table 7 presents the various scenarios for road expansion rates. The low scenario denotes an annual growth of 1 per cent in terms of road network capacity; whereas, the high scenario denotes a growth rate of 3 per cent.

Table 7. Road expansion and awareness program implemented in the model.

Scenario	Parameter Name	Unit	Value			
			Baseline	**Low**	**Neutral**	**High**
Road expansion program	Road expansion rate	Fraction	0%			
				1%		
					2%	
						3%
AV awareness program	AV confidence influence rate	Fraction	40%			
				60%		
					80%	
						100%

Notes: AV, automated vehicle.

4. Results

In this section, we discuss the outcomes under three subsections: baseline scenario, other scenarios, and road expansion and awareness program scenarios.

4.1. Baseline Scenario

Using the data presented in Tables 3–5 and the baseline scenario in Table 6, Figure 8 displays the fluctuation of adoption rates among AVs, CVs, and PT within the 50-year simulation period. In the base year 2018, CV trips accounted for 78 per cent of total trips, while PT trips accounted for 22 per cent. After the trips stabilised in year 30, CV and AV trips by people accounted for 31 per cent and 19 per cent, respectively. The fraction of PT trips remained the same (50 per cent) after year 40, slightly increasing from year 0 (20 per cent). This indicates that after year 30, CV, AV, and PT trips reach equilibrium, and their adoption rates remain stable.

Interestingly, the adoption rates of CVs and PT decrease over time while the adoption rate of AVs increases. This could be due to the technological advancement (lower cost) of AVs and their increased acceptance by the public. As AVs become more affordable and reliable, individuals may switch from CVs to AVs. Similarly, with increasing road network congestion, individuals may shift from CVs and AVs to PT, resulting in a slight increase in PT trips per person until year 40.

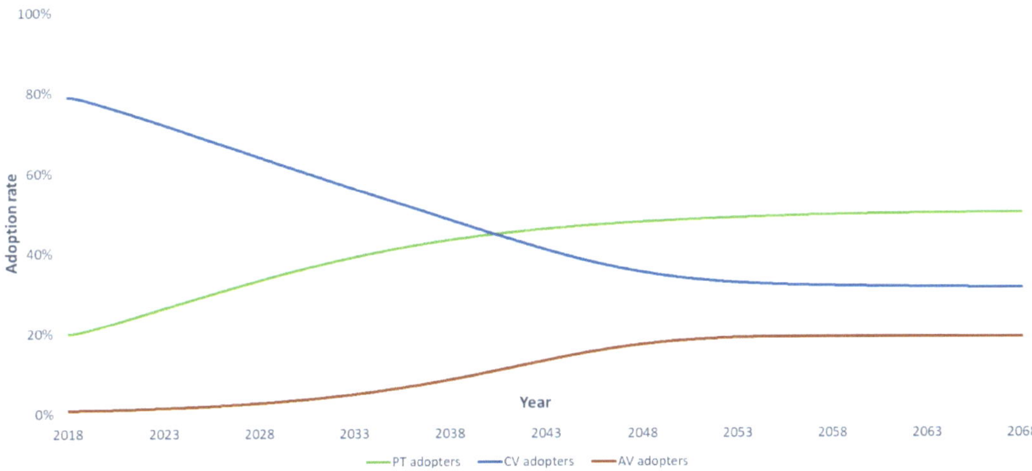

Figure 8. Baseline scenario of AVs, CVs, and PT.

Overall, the findings suggest that AVs will gradually replace CVs as another primary mode of transportation in Melbourne. However, PT will still play a significant role in the transportation system, accounting for 50 per cent of total trips by people after year 40. It is noteworthy that alterations in the "PT capacity max" parameter (currently set at 0.5) will impact the proportion of individuals opting for public transport after convergence. Furthermore, while results can fluctuate due to modifications in road capacity, i.e., the utility function and costs of PT, AVs/CVs, confidence gains, and other factors, the overall trend of these travel choices remains consistent. Figure 8 suggests that while the adoption of AVs is projected to increase over time, PT remains pivotal within the transportation network. Consequently, AVs could be effectively integrated as a solution for first- and last-mile connectivity, contributing to the overall efficiency of the transportation system.

4.2. Other Scenarios

Figure 9 presents the fraction of AV adopters for 10 scenarios from Table 6, including baseline, lower, and upper cases. Due to road network capacity constraints, the graph shows around 16 to 24 per cent of AV adopters among the 10 scenarios, with the lowest adoption rate from the lower case and the highest from the upper case. The adoption rate of AVs started to increase around year 8, due to the assumption that people would start accepting AVs from 2026 (base year 2018), and stabilised around year 34 (2052). Except for the lower and upper scenarios, the lowest adoption rate (17 per cent) was in the scenario 'AV trip cost min (high)', while the highest adoption rate (22 per cent) was in the scenario 'AV trip cost min (low)'. These results show that the cost of AV trips is an important factor in determining the adoption rate of AVs.

Compared with the scenario 'AV trips cost min(high)' and 'PT capacity max(high)', the AV adopters rate of scenario 'AV adopters max(low)' exceeded these two scenarios after year 32. This indirectly proves that the cost of AV trips is the most important factor compared with 'AV adopters max' and 'PT capacity max'. The second- and third-highest AV adopters rate scenarios were 'PT capacity max(low)' and 'AV adopter max(high)', meaning PT demand might reach a high level due to lower capacity, resulting in more people switching to AVs. Therefore, it is essential to consider the balance between PT capacity and AV adoption rate to ensure a smooth transition to AVs. The evaluation of these 10 scenarios, featuring different values for "AV adopters max", "AV trip cost min", and "PT capacity max", reveals a spectrum of adoption rates spanning from 16% to 24%. This implies that these variables introduce relatively minor uncertainties in the results. These findings indicate that cost, whether through financial incentives or subsidies, emerges as the primary determinant influencing the number of individuals who opt for AV adoption once the adoption rate stabilises. These findings can be helpful for policymakers and stakeholders to make informed decisions regarding AV adoption policies and strategies.

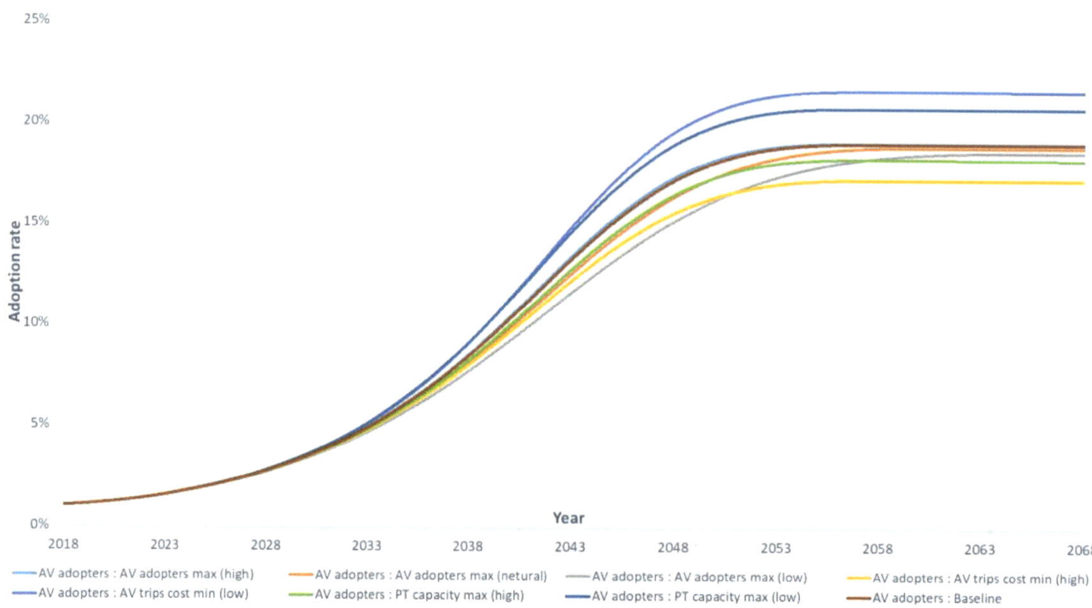

Figure 9. AV adopters in different scenarios.

4.3. Road Expansion and Awareness Program Scenarios

Figure 10 shows the adoption rate of AVs under different road expansion and awareness program scenarios. For the road expansion program, the baseline scenario assumed a 1 per cent annual growth rate in road network capacity, while the high scenario assumed a 3 per cent growth rate. The baseline scenario showed an AV adoption rate of 19 per cent, while the high scenario showed a rate of approximately 24 per cent that continued slightly even after year 50. This suggests that a 1 per cent increase in road network capacity could lead to a 2 per cent increase in the AV adoption rate. However, higher road expansion rates can result in a longer time for AV adoption to stabilise due to increased space on the road. Interestingly, there is little difference in AV adoption rates between road expansion scenarios in the first 20 years. This could be because road network capacity is not a significant determining factor, and cost remains the most critical factor influencing AV adoption.

Further, the study suggests that awareness programs could be more effective than road investment programs in increasing AV adoption rates, particularly between years 4 and 24. However, even with the low and high scenarios for the influence rate, the AV adopter rate remained lower than the low road expansion scenario, at 19.3 per cent and 19.7 per cent, respectively. Therefore, although awareness programs could lead to a more rapid increase in adoption rates, road investment programs are more likely to result in higher adoption rates in the long term.

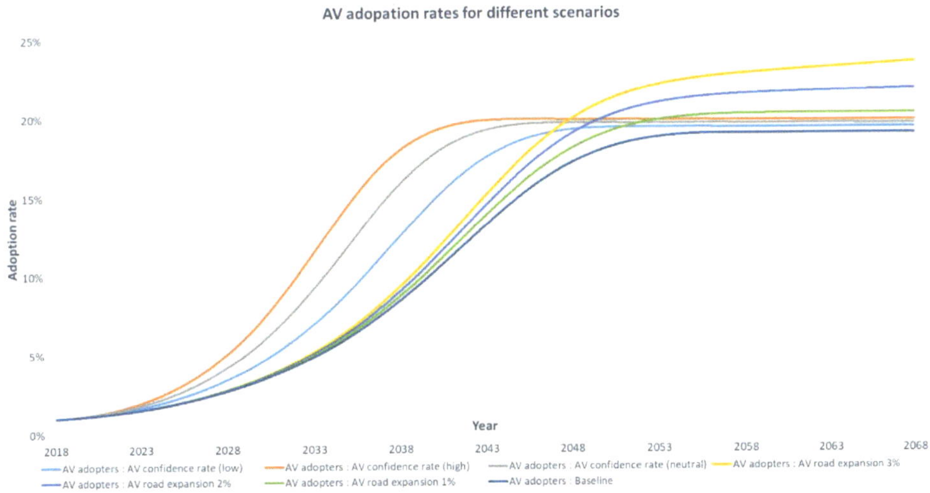

Figure 10. AV adopters in different road expansion and awareness program scenarios.

5. Discussion

In this section, we discuss the following three aspects: AV adoption, awareness program, and cost.

5.1. AV Adoption

A study conducted by [40] developed a framework to forecast the adoption of AVs in Nashville, US. The study projected that AVs would likely capture a 50 per cent market share by year 18 and an 80 per cent market share by year 31, which differs from our study's findings (10 per cent market share by year 18 and 23 per cent after year 30). This variance may be attributed to the difference in the transportation culture between Australia and the US. Unlike Australia, the US has a car-centric culture, which may lead to a more rapid adoption of AVs. Similarly, [41] proposed a simulation-based framework using a multinomial logit model to predict Americans' adoption of CAVs under different scenarios. The authors found that privately owned AVs would be 24.8 per cent in year 30, compared with 18 per cent in the present study, and this result was based on an annual 5 per cent price decrease and the same willingness to pay value.

Additionally, a dynamic approach for designing AV subsidies to accelerate the early deployment of AVs was developed by [42]. The present study also highlights the importance of cost as a critical factor for adoption, which can be addressed through optimal subsidies. Reference [43] used a discrete choice model by incorporating it into the dynamic model with AV subsidies and infrastructure investment as inputs. That study concluded that the optimal subsidy increased from USD 10,000 in year 1 to USD 20,000 in year 60, when AV market penetration was 50 per cent.

In contrast to our study, [44] employed agent-based modelling and considered the reduced value of time of AVs caused by parking restrictions and increased congestion. The study concluded that AVs would decrease transit ridership by 75 per cent, which differs from the present findings that showed an increase in public transit ridership due to congestion and reduced AV costs. In a similar study, [23] used SD modelling to discover that traffic volume would considerably increase, leading to higher congestion equilibrium levels and more VKT. It is, therefore, imperative for the present study to consider the induced traffic volume, as we assumed a relatively stable total number of trips.

Automated on-demand mobility services, such as Uber and taxis, could potentially see a reduction in PT trips by 9–10 per cent in Singapore during peak hours with the introduction of AVs alongside private vehicles [45]. This is because some individuals may shift from PT to AVs due to lower costs compared with existing taxis and on-demand mobility services.

This study's findings suggest that policymakers and stakeholders need to consider the effect of congestion levels on AV adoption rates when developing policies to promote AVs. It is essential to address current congestion levels, as this can influence the adoption rate of AVs. Therefore, road expansion and awareness programs could be a more effective approach to promoting AV adoption. Further, cost is the most important factor in determining the adoption rate of AVs, and optimal subsidies could be used to make AVs more affordable and competitive in the market.

5.2. Awareness Programs

According to [46], social influence and public acceptance are two crucial factors necessary for the widespread adoption of AVs. To encourage the adoption of AVs, governments should work with manufacturers to promote their usefulness and create favourable conditions that foster social influence and public acceptance.

Ref. [47] used SD modelling to find that a lack of customer acceptance was the main barrier to AV adoption. The authors suggested that awareness programs can address this issue, which can help increase the adoption rate. As suggested by [48], the societal dimension of AVs as part of governance processes is important for the transition from CVs to AVs. These recommendations from past studies are consistent with the present study's finding that an awareness program could increase AV adoption rates more quickly than a road investment program.

Although awareness programs can drive an initial surge in AV adoption rates (e.g., from 2028 to 2048), this study indicates that their effect may diminish over time. Therefore, policymakers should consider longer-term strategies, such as investment in AV infrastructure (e.g., charging stations), especially when future cars become electric AVs, to sustain and increase adoption rate.

5.3. Cost

In this study, cost was identified as the most significant factor affecting adoption rates. Similarly, in a study conducted by [49], a lab experiment was carried out in a mixed traffic environment consisting of AVs and CVs to explore the mode preferences of individuals. Participants who received complete information about mode and cost considered perceived cost and inertia during the decision-making process. According to [50], various trade-offs, such as travel time cost, waiting time cost, miles travelled, and operational cost, were captured by considering AVs in private and shared mobility systems. The researchers concluded that technological advancement is necessary to promote AV adoption due to the lower cost of AVs, which aligns with this study's findings.

Additionally, [42] found that optimal subsidies can serve as both an incentive for AV manufacturers to innovate and improve their products and a means of providing competitive pricing to attract potential consumers. Moreover, the research conducted by [51] showed that individuals tended to be more responsive to the cost of the vehicle and the provision of exclusive lanes, which is consistent with the findings of the present study regarding AV trip costs and road expansion programs. The cost factor could also significantly influence individuals' decisions to adopt AVs in Ireland [52].

Therefore, policymakers could consider providing optimal subsidies to AV manufacturers to innovate and improve their products and offer competitive pricing to attract potential consumers. Further, governments could invest in AV infrastructure, such as exclusive lanes, to provide a more seamless and efficient travel experience for AV users. Policymakers should also consider the balance between PT capacity and AV adoption rates to ensure a smooth transition to AVs, which could lead to a reduction in private car usage and, consequently, contribute to reducing greenhouse gas emissions.

6. Conclusions

This study used an SD modelling approach to investigate the effect of AVs on the mode shift between CVs, AVs, and PT in Melbourne, Australia, by systematically considering a range of factors, such as road network, vehicle cost, PT supply, and congestion level. The study also highlights the importance of cost as a critical factor in adoption, which can be addressed through optimal subsidies. Further, the adoption rate of AVs was found to be affected by road network capacity and awareness programs. While higher road expansion rates could result in a longer time for AV adoption rates to stabilise, awareness programs could lead to a more rapid increase in adoption rates. However, road investment programs are more likely to result in higher adoption rates in the long term. Therefore, it is important to facilitate the transition from CVs to AVs in a seamless manner so that road network can accommodate both types of vehicles during the transition period.

The increasing prevalence of AVs in Australia may have significant implications for mobility patterns, particularly in ride-hailing services [53]. This could result in a more congested network, as future travel demand is expected to be primarily carried out via private AVs, with most passengers using shared patterns [6]. Therefore, it is crucial to investigate the adoption rates of AVs in various potential scenarios, including those involving ride-hailing services. Understanding the potential effect of AVs on ride-hailing services will help policymakers develop strategies to manage traffic congestion and ensure a sustainable transportation system.

The SD model developed in this study has the potential to assist planners, policymakers, and researchers to evaluate the potential effect of AVs on the transportation system and plan accordingly to minimise adverse effects and maximise the benefits of AVs. Likewise, analogous to the approach described in reference [54], the adaptation of monitoring strategies in response to evolving conditions holds applicability in making well-informed decisions regarding mode shifts within dynamic transportation systems. However, this study has several limitations that need to be addressed in future research. First, the study only considered a single metropolitan area and assumed that AVs would be available to everyone equally. In reality, AV adoption rates may vary across regions due to factors such as urban design, travel patterns, and demographic characteristics. Thus, future studies should explore the adoption rates of AVs in different regions and the factors that influence them.

The study made three assumptions regarding AV adoption rates in Melbourne, Australia, including a constant 2 per cent growth rate in total trips over time. However, in reality, AVs may induce demand for travel by providing more convenient and accessible transportation, increasing total trips. Thus, future research should explore the potential induced demand due to AV adoption. Additionally, applying equilibrium constraints to traffic assignment models could enhance the model's accuracy by characterising route choice behaviour and vehicle preference, as suggested in the literature [36].

Author Contributions: Y.C.: Conceptualisation, methodology, software, formal analysis, investigation, data curation, writing—original draft, and writing—review and editing. P.S.: Supervision, formal analysis, and funding acquisition. N.S.: Conceptualisation, writing—review and editing, supervision, and funding acquisition. S.K.K.: Formal analysis and writing—original draft. All authors have read and agreed to the published version of the manuscript.

Funding: This research was funded by the Australian Government, Department of Industry, Science, Energy, and Resources, grant number [AEGP000050].

Institutional Review Board Statement: Not applicable.

Informed Consent Statement: Not applicable.

Data Availability Statement: Not applicable.

Conflicts of Interest: The authors declare no conflict of interest.

References

1. Eskandarian, A.; Wu, C.; Sun, C. Research Advances and Challenges of Autonomous and Connected Ground Vehicles. *IEEE Trans. Intell. Transp. Syst.* **2021**, *22*, 683–711. [CrossRef]
2. SAE International. Available online: https://www.sae.org/standards/content/j3016_202104/ (accessed on 6 February 2023).
3. Zou, X.; O'Hern, S.; Ens, B.; Coxon, S.; Mater, P.; Chow, R.; Neylan, M.; Vu, H. On-road virtual reality autonomous vehicle (VRAV) simulator: An empirical study on user experience. *Transp. Res. Part C Emerg. Technol.* **2021**, *126*, 103090. [CrossRef]
4. Chen, Y.; Shiwakoti, N.; Stasinopoulos, P.; Khalid Khan, S. State-of-the-Art of Factors Affecting the Adoption of Automated Vehicles. *Sustainability* **2022**, *14*, 6697. [CrossRef]
5. Zong, F.; Wang, M.; Tang, J.; Zeng, M. Modeling AVs & RVs' car-following behavior by considering impacts of multiple surrounding vehicles and driving characteristics. *Phys. A Stat. Mech. Its Appl.* **2022**, *589*, 126625.
6. Li, J.; Rombaut, E.; Vanhaverbeke, L. A systematic review of agent-based models for autonomous vehicles in urban mobility and logistics: Possibilities for integrated simulation models. *Comput. Environ. Urban Syst.* **2021**, *89*, 101686. [CrossRef]
7. Nassar, R.F.; Ghisolfi, V.; Annema, J.A.; van Binsbergen, A.; Tavasszy, L.A. A system dynamics model for analyzing modal shift policies towards decarbonization in freight transportation. *Res. Transp. Bus. Manag.* **2023**, *48*, 100966. [CrossRef]
8. Gu, Z.; Wang, Z.; Liu, Z.; Saberi, M. Network traffic instability with automated driving and cooperative merging. *Transp. Res. Part C Emerg. Technol.* **2022**, *138*, 103626. [CrossRef]
9. Yu, H.; Jiang, R.; He, Z.; Zheng, Z.; Li, L.; Liu, R.; Chen, X. Automated vehicle-involved traffic flow studies: A survey of assumptions, models, speculations, and perspectives. *Transp. Res. Part C Emerg. Technol.* **2021**, *127*, 103101. [CrossRef]
10. Hu, J.; Kong, L.; Shu, W.; Wu, M.-Y. Scheduling of connected autonomous vehicles on highway lanes. In Proceedings of the 2012 IEEE Global Communications Conference (GLOBECOM), Anaheim, CA, USA, 3–7 December 2012; pp. 5556–5561.
11. Overtoom, I.; Correia, G.; Huang, Y.; Verbraeck, A. Assessing the impacts of shared autonomous vehicles on congestion and curb use: A traffic simulation study in The Hague, Netherlands. *Int. J. Transp. Sci. Technol.* **2020**, *9*, 195–206. [CrossRef]
12. Fakhrmoosavi, F.; Kamjoo, E.; Kavianipour, M.; Zockaie, A.; Talebpour, A.; Mittal, A. A stochastic framework using Bayesian optimization algorithm to assess the network-level societal impacts of connected and autonomous vehicles. *Transp. Res. Part C Emerg. Technol.* **2022**, *139*, 103663. [CrossRef]
13. Chen, B.; Sun, D.; Zhou, J.; Wong, W.; Ding, Z. A future intelligent traffic system with mixed autonomous vehicles and human-driven vehicles. *Inf. Sci.* **2020**, *529*, 59–72. [CrossRef]
14. Ye, L.; Yamamoto, T. Evaluating the impact of connected and autonomous vehicles on traffic safety. *Phys. A Stat. Mech. Its Appl.* **2019**, *526*, 121009. [CrossRef]
15. Hamad, K.; Alozi, A.R. Shared vs. dedicated lanes for automated vehicle deployment: A simulation-based assessment. *Int. J. Transp. Sci. Technol.* **2022**, *11*, 205–215. [CrossRef]
16. Vecchio, P.D.; Secundo, G.; Maruccia, Y.; Passiante, G. A system dynamic approach for the smart mobility of people: Implications in the age of big data. *Technol. Forecast. Soc. Change* **2019**, *149*, 119771. [CrossRef]
17. Feng, B.; Ye, Q.; Collins, B.J. A dynamic model of electric vehicle adoption: The role of social commerce in new transportation. *Inf. Manag.* **2019**, *56*, 196–212. [CrossRef]
18. Wu, D.; Ning, S. Dynamic assessment of urban economy-environment-energy system using system dynamics model: A case study in Beijing. *Environ. Res.* **2018**, *164*, 70–84. [CrossRef]
19. Madadi, B.; van Nes, R.; Snelder, M.; van Arem, B. Multi-stage optimal design of road networks for automated vehicles with elastic multi-class demand. *Comput. Oper. Res.* **2021**, *136*, 105483. [CrossRef]
20. Wang, X.; Lin, X.; Li, M. Aggregate modeling and equilibrium analysis of the crowdsourcing market for autonomous vehicles. *Transp. Res. Part C Emerg. Technol.* **2021**, *132*, 103362. [CrossRef]
21. Hasnat, M.M.; Bardaka, E.; Samandar, M.S. Differential impacts of autonomous and connected-autonomous vehicles on household residential location. *Travel Behav. Soc.* **2023**, *32*, 100570. [CrossRef]
22. SBS News. Available online: https://www.sbs.com.au/news/article/tough-road-ahead-self-driving-cars-are-taking-off-in-the-us-but-australians-might-be-waiting-a-while-yet/a5ebp93on (accessed on 8 February 2023).
23. Gruel, W.; Stanford, J.M. Assessing the Long-Term Effects of Autonomous Vehicles: A speculative approach. *Transp. Res. Procedia* **2016**, *13*, 18–29. [CrossRef]
24. Nieuwenhuijsen, J.; Homem de Almeida Correia, G.; Milakis, D.; Arem, B.; Daalen, E. Towards a quantitative method to analyze the long-term innovation diffusion of automated vehicles technology using system dynamics. *Transp. Res. Part C* **2018**, *86*, 300–327. [CrossRef]
25. Puylaert, S.; Snelder, M.; van Nes, R.; van Arem, B. Mobility impacts of early forms of automated driving—A system dynamic approach. *Transp. Policy* **2018**, *72*, 171–179. [CrossRef]
26. Yang, Y.; Zhang, P.; Ni, S. Assessment of the Impacts of Urban Rail Transit on Metropolitan Regions Using System Dynamics Model. *Transp. Res. Procedia* **2014**, *4*, 521–534. [CrossRef]
27. Stasinopoulos, P.; Shiwakoti, N.; Beining, M. Use-stage life cycle greenhouse gas emissions of the transition to an autonomous vehicle fleet: A System Dynamics approach. *J. Clean. Prod.* **2021**, *278*, 123447. [CrossRef]
28. Bhardwaj, C.; Axsen, J.; Crawford, C. Simulating long-term emissions from private automated vehicles under climate policies. *Transp. Res. Part D Transp. Environ.* **2023**, *118*, 103665. [CrossRef]

29. Laufer, J.; Shahhoseini, Z.; Yue, W.L.; Javanshour, F. The Development of a Mesoscopic Melbourne Model. In Proceedings of the Australasian Transport Research Forum 2019 Proceedings, Canberra, Australia, 30 September–2 October 2019.
30. Currie, G.; Delbosc, A. Exploring the trip chaining behaviour of public transport users in Melbourne. *Transp. Policy* **2011**, *18*, 204–210. [CrossRef]
31. Nguyen-Phuoc, D.Q.; Currie, G.; De Gruyter, C.; Young, W. Congestion relief and public transport: An enhanced method using disaggregate mode shift evidence. *Case Stud. Transp. Policy* **2018**, *6*, 518–528. [CrossRef]
32. Chen, Y. Modelling the Adoption and Deployment of Automated Vehicles. Ph.D. Thesis, RMIT University, Melbourne, Australia, 2023.
33. Li, Z.; Hensher, D.; Rose, J. Willingness to Pay for Travel Time Reliability in Passenger Transport: A Review and Some New Empirical Evidence. *Transp. Res. Part E Logist. Transp. Rev.* **2010**, *46*, 384–403. [CrossRef]
34. Australian Transport Assessment and Planning. Available online: https://www.atap.gov.au/tools-techniques/travel-demand-modelling/appendix-b (accessed on 30 March 2023).
35. Miller, E.J. *Simulating Autonomous Vehicles: A Discussion Paper and Research Proposal*; University of Toronto Transportation Research Institute: Toronto, ON, Canada, 2016.
36. Chen, S.; Wang, H.; Meng, Q. Designing autonomous vehicle incentive program with uncertain vehicle purchase price. *Transp. Res. Part C Emerg. Technol.* **2019**, *103*, 226–245. [CrossRef]
37. Sharma, I.; Mishra, S. Quantifying the consumer's dependence on different information sources on acceptance of autonomous vehicles. *Transp. Res. Part A Policy Pract.* **2022**, *160*, 179–203. [CrossRef]
38. Sterman, J.D. System dynamics modeling: Tools for learning in a complex world. *Calif. Manag. Rev.* **2001**, *43*, 8–25. [CrossRef]
39. Mo, B.; Wang, Q.; Moody, J.; Shen, Y.; Zhao, J. Impacts of subjective evaluations and inertia from existing travel modes on adoption of autonomous mobility-on-demand. *Transp. Res. Part C Emerg. Technol.* **2021**, *130*, 103281. [CrossRef]
40. Dubey, S.; Sharma, I.; Mishra, S.; Cats, O.; Bansal, P. A General Framework to Forecast the Adoption of Novel Products: A Case of Autonomous Vehicles. *Transp. Res. Part B Methodol.* **2022**, *165*, 63–95. [CrossRef]
41. Bansal, P.; Kockelman, K.M. Forecasting Americans' long-term adoption of connected and autonomous vehicle technologies. *Transp. Res. Part A Policy Pract.* **2017**, *95*, 49–63. [CrossRef]
42. Luo, Q.; Saigal, R.; Chen, Z.; Yin, Y. Accelerating the adoption of automated vehicles by subsidies: A dynamic games approach. *Transp. Res. Part B Methodol.* **2019**, *129*, 226–243. [CrossRef]
43. Wang, S.; Li, Z.; Levin, M.W. Optimal policy for integrating Autonomous Vehicles into the auto market. *Transp. Res. Part C Emerg. Technol.* **2022**, *143*, 103821. [CrossRef]
44. Llorca, C.; Moreno, A.; Ammar, G.; Moeckel, R. Impact of autonomous vehicles on household relocation: An agent-based simulation. *Cities* **2022**, *126*, 103692. [CrossRef]
45. Nguyen-Phuoc, D.; Zhou, M.; Chua, M.; Alho, A.; Oh, S.; Seshadri, R.; Le-Klähn, D. Examining the effects of Automated Mobility-on-Demand services on public transport systems using an agent-based simulation approach. *Transp. Res. Part A Policy Pract.* **2023**, *169*, 103583. [CrossRef]
46. Gopinath, K.; Narayanamurthy, G. Early bird catches the worm! Meta-analysis of autonomous vehicles adoption—Moderating role of automation level, ownership and culture. *Int. J. Inf. Manag.* **2022**, *66*, 102536. [CrossRef]
47. Raj, A.; Kumar, J.A.; Bansal, P. A multicriteria decision making approach to study barriers to the adoption of autonomous vehicles. *Transp. Res. Part A Policy Pract.* **2020**, *133*, 122–137. [CrossRef]
48. Milakis, D.; Muller, S. The societal dimension of the automated vehicles transition: Towards a research agenda. *Cities* **2021**, *113*, 103144. [CrossRef]
49. Zhang, Q.; Ma, S.; Tian, J.; Rose, J.; Ning, J. Mode choice between autonomous vehicles and manually-driven vehicles: An experimental study of information and reward. *Transp. Res. Part A Policy Pract.* **2022**, *157*, 24–39. [CrossRef]
50. Singh, H.; Kavianipour, M.; Ghamami, M. Adoption of autonomous and electric vehicles in private and shared mobility systems. *Transp. Res. Part D Transp. Environ.* **2023**, *115*, 103561. [CrossRef]
51. Shabanpour, R.; Golshani, N.; Shamshiripour, A.; Mohammadian, A. Eliciting preferences for adoption of fully automated vehicles using best-worst analysis. *Transp. Res. Part C Emerg. Technol.* **2018**, *93*, 463–478. [CrossRef]
52. Rezaei, A.; Caulfield, B. Examining public acceptance of autonomous mobility. *Travel Behav. Soc.* **2020**, *21*, 235–246. [CrossRef]
53. Irannezhad, E.; Mahadevan, R. Examining factors influencing the adoption of solo, pooling and autonomous ride-hailing services in Australia. *Transp. Res. Part C Emerg. Technol.* **2022**, *134*, 103524. [CrossRef]
54. Duan, C. Dynamic Bayesian monitoring and detection for partially observable machines under multivariate observations. *Mech. Syst. Signal Process* **2021**, *158*, 107714. [CrossRef]

sensors

MDPI

Article

Research on the Prediction Model of Engine Output Torque and Real-Time Estimation of the Road Rolling Resistance Coefficient in Tracked Vehicles

Weijian Jia [1], Xixia Liu [1,*], Guodong Jia [2], Chuanqing Zhang [1] and Bin Sun [2]

1 Army Academy of Armored Forces, Beijing 100072, China; jwjyxsl@163.com (W.J.)
2 CITIC Machinery Manufacturing Inc., Linfen 043000, China
* Correspondence: zgyflw@163.com

Abstract: Road parameter identification is of great significance for the active safety control of tracked vehicles and the improvement of vehicle driving safety. In this study, a method for establishing a prediction model of the engine output torques in tracked vehicles based on vehicle driving data was proposed, and the road rolling resistance coefficient f was further estimated using the model. First, the driving data from the tracked vehicle were collected and then screened by setting the driving conditions of the tracked vehicle. Then, the mapping relationship between the engine torque T_e, the engine speed n_e, and the accelerator pedal position β was obtained by a genetic algorithm–backpropagation (GA–BP) neural network algorithm, and an engine output torque prediction model was established. Finally, based on the vehicle longitudinal dynamics model, the recursive least squares (RLS) algorithm was used to estimate the f. The experimental results showed that when the driving state of the tracked vehicle satisfied the set driving conditions, the engine output torque prediction model could predict the engine output torque $\hat{T}_e$ in real time based on the changes in the n_e and β, and then the RLS algorithm was used to estimate the road rolling resistance coefficient $\hat{f}$. The average coefficient of determination R of the $\hat{T}_e$ was 0.91, and the estimation accuracy of the $\hat{f}$ was 98.421%. This method could adequately meet the requirements for engine output torque prediction and real-time estimation of the road rolling resistance coefficient during tracked vehicle driving.

Keywords: tracked vehicles; engine output torque prediction model; GA–BP neural network; estimation of rolling resistance coefficient

Citation: Jia, W.; Liu, X.; Jia, G.; Zhang, C.; Sun, B. Research on the Prediction Model of Engine Output Torque and Real-Time Estimation of the Road Rolling Resistance Coefficient in Tracked Vehicles. *Sensors* **2023**, *23*, 7549. https://doi.org/10.3390/s23177549

Academic Editors: Yafei Wang, Chao Huang, Hailong Huang, Zhiqiang Zuo, Bo Leng and Peng Hang

Received: 10 August 2023
Revised: 25 August 2023
Accepted: 28 August 2023
Published: 31 August 2023

1. Introduction

Tracked vehicles are mostly used in agricultural, fire protection, and military fields due to their good trafficability. A driving road is complex and changeable, and the demands on the dynamics and safety of tracked vehicles are high [1]. The road parameters affect the acceleration, braking, and steering performance of vehicles, and they are important parameters for risk assessment and active safety control during tracked vehicle driving [2]. Road parameter estimation is of great significance for improving the safety and dynamic performance of vehicles [3]. Road identification is generally achieved by estimating a parameter that reflects the characteristics of the road surface, such as the road adhesion coefficient, roughness, rolling resistance coefficient, or slope. To realize road surface recognition and improve the safety of vehicle driving, researchers have conducted in-depth studies. For tracked vehicles, researchers have mostly focused on the dynamic characteristics of tracked vehicles [4] and the coupling between the track and ground during vehicle driving. There have been few studies on the further application of research results to road recognition.

At present, there are two main methods to realize road recognition. One is to directly measure the road through environmental sensors or vehicle state sensors [5,6]. Abhinav et al. [7] proposed a terrain recognition method based on a deep learning long short-term memory model, using the acoustic waves generated by the interactions between the vehicle and the terrain as the terrain feature variable. Huang et al. [8] proposed a road boundary monitoring method based on a deep learning network. The experimental results showed that the method had high accuracy and robustness for lane boundary line monitoring in various scenarios. Based on road images, laser radar point cloud data, and vehicle state information, Zhao et al. [9] realized road recognition by parameter estimation and state fusion. The direct measurement method using environmental sensors for road recognition has the advantages of real-time measurement capabilities and high estimation accuracy. However, to achieve large-scale commercial applications, it is necessary to further reduce the cost of environmental sensors and their reliability in harsh environments [10].

Another method is to estimate the road feature parameters through a vehicle model and then perform road recognition. The method of estimating the road adhesion coefficients of wheeled vehicles based on the slip-slope method [11] using the $\mu - s$ curve is comparatively mature, but the implementation of this method is based on an accurate tire model. Since the vehicle longitudinal dynamics model does not include the tire model, the method of estimating the road characteristics parameters based on this model is theoretically more suitable for tracked vehicles. The application of this method to wheeled vehicles could also be used as a reference. The estimation of the engine output torque is a difficult task for road parameter estimation based on the vehicle longitudinal dynamics model. Chu et al. [12] used the vehicle longitudinal dynamics model to estimate the road slope by making full use of the accurate driving force information from an electric-drive vehicle. Liu et al. [13] established an engine output torque prediction model, including a fuel supply system model, an in-cylinder combustion model, and a crankshaft dynamics model. Based on the vehicle longitudinal kinematics model, the Kalman filter algorithm was used to estimate the road slope. Because the established engine output torque prediction model was too complex, this method was difficult to use for real vehicle control. Cong et al. [14] established a look-up table model of the engine output torque by fitting the corresponding relationship between the engine output torque, engine speed, and accelerator pedal position and estimated the road slope based on the Kalman filter algorithm. This method required a lot of manual calibration work in the bench test stage of the vehicle, and it was difficult to update after the calibration was complete. The predicted torque error of the model would increase with the engine performance degradation. The development of intelligent algorithms provides a new idea for establishing an engine model with nonlinearity, multiple disturbances, and time lag [15].

In this study, the tracked vehicle was the research object, and a method for establishing a prediction model of the engine output torque in the tracked vehicle, based on the vehicle driving data, was proposed. This method saves a lot of manual calibration work during the vehicle bench test stage and has the advantage of enabling real-time updates. The rolling resistance coefficient estimation for the road was realized using the model. The vehicle driving data, during the driving process, from the tracked vehicle were collected. The vehicle driving data were screened by setting the driving conditions of the vehicle to estimate the f. The Kalman filter algorithm was used to filter the longitudinal acceleration a in the selected data segments, and the engine output torque T_e of each data segment was calculated based on the filtered longitudinal acceleration $\hat{a}$. The engine speed n_e, the engine speed variation rate n_e', the accelerator pedal position β, and the accelerator pedal position variation rate β' were the inputs, and the engine output torque T_e was the output. A genetic algorithm–backpropagation (GA–BP) neural network algorithm was used to fit the mapping relationship between the inputs and the output, $T_e = f(n_e, n_e', \beta, \beta')$. Based on the longitudinal dynamics model of tracked vehicles, the RLS algorithm with the forgetting factor λ was used to estimate the $\hat{f}$. The implementation process is shown in Figure 1. The experimental results showed that the sensor information could be used to automatically judge the driving conditions during the driving process of the tracked vehicle. When the driving conditions of the vehicle satisfied the set driving conditions, the engine output torque prediction model predicted the $\hat{T}_e$ in realtime based on the n_e and β, and then it further estimated the $\hat{f}$. The estimation results had high accuracy and could better meet the requirements for the real-time estimation of the road parameters for tracked vehicles.

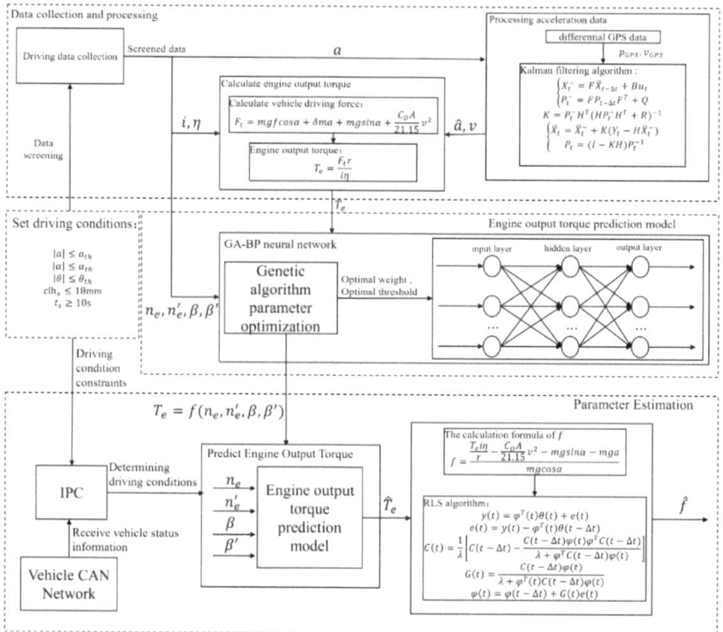

Figure 1. Process of engine output torque prediction and rolling resistance coefficient estimation.

2. Estimation of *f* Based on Recursive Least Squares (RLS) Algorithm

The longitudinal driving forces on the tracked vehicle are shown in Figure 2. The vehicle driving force F_t can be expressed as follows:

$$F_t = mgf\cos\alpha + \delta ma + mg\sin\alpha + \frac{C_D A}{21.15}v^2, \tag{1}$$

where m is the mass of the vehicle, g is the acceleration of gravity, f is the rolling resistance coefficient of the road, α is the road slope, δ is the rotating mass scaling factor, a is the longitudinal acceleration of the tracked vehicle, C_D is the air resistance coefficient, A is the windward area of the tracked vehicle, and v is the speed of the tracked vehicle.

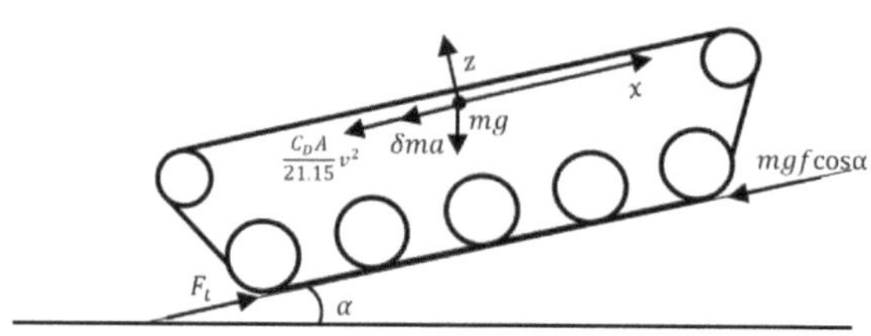

Figure 2. Forces on the tracked vehicle during longitudinal driving.

The relationship between the T_e and vehicle driving force F_t can be expressed as follows:

$$T_e = \frac{F_t r}{i\eta}, \tag{2}$$

where i is the transmission ratio from the engine to the driving wheel, η is the transmission efficiency, and r is the radius of the sprocket.

The formula for f can be obtained by combining (1) and (2):

$$f = \frac{\frac{T_e i\eta}{r} - \frac{C_D A}{21.15}v^2 - mg\sin\alpha - \delta ma}{mg\cos\alpha}. \quad (3)$$

From Formula (3), it can be seen that f can be obtained when T_e, a, and α are known.

The RLS algorithm with the forgetting factor λ is used to estimate the f. The RLS algorithm can be expressed as follows:

$$y(t) = \varphi^T(t)\theta(t) + e(t), \quad (4)$$

where $\varphi(t)$ is the estimated parameter vector at time t, $\theta(t)$ is the regression vector at time t, and $e(t)$ is the deviation between the measured value $y(t)$ and the estimated value $\varphi^T(t)\theta(t)$ at time t.

The RLS algorithm iteratively updates the position parameter vector $\varphi(t)$ at each sampling time by making the regression vector $\theta(t)$ contain the input and output data from a previous time. The RLS algorithm minimizes the estimation bias for each iteration period by updating the vector regression $\theta(t)$. In this paper, $y(t) = T_e i\eta/r - \frac{C_D A}{21.15}v^2 - mg\sin\alpha(t) - \delta ma(t)$, $\theta(t) = mg\cos\alpha(t)$, and $\lambda = 0.98$.

The calculation steps for the RLS algorithm at each time t are as follows:

(1) The system output $y(t)$ is measured, and the regression vector $\theta(t)$ is calculated.

(2) The difference $e(t)$ between the actual output of the system $y(t)$ at time t and the output of the prediction model obtained by estimating the parameters $\varphi^T(t)\theta(t-\Delta t)$ is calculated. Δt is the time interval. The difference $e(t)$ can be expressed as follows:

$$e(t) = y(t) - \varphi^T(t)\theta(t-\Delta t). \quad (5)$$

(3) The updated gain vector $G(t)$ and the covariance matrix $C(t)$ are calculated. These can be expressed as follows:

$$C(t) = \frac{1}{\lambda}\left[C(t-\Delta t) - \frac{C(t-\Delta t)\varphi(t)\varphi^T(t)C(t-\Delta t)}{\lambda + \varphi^T(t)P(t-\Delta t)\varphi(t)}\right], \quad (6)$$

$$G(t) = \frac{C(t-\Delta t)\varphi(t)}{\lambda + \varphi^T(t)C(t-\Delta t)\varphi(t)}. \quad (7)$$

(4) The parameter estimation vector $\varphi(t)$ is updated as follows:

$$\varphi(t) = \varphi(t-\Delta t) + Ge(t). \quad (8)$$

3. Tracked Vehicle Driving Data Acquisition and Processing

The tracked vehicle examined in this study was equipped with a diesel engine, a dry clutch, and a fixed shaft gearbox. The tracked vehicle was equipped with a combined inertial navigation module, including an acceleration sensor and a gyroscope. The positioning data were processed by differential processing, and the accuracy reached the centimeter level. The acceleration sensor was used to measure the longitudinal acceleration value of the tracked vehicle in real time. The gyroscope was used to measure the vehicle pitch angle, and the vehicle pitch angle was assumed to be equal to the road slope value. The vehicle controller received the driver's control instructions to control the vehicle, and the driving data recorder recorded the vehicle's driving data, which was convenient for the data analysis and control optimization of the vehicle. The communication structure of each module is shown in Figure 3.

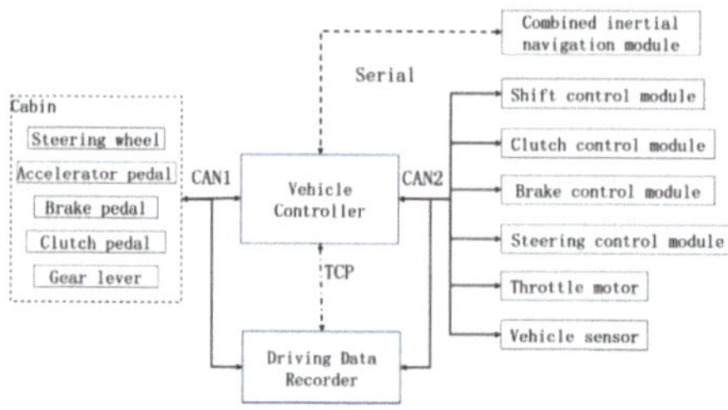

Figure 3. Communication structure of each module.

The test site was a vehicle driving test site in Shanxi, China. The total length of the test site route was 10 km, including a sand road and a cement road. The rolling resistance coefficients of the two roads were measured to be 0.06 and 0.045. Figure 4 shows the satellite map of the experimental site. A total of 119.67 h driving data were collected by the driving data recorder. The collected data included GPS coordinates, vehicle speed, longitudinal acceleration, pitch angle, heading angle, and gear and clutch displacement.

Figure 4. Satellite map of the experimental site.

To ensure the accuracy of the engine output torque model, the driving data were screened by setting the driving conditions for the tracked vehicle. The vehicle driving data that satisfied the driving conditions were used as the effective data to establish the engine output torque prediction model.

(1) The measurement of the vehicle pitch angle by the gyroscope was affected not only by the road slope but also by installation error, the suspension state, and other factors. Under some conditions, for example, the clutch was engaged too fast when shifting, which caused the vehicle to pitch in a short time even if it was driving on a flat road, resulting in measurement errors. At the same time, a large change in the road slope also increased the measurement error of the acceleration sensor. To improve the prediction accuracy of the model, the driving data with a large angle measured by the gyroscope were eliminated by setting the ramp threshold to $\alpha_{th} = 3^{\circ}$, so that the tracked vehicle could drive on a flat road, which was approximately level, as far as possible.

(2) The selected vehicle driving data did not include the clutch separation process, and the driving force during the vehicle driving process was only provided by the engine. The engagement and separation state of the clutch was judged by the displacement of the clutch control cylinder clh_x. When the clutch combination displacement $clh_x \leq 18$ mm, the clutch was considered engaged. Setting the vehicle acceleration threshold a_{th} and the minimum stable driving time threshold t_s ensured that the stable driving data were screened after the vehicle shift was complete. By analyzing the driving data, we set $a_{th} = 0.4\ \mathrm{m/s^2}$ and $t_s = 10$ s. The driving data when the vehicle acceleration $|a| \leq 0.4\ \mathrm{m/s^2}$ for more than 10 s after the clutch was engaged was considered stable and valid data.

(3) It was necessary to limit the heading angle γ in the screened vehicle driving data to ensure that the tracked vehicle was in a straight-line driving state. Considering the influence of sensor measurement error and random road disturbances, we set $\gamma_{th} = 5°$. In the selected driving data, the change in the heading angle of the vehicle between the initial moment and the final moment could not exceed 5°.

Due to the body vibration and acceleration sensor measurement bias during the running of the tracked vehicle, the acceleration measurement data had a larger error than the real value. The Kalman filter algorithm was used to filter a to obtain an accurate longitudinal acceleration $\hat{a}$. The vehicle displacement p, velocity v, and acceleration a were the state variables, and u was the measured value of the acceleration sensor. The driving displacement p_{GPS} and the driving speed v_{GPS} obtained by the vehicle's integrated inertial navigation module through the differential positioning system were taken as the observed quantities. The vehicle state equation at time t can be expressed as follows:

$$\begin{bmatrix} p_t \\ v_t \\ a_t \end{bmatrix} = \begin{bmatrix} 1 & \Delta t & \frac{1}{2}\Delta t^2 \\ 0 & 1 & \Delta t \\ 0 & 0 & 0 \end{bmatrix} \begin{bmatrix} p_{t-\Delta t} \\ v_{t-\Delta t} \\ a_{t-\Delta t} \end{bmatrix} + \begin{bmatrix} 0 \\ 0 \\ 1 \end{bmatrix} u_{t-\Delta t}, \tag{9}$$

$$\begin{bmatrix} p_{GPS_{t-\Delta t}} \\ v_{GPS_{t-\Delta t}} \\ \frac{v_{GPS_t} - v_{GPS_{t-\Delta t}}}{\Delta t} \end{bmatrix} = \begin{bmatrix} 1 & 0 & 0 \\ 0 & 1 & 0 \\ 0 & 0 & 1 \end{bmatrix} \begin{bmatrix} p_{t-\Delta t} \\ v_{t-\Delta t} \\ a_{t-\Delta t} \end{bmatrix}, \tag{10}$$

where $X_k = \begin{bmatrix} p_t \\ v_t \\ a_t \end{bmatrix}$, $Y_k = \begin{bmatrix} p_{GPS_{t-\Delta t}} \\ v_{GPS_{t-\Delta t}} \\ \frac{v_{GPS_t} - v_{GPS_{t-\Delta t}}}{\Delta t} \end{bmatrix}$, $F = \begin{bmatrix} 1 & \Delta t & \frac{1}{2}\Delta t^2 \\ 0 & 1 & \Delta t \\ 0 & 0 & 0 \end{bmatrix}$, $B = \begin{bmatrix} 0 \\ 0 \\ 1 \end{bmatrix}$, $H = \begin{bmatrix} 1 & 0 & 0 \\ 0 & 1 & 0 \\ 0 & 0 & 1 \end{bmatrix}$, and Δt is the calculation time step.

The Kalman filter algorithm uses a recursive method to solve the filtering problem of discrete linear data [16]. The steps are as follows:

(1) Update the prediction equation:

$$\begin{cases} \overline{X}_t^- = F\overline{X}_{t-\Delta t} + Bu_t \\ P_t^- = FP_{t-\Delta t}F^T + Q \end{cases}. \tag{11}$$

(2) Update the Kalman gain coefficient:

$$K = P_t^- H^T (HP^- H^T + R)^{-1}. \tag{12}$$

(3) Update the measurement equation:

$$\begin{cases} \overline{X}_t = \overline{X}_t^- + K(Y_t - H\overline{X}_t^-) \\ P_t = (I - KH)P_t^- \end{cases}. \tag{13}$$

In these formulas, $\overline{X}_t^-$ is the prior state estimation at time t, $\overline{X}_t$ is the posterior state estimation at time t, P_t^- is the prior covariance matrix at time t, P_t is the covariance matrix at time t, Q is the process noise covariance matrix, R is the observation noise covariance matrix, and K is the Kalman gain coefficient. In this study, $Q = \begin{bmatrix} 0.5 & 0 & 0 \\ 0 & 0.5 & 0 \\ 0 & 0 & 0.5 \end{bmatrix}$, $R = \begin{bmatrix} 0.2 & 0 & 0 \\ 0 & 0.2 & 0 \\ 0 & 0 & 0.2 \end{bmatrix}$, and $P_0 = \begin{bmatrix} 0.1 & 0 & 0 \\ 0 & 0.1 & 0 \\ 0 & 0 & 0.1 \end{bmatrix}$. The acceleration data measured by some accelerometers are filtered, and the filtering effect is shown in Figure 5.

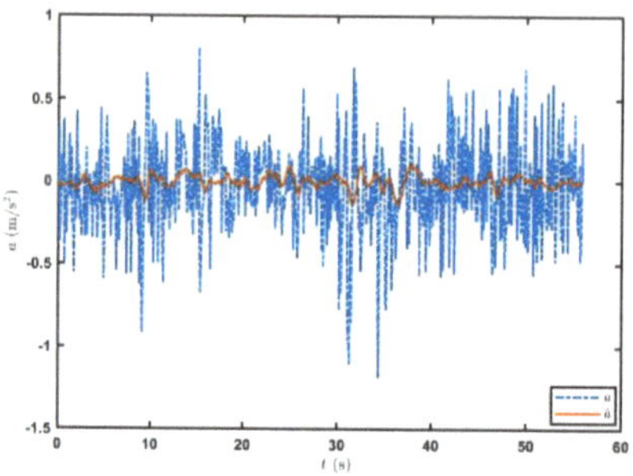

Figure 5. Filtering effect on the acceleration signal.

4. Engine Output Torque Prediction Model

After obtaining $\hat{a}$ using the Kalman filter algorithm, the driving force F_t of the vehicle was calculated according to Formula (1), and the T_e was further obtained. The n_e, n_e', β, and β' were used as the inputs in the engine output torque prediction model, and the corresponding engine output torque T_e, calculated by $\hat{a}$, was used as the model output. The GA–BP neural network algorithm was trained on the inputs and output, and the mapping relationship between the T_e and n_e, n_e', β, and β' was $T_e = f(n_e, n_e', \beta, \beta')$. Thus, the engine output torque prediction model was established.

The GA–BP neural network algorithm makes use of the global optimization ability of the genetic algorithm to make up for the shortcomings of BP neural networks, such as the slow learning convergence speeds, uncertain network structures, and ease of falling into the local minimum. The initial weights and thresholds in the BP neural network were used as genes in the genetic algorithm. The values on the genes represented the connection weights or thresholds in the BP neural network and formed the chromosomes of the genetic algorithm. A certain number of chromosomes were used as the initial population of the genetic algorithm. After selection, crossover, and mutation iterations, the initial weights and thresholds of the optimal BP neural network were obtained. The GA–BP neural network algorithm structure diagram is shown in Figure 6. After the simulation test, the number of genetic iterations in the genetic algorithm is set to 30, the number of populations is 5, the probability of crossover is 0.7, and the probability of mutation is 0.1.

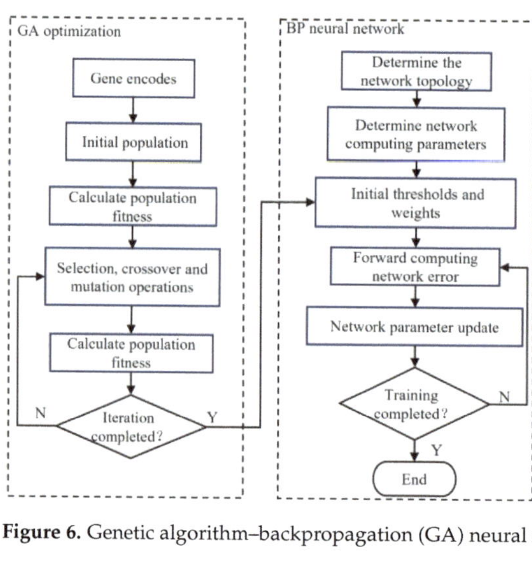

Figure 6. Genetic algorithm–backpropagation (GA) neural network algorithm structure diagram.

The BP neural network continuously corrected the weights and thresholds of each neural network layer through error backpropagation. The number of input nodes in the BP neural network is four, the number of output nodes is one, and the hidden layer is ten. When the training results met the set requirements, the training was stopped and the prediction results were output. The network structure is shown in Figure 7.

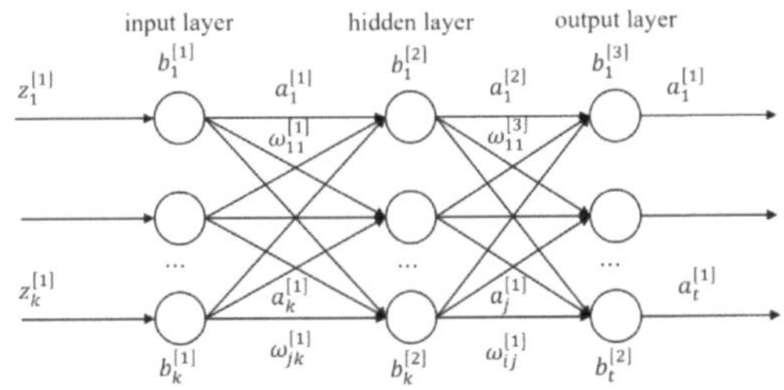

Figure 7. Structure of the backpropagation (BP) network.

In the graph, $\omega_{jk}^{[n]}$ is the weight value from the k-th node to the j-th node of the $(n-1)$-th layer in the neural network, $b_j^{[n]}$ represents the threshold of the j-th node of the n-th layer neural network, $z_j^{[n]}$ is the linear result of the j-th node added to the n-th layer neural network, and $a_j^{[n]}$ represents the output value of the j-th node of the n-th layer neural network.

The initial weights and thresholds for each layer of the neural network were calculated by the genetic algorithm to obtain the optimal solution. σ denotes the activation function, and χ donates the learning rate. The input signal fitting process is as follows:

Gradient of output layer:

$$\sigma_j^{[n]} = \frac{\partial s}{\partial \alpha_j^{[n]}} \sigma'\left(z_j^{[s]}\right); \tag{14}$$

Gradient of hidden layer:

$$\sigma_j^{[n]} = \sum \omega_{kj}^{[n]} \sigma_k^{[n+1]} \sigma'\left(z_j^{[n]}\right); \tag{15}$$

s-th iteration threshold:

$$b_j^{[n]}(s) = b_j^{[n]}(s-1) - \eta \sigma_j^{[n]}; \tag{16}$$

s-th iteration weight:

$$\omega_{jk}^{[n]}(s) = \omega_{jk}^{[n]}(s-1) - \eta \sigma_j^{[n]} a_k^{[n-1]}. \tag{17}$$

The learning rate χ was adaptively adjusted according to the error change $e(s)$, which can be expressed as:

$$\chi(s) = \begin{cases} 1.05\chi(s-1) & e(s-1) < e(s-2) \\ 0.5\chi(s-1) & e(s-1) < 1.04e(s-2) \\ \chi(s-1) & other \end{cases}. \tag{18}$$

5. Experimental Results and Analysis

The accuracy of the tracked vehicle engine output torque prediction model and the effectiveness of the f estimation method were verified by experiments. The experimental pavement was a sand road and a cement road. The structural parameters of the experimental vehicle are shown in Table 1.

Table 1. Tracked vehicle structural parameters.

Parameter	Value
m (kg)	31,000
r (m)	0.283
A (m^2)	6
C_D	0.45
δ	1.24
i (1st gear to 5th gear)	28.35/13.23/9.45/6.71/4.3
η (1st gear to 5th gear)	0.79/0.77/0.76/0.75/0.73

To improve the computational efficiency, the engine output torque prediction model was established through offline updates and online prediction. In this study, MATLAB 2021b was used to train the engine output torque prediction model offline through the selected vehicle driving data, and the generated model was converted into C code and imported into an industrial personal computer (IPC).

The IPC received the vehicle state in real time through the controller area networks (CAN) bus. When the vehicle state was determined to meet the set working conditions, a data storage container was established to store the vehicle state data. When it was determined that the current vehicle driving state did not meet the set conditions, the data stored in the container were emptied and the vehicle state was continuously monitored. When the container stored data for more than 10 s, i.e., the vehicle had been running in a specific state for 10 s, the engine output torque prediction model began to predict the $\hat{T}_e$ based on the n_e and β, and the $\hat{f}$ was further estimated. The data container was used to store data to estimate the $\hat{f}$ and update the $\hat{f}$ in real time, based on the current vehicle state. The process through which the IPC processed the tracked vehicle driving data is shown in Figure 8.

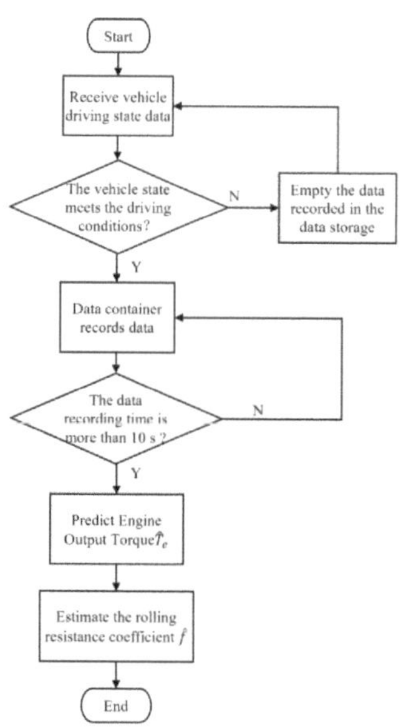

Figure 8. Process through which the industrial personal computer (IPC) processed the tracked vehicle driving data.

5.1. Estimation of $\hat{f}$ for Tracked Vehicles Driving on a Sand Road

The annular sand road in the vehicle driving test field was selected as the experimental test road. According to the driving habits and environmental conditions, the driver determined the gear and speed to maintain driving safety and speed. The IPC received the driver's control information and the vehicle driving state data in real time, and automatically determined whether the vehicle's state satisfied the set driving conditions. When the driving conditions were satisfactory, the engine output torque prediction model predicted the $\hat{T}_e$ in real time and estimated the $\hat{f}$.

Figure 9 shows the trajectory of the tracked vehicle. The vehicle started from the starting point and traveled around the circular runway. The driving distance was 3694 m. The red solid line section marked in Figure 9 indicates that the state of the tracked vehicle on this section satisfied the set driving conditions. The total length of the red solid line segment was 2054.5 m. Figure 10 shows the change in the speed, and the gear and clutch cylinder displacement in the tracked vehicle during the whole driving process. As can be seen from Figure 10, the total driving time of the vehicle was 622.3 s. The vehicle starts in second gear, the highest gear was fifth gear, the highest speed was 32.5 km/h, the commonly used gear during the vehicle driving was fourth gear, and the average driving speed was 20.55 km/h. When $flag_{state} = 1$, the driving state of the vehicle meets the set conditions. The entire driving process satisfied the set driving conditions during 10 periods, and the total time was 313.8 s. Figure 11 shows the changes in the acceleration, pitch angle, and heading angle of the tracked vehicle. It can be seen from Figure 11 that the acceleration of the vehicle increased rapidly in a short time during the engagement of the vehicle's clutch, and the impact of shifting the vehicle was large. After the clutch engagement was completed, the vehicle acceleration changed relatively smoothly when the vehicle was accelerating and decelerating. The acceleration measurement value was more credible, and it was reasonable to screen the vehicle driving data by setting the acceleration threshold. It can be seen from the change in the heading angle that when the tracked vehicle was under the set driving conditions, the heading angle of the vehicle was almost unchanged, and the vehicle could be considered to have maintained, approximately, a straight driving state. Under the set driving conditions, the change in the pitch angle of the tracked vehicle was in the range of the set pitch angle. Figure 12 shows the changes in the engine speed and accelerator pedal angle during the driving process of the vehicle. The selected driving information data excluded the rapid variation stage of the engine speed during the shifting process. The accelerator pedal angle and engine speed varied smoothly, and the vehicle ran stably.

Figure 13 shows the $\hat{T}_e$ prediction of the engine output torque prediction model when the tracked vehicle satisfied the driving conditions for the first time on the sand road. The T_e, calculated based on $\hat{a}$ using Equation (1), was the real value, and the engine output torque predictions of the BP neural network were used as the control data. The engine output torque value $\hat{T}_{e_{GA-BP}}$ predicted by the GA–BP neural network was closer to the T_e. The root mean square error σ_e and the coefficient of determination R were used as the evaluation indices on the accuracy of the engine output torque estimation.

$$\sigma_e = \sqrt{(\hat{T}_e - T_e)^2/n}, \tag{19}$$

$$R = 1 - \frac{\sum(T_e - \hat{T}_e)^2}{\sum(T_e - \overline{T}_e)^2}, \tag{20}$$

where n is the number of sample data, and $\overline{T}_e$ is the average value of the true values on the output torque of the transmitter calculated from the sample data.

The root mean square error of the engine output torque obtained by the engine output torque prediction model established by the BP neural network was 78.65, and the coefficient of determination was 0.768. The root mean square error of the engine output torque calculated by the GA–BP was 45.06, and the coefficient of determination was 0.924, which was 42.71% and 20.31% higher than those of the BP neural network, respectively. Figure 14 shows the result on the further f estimation by the RLS algorithm. The rolling resistance coefficient estimated by the GA–BP neural network method was more convergent.

The average error of the $\hat{f}$ value estimated by the RLS algorithm was 0.00093, while it was 0.00117 for the BP neural network. Thus, the estimation accuracy was improved by 20.51%.

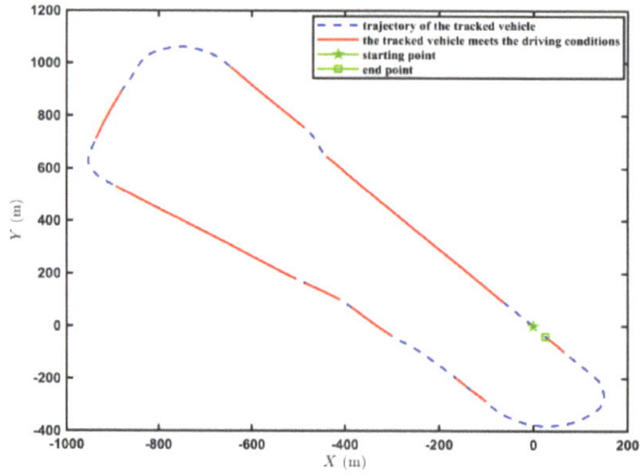

Figure 9. Trajectory of the tracked vehicle driving on the sand road.

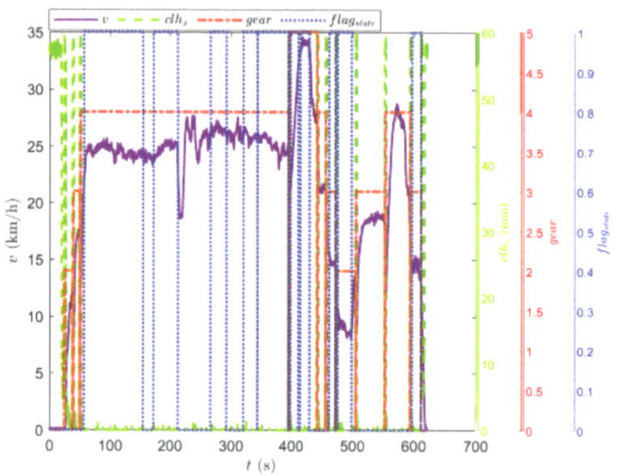

Figure 10. Changes in the speed, and gear and clutch cylinder displacement in the tracked vehicle on the sand road.

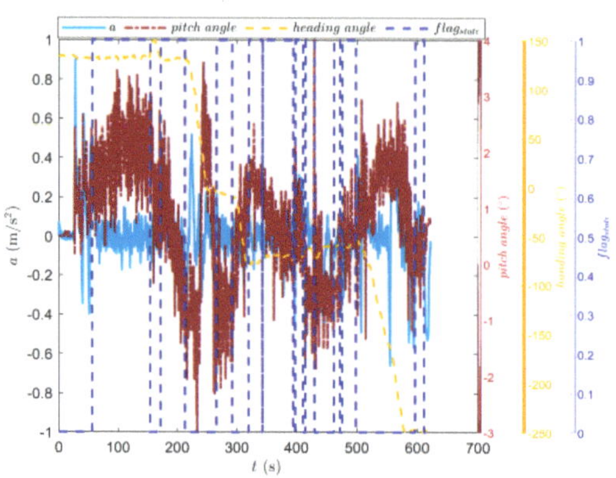

Figure 11. Changes in the longitudinal acceleration, heading angle, and pitch angle of the tracked vehicle running on the sand road.

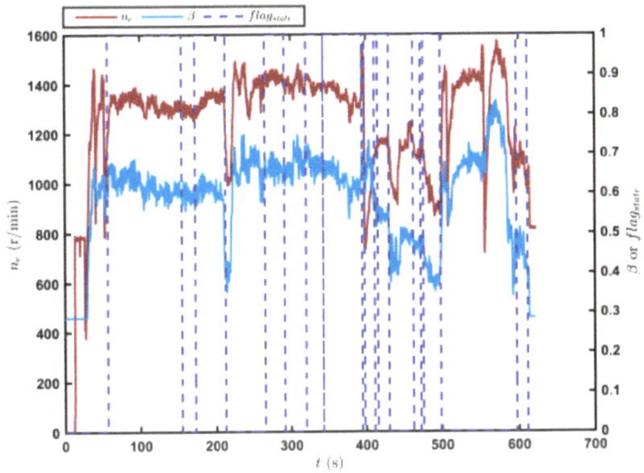

Figure 12. Changes in the engine speed and accelerator pedal position in the tracked vehicle running on the sand road.

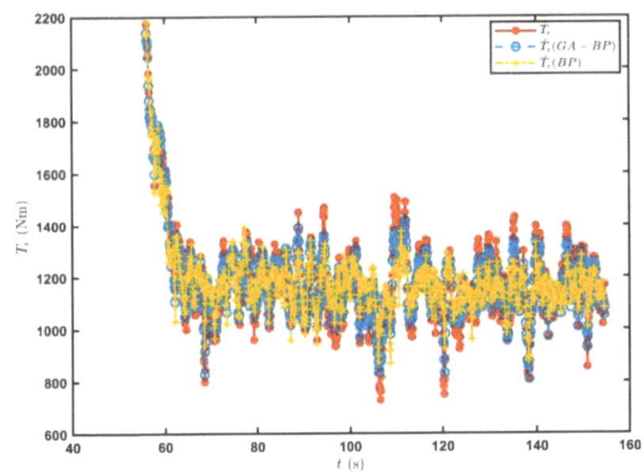

Figure 13. Predicted $\hat{T}_e$ values when the tracked vehicle satisfied the driving conditions for the first time on the sand road.

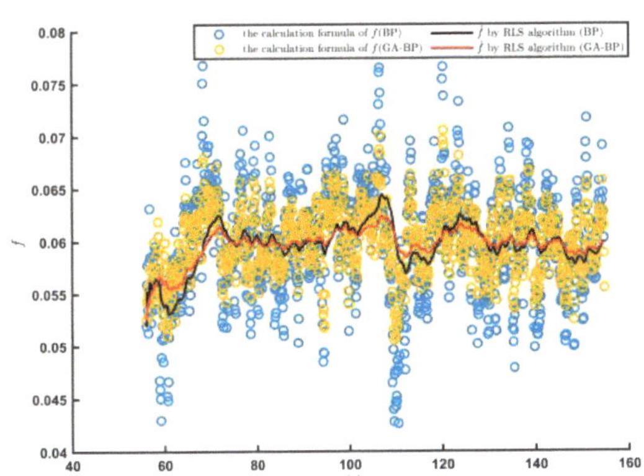

Figure 14. Predicted $\hat{f}$ values when the tracked vehicle satisfied the driving conditions for the first time on the sand road.

Table 2 shows the vehicle state and $\hat{T}_e$ prediction for the 10 periods when the tracked vehicle satisfied the driving conditions. The average root mean square error of the engine output torque estimated by the GA–BP neural network was 42.24, and the average root mean square error calculated by the BP neural network was 73.95. The average root mean square error of the engine output torque calculated by the GA–BP neural network on the sand road was 42.87% higher than that of the BP neural network. Similarly, the average coefficient of determination of the engine output torque estimated by the engine output torque model established by the GA–BP neural network was 0.918, which was 27.73% higher than that of the BP neural network.

Table 2. Vehicle state, $\hat{T}_e$, and estimated $\hat{f}$ when the tracked vehicle satisfied the driving conditions on the sand road.

	Time (s)	Distance (m)	Gear	Average Speed (km/h)	σ_e (GA–BP)	σ_e (BP)	R (GA–BP)	R (BP)
1	56.1–154.6	671.13	4	24.48	43.6	79.69	0.9287	0.7624
2	172.2–212.8	279.84	4	24.8	32.34	60.07	0.9305	0.7604
3	266–291.8	189.63	4	26.46	43.47	75.6	0.8727	0.6452
4	319.8–343.1	167.53	4	25.88	35.83	75.93	0.903	0.5640
5	343.3–394.3	356.49	4	25.11	40.59	76.04	0.9118	0.6907
6	398.1–411.1	96.98	5	26.82	42.59	79.97	0.9498	0.7437
7	415.2–429.6	136.84	5	33.99	52.66	79.26	0.9387	0.7832
8	461.6–471.7	41.89	3	14.78	51.73	69.84	0.9184	0.7142
9	475.5–498.3	57.39	2	9.04	42.95	64.35	0.8942	0.7627
10	596.9–611.2	56.83	3	14.3	36.67	78.73	0.9215	0.7608

Figure 15 shows the change in the average absolute error e_{mb} of the $\hat{f}$ estimation results obtained by the two methods, which was calculated as

$$e_{mb} = \frac{\sum \left| \hat{T}_e - T_e \right|}{n}. \quad (21)$$

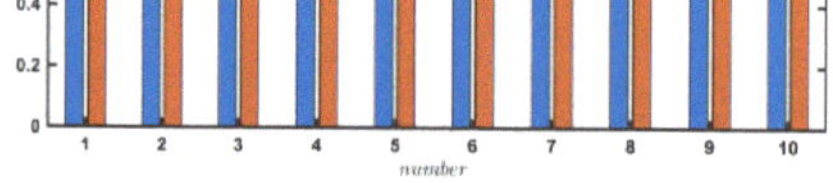

Figure 15. Engine output torque predictions when the tracked vehicle satisfied the driving conditions during 10 periods on the sand road.

The e_{mb} value of the road rolling resistance coefficient estimated by the GA–BP neural network was 0.00108, which was 24.44% higher than that of the BP neural network. The engine output torque model was established by the GA–BP neural network on the sand road, and the engine output torque was estimated. The estimation accuracy of the rolling resistance coefficient of the road was significantly improved compared with that of the BP neural network. Thus, the GA–BP neural network can better meet the real-time estimation requirements of the rolling resistance coefficient during the driving process of a vehicle on a sand road.

5.2. Estimation of $\hat{f}$ for Tracked Vehicles Running on a Cement Road

The driver drove the tracked vehicle by starting in second gear on the cement road. During the driving process of the vehicle, the IPC received the state information on the vehicle, predicted the engine output torque, and estimated the road rolling resistance coefficient in real time, when the vehicle state satisfied the driving conditions.

Figure 16 shows the driving route of the tracked vehicle on the cement road, with a driving distance of 776.5 m. The red solid line in the figure represents the road section where the tracked vehicle predicted the $\hat{T}_e$ and estimated the $\hat{f}$. During the whole driving process of the tracked vehicle, the engine output torque was predicted and the rolling resistance coefficient was estimated four times. The total length of the driving section was 554.3 m. Figure 17 shows the speed, the displacement of the clutch cylinder, and the change in the gear when the tracked vehicle was driving on the cement road. The tracked vehicle did not shift again after starting in second gear, and it ran in second gear to complete the trip. The maximum speed was 14.36 km/h and the time was 260 s. The time for the tracked vehicle to meet the set driving conditions was 154.4 s. Figure 18 shows the changes in the acceleration, pitch angle, and heading angle of the tracked vehicle. The acceleration changed smoothly when the tracked vehicle was under the set working conditions, the pitch angle change was less than the set threshold, and the heading angle change was small. The vehicle could be considered to be in a straight driving state. Figure 19 shows the relationship between the engine speed and the percentage change in the accelerator pedal. The engine speed and the driver's control accelerator pedal changed smoothly during the selected driving data.

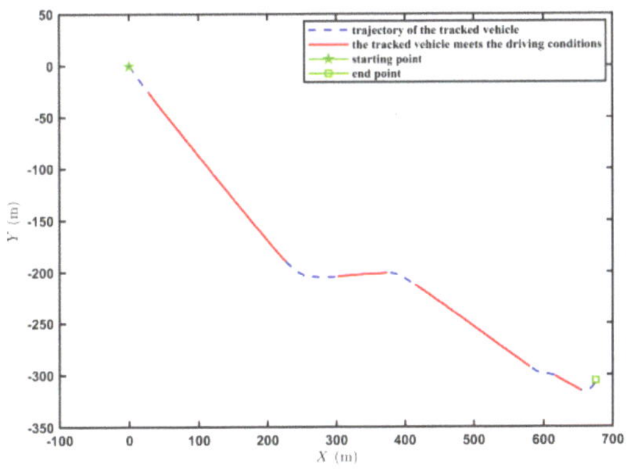

Figure 16. Trajectory of the tracked vehicle driving on the cement road.

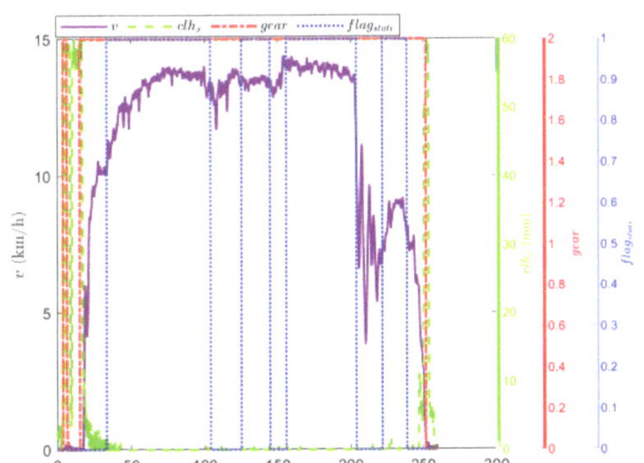

Figure 17. Changes in the speed, and gear and clutch cylinder displacement in the tracked vehicle on the cement road.

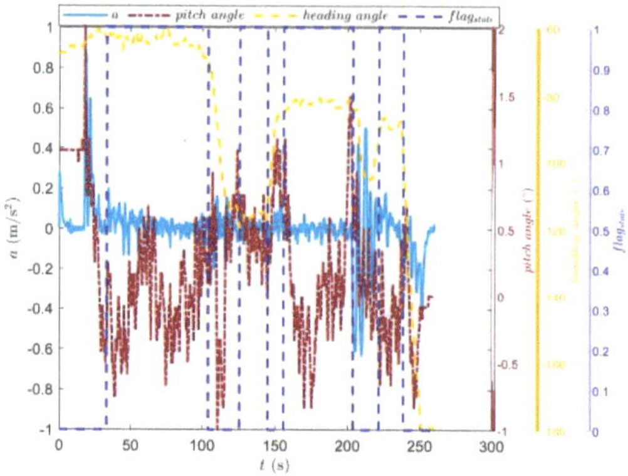

Figure 18. Changes in the longitudinal acceleration, heading, and angle of the tracked vehicle running on the cement road.

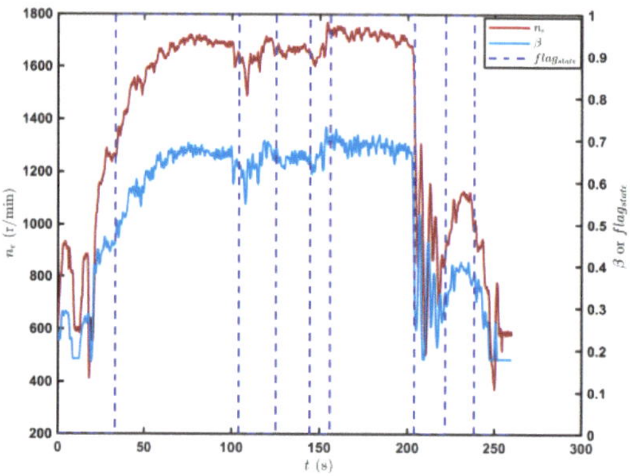

Figure 19. Changes in the engine speed and accelerator pedal position in the tracked vehicle running on the cement road.

Figure 20 shows the predicted values of the $\hat{T}_e$ and the estimated $\hat{f}$ values when the tracked vehicle ran on the cement road and met the set driving conditions for the first time. The $\hat{T}_e$ predicted by the engine output torque prediction model established by the GA–BP neural network was closer to the real value of the engine output torque. Figure 21 shows the estimation of the road rolling resistance coefficient $\hat{f}$ by the RLS algorithm. The GA–BP neural network was used to estimate the engine output torque and further estimated the road rolling resistance coefficient with better accuracy.

Table 3 shows the estimated values on the vehicle state, $\hat{T}_e$, and $\hat{f}$ of the tracked vehicle on the cement road under the set conditions. The σ_e value in the results estimated by the GA–BP neural network was 13.09, and the value for the BP neural network method was 42.4% greater. The R from the GA–BP method was 0.895, which was 12.8% higher than that of the BP neural network. Figure 22 shows the change in the e_{mb} values for the $\hat{f}$ estimation results obtained by the two methods. The e_{mb} value by the GA–BP neural network predictions was 0.00061, which was 38.1% higher than that of the BP neural network predictions.

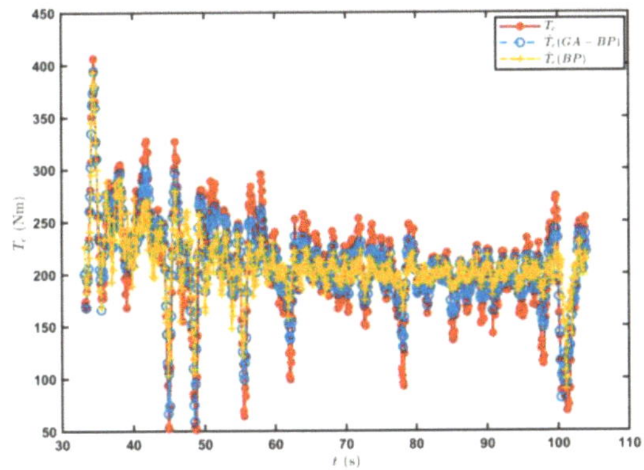

Figure 20. Prediction of the $\hat{T}_e$ values when the tracked vehicle satisfied the driving conditions for the first time on the cement road.

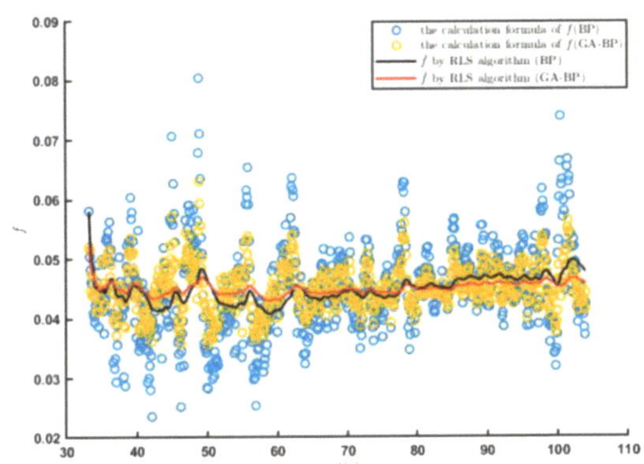

Figure 21. Prediction of the $\hat{f}$ values when the tracked vehicle satisfied the driving conditions for the first time on the cement road.

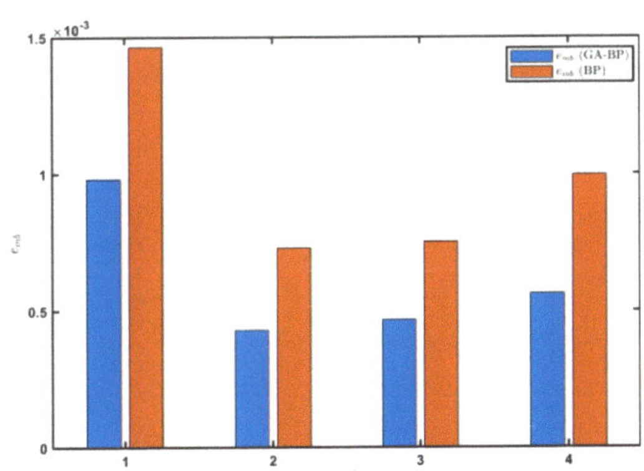

Figure 22. Engine output torque prediction when the tracked vehicle satisfied the driving conditions during 10 periods on the cement road.

Table 3. Vehicle state, $\hat{T}_e$, and estimated $\hat{f}$ when the tracked vehicle satisfied the driving conditions on the cement road.

	Time (s)	Distance (m)	Gear	Average Speed (km/h)	σ_e (GA–BP)	σ_e (BP)	R (GA–BP)	R (BP)
1	33.3–104	257.9	2	13.12	19.01	33.38	0.86	0.75
2	125.3–144.7	75.5	2	13.45	11.2	18.12	0.83	0.748
3	156.1–203.7	184.1	2	13.89	10.49	18.54	0.92	0.76
4	221.7–238.4	39.8	2	8.51	11.66	20.88	0.89	0.81

6. Conclusions

In this paper, a method for establishing an engine output torque prediction model based on vehicle driving data was proposed for tracked vehicles, and the model was used to further estimate the rolling resistance coefficient of the road. The following conclusions can be drawn from the experimental results:

(1) The engine output torque prediction model obtained by fitting the vehicle driving data with the GA–BP neural network had a high level of engine output torque prediction accuracy. The engine output torque prediction model was established using vehicle driving data, which reduced the calibration work in the engine bench test stage significantly and had real-time updating capabilities. This method provides a new option for the establishment of an engine output torque model.

(2) In this study, a prediction model of the engine output torque was established, and the RLS algorithm was used to estimate the road rolling resistance coefficients of tracked vehicles under certain driving conditions. The experimental results showed that when the tracked vehicle was driving on a sand road and a cement road, the rolling resistance coefficient of the road could be automatically estimated and had high accuracy when the vehicle driving state satisfied the set driving conditions. To a certain extent, this method meets the requirements for the real-time estimation of the rolling resistance coefficient of a road when a tracked vehicle drives longitudinally.

(3) Limited by the system structure of the tracked vehicle and the measurement error of the sensor, to ensure the prediction accuracy of the engine output torque prediction model and the estimation accuracy of the road rolling resistance coefficient, it is necessary to limit the driving conditions of the tracked vehicle, which makes it difficult to apply this model throughout the whole driving process. Determining how to make the tracked vehicle estimate the road parameters over the whole driving process will be the focus of future research.

Author Contributions: Conceptualization, X.L. and G.J.; Methodology, X.L. and W.J.; Software, W.J. and C.Z.; Validation, W.J. and B.S.; Formal analysis, G.J.; Investigation, C.Z.; Data curation, B.S.; Writing—original draft, X.L. and W.J.; Writing—review & editing, W.J.; Project administration, G.J. All authors have read and agreed to the published version of the manuscript.

Funding: This research received no external funding.

Institutional Review Board Statement: Not applicable.

Informed Consent Statement: Not applicable.

Data Availability Statement: No new data was created.

Conflicts of Interest: The authors declare no conflict of interest.

References

1. Purdy, D.J.; Simner, D.; Diskett, D.; Duncan, A.; Wormell, P.J.; Stonier, C. An experimental and theoretical investigation into the roll-over of tracked vehicles. *Proc. Inst. Mech. Eng. Part D J. Automob. Eng.* **2016**, *230*, 291–307. [CrossRef]
2. Lu, E.; Li, W.; Jiang, S.; Liu, Y. Anti-disturbance speed control of permanent magnet synchronous motor based on fractional order sliding mode load observer. *IEEE Access* **2022**. [CrossRef]

3. Komnos, D.; Broekaert, S.; Grigoratos, T.; Ntziachristos, L.; Fontaras, G. In use determination of aerodynamic and rolling resistances of heavy-duty vehicles. *Sustainability* **2021**, *13*, 974. [CrossRef]
4. Wang, H.; Wang, Q.; Rui, Q.; Gai, J.; Zhou, G. Analyzing and testing verification the performance about high-speed tracked vehicles in steering process. *Chin. J. Mech.* **2014**, *50*, 162–172. [CrossRef]
5. Dewangan, D.K.; Sahu, S.P. RCNet: Road classification convolutional neural networks for intelligent vehicle system. *Intell. Serv. Robot.* **2021**, *14*, 199–214. [CrossRef]
6. Selvathai, T.; Varadhan, J.; Ramesh, S. Road and off road terrain classification for autonomous ground vehicle. In Proceedings of the 2017 International Conference on Information Communication and Embedded Systems (ICICES), Chennai, India, 23–24 February 2017; pp. 1–3.
7. Valada, A.; Burgard, W. Deep spatiotemporal models for robust proprioceptive terrain classification. *Int. J. Robot. Res.* **2017**, *36*, 1521–1539. [CrossRef]
8. Huang, Y.; Chen, S.; Chen, Y.; Jian, Z.; Zheng, N. Spatial-temproal based lane detection using deep learning. In Proceedings of the Artificial Intelligence Applications and Innovations: 14th IFIP WG 12.5 International Conference, AIAI 2018, Rhodes, Greece, 25–27 May 2018; pp. 143–154.
9. Zhao, J.; Li, Y.; Zhu, B.; Deng, W.; Sun, B. Method and applications of LiDAR modeling for virtual testing of intelligent vehicles. *IEEE Trans. Intell. Transp. Syst.* **2020**, *22*, 2990–3000. [CrossRef]
10. Zhu, Y.; Luo, K.; Ma, C.; Liu, Q.; Jin, B. Superpixel segmentation based synthetic classifications with clear boundary information for a legged robot. *Sensors* **2018**, *18*, 2808. [CrossRef] [PubMed]
11. Rajendran, S.; Spurgeon, S.K.; Tsampardoukas, G.; Hampson, R. Estimation of road frictional force and wheel slip for effective antilock braking system (ABS) control. *Int. J. Robust Nonlinear Control.* **2019**, *29*, 736–765. [CrossRef]
12. Wenbo, C.; Yugong, L.; Jian, L.; Keqiang, L. Vehicle mass and road slope estimates for electric vehicles. *J. Tsinghua Univ. (Sci. Technol.)* **2014**, *54*, 724–728.
13. Liu, P.; Liu, B.; Dong, T.; Zhang, C. Road Roughness Identification and Shift Control Study for a Heavy-duty Powertrain. *Energy Procedia* **2017**, *105*, 2885–2890. [CrossRef]
14. Cong, X.-Y.; Zhang, Y.-G.; Wang, C.-P.; Zhu, H.-X.; Wang, Z.-C. Look-ahead gear-shifting strategy on ramps for heavy trucks with automated mechanical transmission. *Adv. Mech. Eng.* **2019**, *11*, 1687814018822911. [CrossRef]
15. Qian, K.; Hou, Z.; Sun, D. Sound quality estimation of electric vehicles based on GA-BP artificial neural networks. *Appl. Sci.* **2020**, *10*, 5567. [CrossRef]
16. Wang, J.; Alexander, L.; Rajamani, R. Friction estimation on highway vehicles using longitudinal measurements. *J. Dyn. Sys. Meas. Control* **2004**, *126*, 265–275. [CrossRef]

sensors

MDPI

Article

Full-Field Vibration Response Estimation from Sparse Multi-Agent Automatic Mobile Sensors Using Formation Control Algorithm

Debasish Jana [1,2] and Satish Nagarajaiah [2,3,*]

[1] Samueli Civil and Environmental Engineering, University of California, Los Angeles, CA 90095, USA; dj93@ucla.edu
[2] Civil and Environmental Engineering, Rice University, Houston, TX 77005, USA
[3] Mechnanical Engineering, Rice University, Houston, TX 77005, USA
* Correspondence: satish.nagarajaiah@rice.edu

Abstract: In structural vibration response sensing, mobile sensors offer outstanding benefits as they are not dedicated to a certain structure; they also possess the ability to acquire dense spatial information. Currently, most of the existing literature concerning mobile sensing involves human drivers manually driving through the bridges multiple times. While self-driving automated vehicles could serve for such studies, they might entail substantial costs when applied to structural health monitoring tasks. Therefore, in order to tackle this challenge, we introduce a formation control framework that facilitates automatic multi-agent mobile sensing. Notably, our findings demonstrate that the proposed formation control algorithm can effectively control the behavior of the multi-agent systems for structural response sensing purposes based on user choice. We leverage vibration data collected by these mobile sensors to estimate the full-field vibration response of the structure, utilizing a compressive sensing algorithm in the spatial domain. The task of estimating the full-field response can be represented as a spatiotemporal response matrix completion task, wherein the suite of multi-agent mobile sensors sparsely populates some of the matrix's elements. Subsequently, we deploy the compressive sensing technique to obtain the dense full-field vibration complete response of the structure and estimate the reconstruction accuracy. Results obtained from two different formations on a simply supported bridge are presented in this paper, and the high level of accuracy in reconstruction underscores the efficacy of our proposed framework. This multi-agent mobile sensing approach showcases the significant potential for automated structural response measurement, directly applicable to health monitoring and resilience assessment objectives.

Keywords: full-field sensing; compressive sensing; multi-agent system; mobile sensors; formation control; structural health monitoring

Citation: Jana, D.; Nagarajaiah, S. Full-Field Vibration Response Estimation from Sparse Multi-Agent Automatic Mobile Sensors Using Formation Control Algorithm. *Sensors* **2023**, *23*, 7848. https://doi.org/10.3390/s23187848

Academic Editors: Yafei Wang, Chao Huang, Hailong Huang, Zhiqiang Zuo, Bo Leng and Peng Hang

Received: 8 August 2023
Revised: 1 September 2023
Accepted: 11 September 2023
Published: 13 September 2023

1. Introduction

Bridge health monitoring is essential to ensure public safety, prolong infrastructure lifespan, and mitigate potential risks through continuous assessment of structural integrity and performance. Although fixed sensors placed on the structure are commonly used for vibration-based bridge health monitoring [1], they require ongoing monitoring of sensor health to ensure data reliability [2,3]. Mobile sensing presents an alternative approach, involving the installation of vibration sensors on mobile vehicles or carriers [4]. These mobile units traverse the structure, collecting vibration response data in relation to spatial and temporal variations. Mobile sensing offers distinct benefits compared to traditional fixed sensors, including increased spatial information, scalability, and reduced maintenance costs [5]. The advancements in wireless sensing technologies enable mobile sensor networks to complement wired counterparts, facilitating expanded usage of mobile sensors [6]. Modern smartphones, equipped with motion sensing chips like accelerometers and gyroscopes, enhance the potential for crowd-sourced data collection [7–9]. Nevertheless, these mobile sensing techniques often require multiple passes on bridges to record vibration response data. Notably, Matarazzo et al. [10] demonstrated that controlled field experiments and UBER trips on the Golden Gate Bridge enable continuous modal information extraction from smartphone-recorded vibration data. In their study, the researchers traversed the bridge 102 times and utilized 72 UBER trips, capturing acceleration vibration data with smartphones. The collected data were then analyzed to determine the most probable modal frequencies (MPMFs) using the synchro-squeezed wavelet transform [11]. Importantly, multiple vehicles were employed simultaneously for data collection. Additionally, the effectiveness of this mobile sensing approach was verified on the Harvard Bridge by Matarazzo et al. [12]. In a similar vein, Eshkevari et al. [13] conducted an experimental study validating crowd-sourced modal identification using continuous wavelets (CMICW), utilizing a collection of smartphone-mounted sensors on vehicles that moved back and forth across the bridge using a motored pulley system. Therefore, this paper introduces a multi-agent formation control framework to automate the mobile sensing procedure, eliminating the need for manual driving stages, particularly in the context of structural vibration response sensing.

The concept of capturing bridge vibration data through sensors on moving vehicles was first introduced by Yang et al. [14]. Leveraging the pure structural responses recorded by these mobile sensors, system properties have been deduced using input–output balance [15]. Over the past decade, extensive studies have explored mobile sensing through diverse avenues, encompassing analytical and numerical analyses [16–19], laboratory-scale experiments [20–23], and real-life scenarios [24,25]. The literature predominantly emphasizes bridge modal identification via mobile sensing. Oshima et al. [26] detected mode shape-based support damage by mapping mobile sensor data to fixed sensor data. Mode shape-based bridge damage detection was achieved by Malekjafarian and O'Brien [27] using the short time-frequency domain decomposition (STFDD) method. High-resolution mode shapes were obtained via laser vibrometer and accelerometers mounted on vehicles as proposed by O'Brien and Malekjafarian [28]. Additional signal processing techniques like Short-Time Fast Fourier Transform (STFFT) [29], Empirical mode decomposition (EMD) [30,31], and Hilbert transform [32] are employed for estimating mode shapes from data collected by mobile sensors. Matarazzo et al. [33,34] introduced the "structural identification using expectation maximization (STRIDE)" method for mode shape identification from mobile sensors. Eshkevari et al. [35–37] formulated mobile sensing data as a sparse matrix with missing values. They employed alternating least-square (ALS) for matrix completion, followed by principal component analysis (PCA) and structured optimization analysis (SOA) for modal identification. Matrix completion approaches have gained traction in recent years for health monitoring due to their data-driven nature, applicable to both fixed sensors [38] and mobile sensors [19,36,37]. Yang and Nagarajaiah [38] utilized nuclear norm minimization for matrix completion, and a comprehensive overview of such methods is presented in Nagarajaiah and Yang [39].

Throughout the aforementioned research, instances involving multiple mobile sensors for structural sensing or system identification typically involve independent manual control of each vehicle by humans. In certain cases, trains or vehicles with multiple trailers [26] have been employed, attaching sensors to each axle. The evolution of self-driving cars [40] presents the potential to streamline and enhance the mobile sensing process for structural health monitoring (SHM). Multiple self-driving cars could be useful for this purpose; however, autonomous vehicles are optimized for individual operations, often proving expensive for structural vibration response measurement work. Thus, an alternative approach is imperative to automate the mobile sensing procedure without relying on costly self-driving vehicles.

The primary objective of this paper is to automate the mobile sensing process instead of manually driving the vehicles or deploying fully self-driving cars. As a novel contribution, this paper introduces a formation control-based framework to collect the bridge vibration response data through multi-agent systems. The vibration measurement sensors installed on the multi-agent can capture structural responses at the corresponding position of their movement on the system, which can create a sparse space–time response matrix. In recent work, we proposed full-field structural vibration response estimation from a limited number of fixed sensors using data-driven [41] and physics-based [42] approaches. In this study, we deploy this concept to acquire full-field structural vibration responses but utilizing a limited number of automatic multi-agent mobile sensors.

In this paper, we first introduce a framework designed for acquiring full-field responses using mobile sensors and underscore its significance in Structural Health Monitoring (SHM) for bridges. Within this proposed framework, two primary components take center stage: the formation control strategy and compressive sensing. We provide a brief overview of these fundamental concepts. The formation control strategy streamlines the process by harnessing mobile sensors, while the compressive sensing algorithm is employed to estimate full-field responses with a limited sensor setup. This becomes especially pertinent as we leverage data from a network of multi-agent sensors to achieve this objective. Subsequently, we validate the effectiveness of this proposed framework through a numerical study involving a simply supported bridge. We explore two distinct scenarios for formation control. Additionally, we outline practical implementation recommendations that should be taken into consideration. Finally, we delve into the results, draw conclusions, and outline potential avenues for future research.

2. Proposed Framework and Its Significance in Bridge SHM

In this section, we initially outline the overarching process for vibration response sensing for bridge health monitoring. Subsequently, we delineate where our proposed automatic formation control-based mobile sensing framework fits in the domain of bridge SHM. Following this, we introduce the details of the formation control-based mobile sensing framework.

The overall sensing pipeline for the bridge structural health monitoring is illustrated in Figure 1. Generally, such sensing can be performed using either fixed or moving/mobile sensors. For fixed sensing cases, sensors are strategically placed at specific locations on the bridge. This approach facilitates the measurement of vibration responses at those particular bridge locations. By analyzing these responses, the health of the bridge can be monitored by estimating system parameters, detecting the presence of damage, and deriving a full-field response from a limited array of sensors. This full-field response can then be utilized for tasks such as damage localization, full-state estimation, or control purposes. An alternative approach to obtaining similar information involves mobile sensing. In this setup, sensors are mounted on vehicles, which can be operated manually or autonomously. Given that the aim of this paper revolves around achieving a fully automated sensing process with minimal human intervention, we refrain from addressing manual driving. Also, achieving coordination through manual driving is complex. In the context of automated driving, two possibilities emerge: (a) autonomous vehicles like Tesla or Waymo, and (b) multi-agent systems wherein vehicles interact with each other. Self-driving cars entail substantial costs, and using multiple self-driving cars for sensing purposes might prove economically unfeasible, as a single sensor might not suffice for bridge health monitoring. Thus, this paper concentrates on multi-agent systems to perform such tasks, as they offer a cost-effective and automated approach to the sensing process.

Nevertheless, within all mobile sensing strategies—whether manual driving or automatic driving—the responses measured by the sensor installed within the vehicle comprise bridge responses, superimposed with road roughness and vehicle dynamics [43], attributed to vehicle–bridge interaction [44,45]. When the recorded sensor data are amalgamated with the vibration from other sources mentioned above, conventional SHM methods cannot be directly applied as they are designed to work with pure structural vibration responses [46]. To circumvent this problem, the existing literature addresses primary approaches: (1) controlling the sensing conditions like vehicle speed and road roughness intensity so that the recorded response mainly contains the bridge vibration [14,27], (2) modeling the vehicle–bridge interaction in a closed form to eliminate the uncertainties due to vibration sources other than the bridge itself [47,48], (3) to use blind source separation (BSS) techniques to extract the different sources of the recorded response [35]. The BSS technique is capable of estimating pure bridge vibration response.

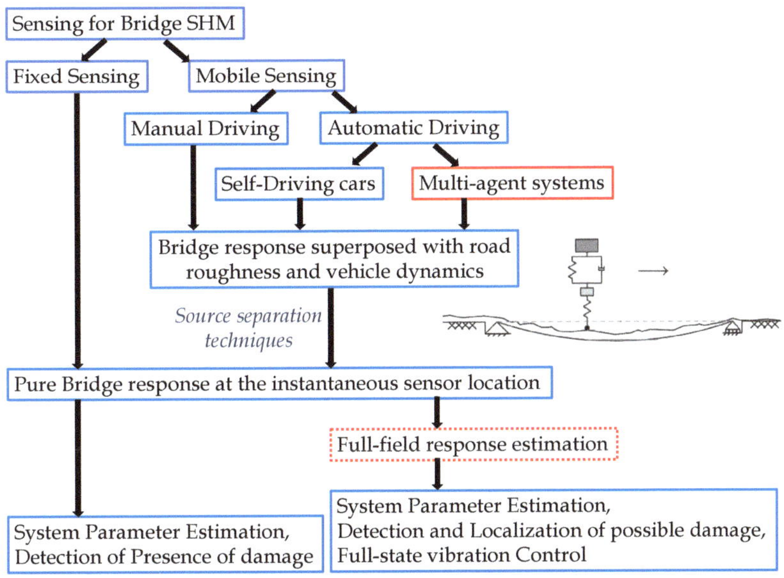

Figure 1. Framework for bridge vibration response sensing and structural health monitoring. This paper proposes a multi-agent formation control strategy for automatic driving in mobile sensing (marked with a red solid border). The result is numerically validated in the full-field response estimation (marked with a red dotted border). All the other framework components are marked with blue solid border. The image on the right shows that the acquired vibration response in the sensor contains the bridge vibration, road roughness, and vehicle dynamics.

In this paper, we focus on the multi-agent-system-based sensing framework and assume that the recorded bridge vibration response has undergone prior processing to eliminate the unintended road roughness and vehicle–bridge-interaction-related motion to work with pure structural vibration. This is a valid assumption as we explore only numerical scenarios in this study.

In the formation-control-based framework, the multi-agent system is designed to traverse the bridge in a user-defined formation. At a given instantaneous position of all the mobile agents, they collect the vertical vibration response of the bridge. The ultimate goal of this paper is to estimate full-field vibration response time history. In terms of matrix terminology, the objective is to complete the spatiotemporal response matrix. The mobile agent sparsely populates some of the elements of this spatiotemporal response matrix. To better elucidate the matrix completion process, drawing a comparison with fixed sensors could provide enhanced clarity.

The distinction between fixed and mobile sensing for a simply supported bridge is depicted in Figure 2. In scenarios involving fixed sensors, the spatiotemporal response matrix is populated along a particular column based on the sensor's position, as illustrated in Figure 2a. Conversely, with mobile sensors, the spatiotemporal response matrix is populated diagonally in accordance with the movement and instantaneous positions of the vehicles, as shown in Figure 2b. The slope of these diagonals depends on the vehicle speeds. Now, considering one particular time instance marked by the red arrow in Figure 2b, only four elements are occupied (as four sensors are considered for demonstration purposes). The spatial compressive sensing algorithm's task is to estimate other values from the measured ones. Consequently, executing this procedure in real time for all rows leads to a complete spatiotemporal matrix, essentially constituting the full-field vibration response.

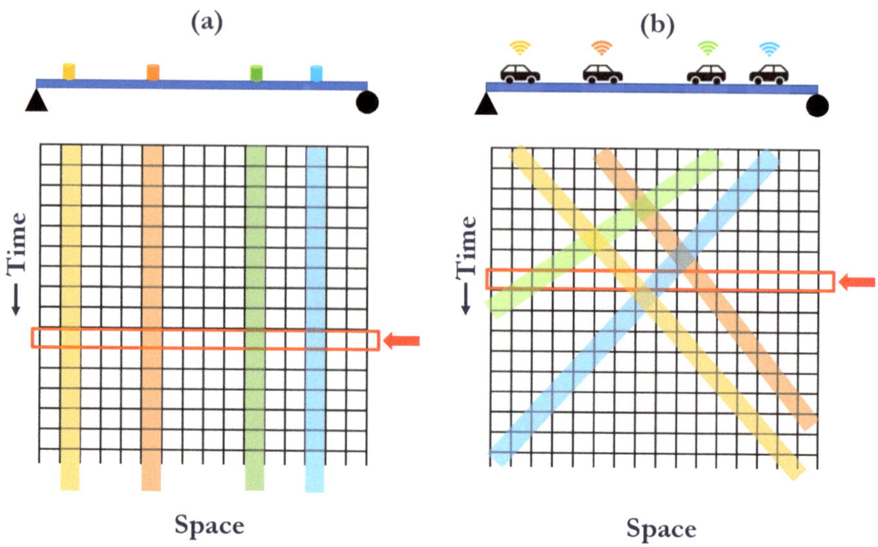

Figure 2. Vibration response sensing and the sensed entries of the spatiotemporal response matrix using (**a**) fixed sensors, (**b**) mobile sensors.

The authors recently published studies to complete the spatiotemporal matrix using the compressive sensing algorithm for fixed sensor cases [41,42], addressing the problem in Figure 2a. In the present paper, we intend to employ the same algorithm to accomplish spatiotemporal matrix completion for mobile sensor cases, e.g., addressing the problem in Figure 2b. In essence, the core proposition of this paper revolves around introducing a formation control algorithm to efficiently and autonomously populate the elements of the spatiotemporal matrix using a multi-agent system. Subsequently, the capability of compressive sensing is leveraged to complete the spatiotemporal response matrix utilizing response data collected through the multi-agent system, thus estimating the full-field vibration response of the structure.

3. Formation Control of Multi-Agent System Formulation

In this section, we introduce the formation control algorithm intended for automated sensing within the multi-agent system. We adopt a graphical model to depict interactions among these multi-agent systems, a prevalent approach in the state-of-the-art study. This framework often utilizes a graph, wherein agents represent graph nodes, while the graph edges symbolize communication and sensing exchanges between these agents, as emphasized by Godsil and Royle [49].

Graph $\mathcal{G}$ is defined as $(\mathcal{V}, \mathcal{E})$; here, $\mathcal{V}$ denotes the set of nodes or vertices of graph and $\mathcal{E} \subseteq \mathcal{V} \times \mathcal{V}$ symbolizes the edges. Here, it is assumed that there are no self-edges, viz., $(i,i) \notin \mathcal{E}$ for any $i \in \mathcal{V}$, which is a valid assumption. The neighbor set of node $i \in \mathcal{V}$ is defined as $\mathcal{N}_i := \{j \in \mathcal{V} : (i,j) \in \mathcal{E}\}$. The graph edges are weighted by w_{ij} which are associated with (i,j) for $i,j \in \mathcal{V}$; here, $w_{ij} > 0$ if $(i,j) \in \mathcal{E}$ and $w_{ij} = 0$ otherwise. The Laplacian Matrix $\mathcal{L} = [l_{ij}] \in \mathbb{R}^{|\mathcal{V}| \times |\mathcal{V}|}$ of $\mathcal{G} = (\mathcal{V}, \mathcal{E})$ is defined as

$$l_{ij} = \begin{cases} \sum_{k \in \mathcal{N}_i} w_{ik}, & \text{if } i = j \\ -w_{ij}, & \text{if } i \neq j \end{cases}. \quad (1)$$

The majority of dynamic models for automated mobile sensors adhere to the double-integrator dynamic models, given that the control feedback can be linearized in this form, as outlined in Ren and Beard [50]. In the context of graph $\mathcal{G}$, a collection of n agents, each modeled as a double-integrator system, residing within an m-dimensional space, can be succinctly expressed:

$$\begin{cases} \dot{p}_i = v_i, \\ \dot{v}_i = u_i, \end{cases} \qquad i = 1, \cdots, n. \tag{2}$$

Here, $p_i \in \mathbb{R}^m$, $v_i \in \mathbb{R}^m$, and $u_i \in \mathbb{R}^m$ indicate the position, velocity, and control input of agent i in relation to the global coordinate system. The agents have the capacity to sense the relative positions and velocities of neighboring agents within the global coordinate system. The overarching aim of the agents is to maneuver in a manner that controls a formation shape, where the designated desired position $p^* \in \mathbb{R}^{mn}$ and desired velocity $v^* \in \mathbb{R}^{mn}$ are pre-defined.

A formation comprises agents that move to acquire and maintain a specific geometric configuration based on the relative positions of neighboring agents. A formation can be denoted by specifying the desired position $p_i^*(t)$ and the desired velocity $v_i^*(t)$ for all $i = 1, \cdots, n$ at any given time instant $t \geq 0$. The primary goal is to formulate a control strategy u_i in a manner such that,

$$\begin{cases} p_i(t) \longrightarrow p_i^*(t), \\ v_i(t) \longrightarrow v_i^*(t), \end{cases} \qquad i = 1, \cdots, n. \tag{3}$$

Here, the symbol, "$\longrightarrow$", indicates the desirability, e.g., the actual position and actual velocity of agent i at time instant t which are $p_i(t)$ and $v_i(t)$, but desired to be $p_i^*(t)$ and $v_i^*(t)$.

If the agents can precisely measure their positions and velocities in the global coordinate system in which $p_i^*(t)$ and $v_i^*(t)$ are specified, then the Equation (3) could be solved in a straightforward manner using classical control. In this case, the controller follows the subsequent equation:

$$u_i(t) = g_p(p_i^*(t) - p_i(t)) + g_v(v_i^*(t) - v_i(t)); \qquad i, j = 1, \cdots, n; \qquad g_p, g_v > 0. \tag{4}$$

In this context, parameters g_p and g_v function as scaling factors linked to the position and velocity components of the control force. The formulation of formation control presented in Equation (4) corresponds to the concept of position-based formation control, a framework previously explored in works such as Ren and Beard [51] and Oh et al. [52]. However, within this position-based control scheme, every agent must possess sophisticated sensors capable of precisely measuring position and velocity with respect to global coordinates. Implementing this control strategy could prove challenging due to the associated financial costs tied to the requirement for advanced sensors, especially only for structural health monitoring purposes. Nevertheless, if agents are restricted to sensing their neighboring agents' relative positions and velocities solely, the goal outlined in Equation (4) becomes notably more intricate to attain. A more lenient objective emerges, revolving around maintaining relative positions and velocities amongst the agents. This approach is termed displacement-based control [52]. In this context, agents actively regulate their neighboring counterparts to realize the intended formation, with most agents operating without knowledge of the global coordinate system. Consequently, a less rigid objective is to devise a control law u_i such that

$$\begin{cases} p_i(t) - p_j(t) \longrightarrow p_i^*(t) - p_j^*(t), \\ v_i(t) - v_j(t) \longrightarrow v_i^*(t) - v_j^*(t), \end{cases} \qquad i, j = 1, \cdots, n. \tag{5}$$

To satisfy the objective in Equation (5), the control law can be written as

$$u_i(t) = g_p \sum_{j\in\mathcal{N}_i} (p_j(t) - p_i(t) - p_j^*(t) + p_i^*(t)) \\ + g_v \sum_{j\in\mathcal{N}_i} (v_j(t) - v_i(t) - v_j^*(t) + v_i^*(t)); \qquad g_p, g_v > 0; \qquad i = 1, \cdots, n. \tag{6}$$

Considering the formation moves in a constant velocity, $\dot{p}^* = v^*$ and $\dot{v}^* = 0$. We assume

$$p = \begin{Bmatrix} p_1 \\ \vdots \\ p_n \end{Bmatrix}; \qquad v = \begin{Bmatrix} v_1 \\ \vdots \\ v_n \end{Bmatrix}; \qquad p^* = \begin{Bmatrix} p_1^* \\ \vdots \\ p_n^* \end{Bmatrix}; \qquad v^* = \begin{Bmatrix} v_1^* \\ \vdots \\ v_n^* \end{Bmatrix}.$$

From Equation (2), the system dynamics are

$$\begin{Bmatrix} \dot{p} \\ \dot{v} \end{Bmatrix} = \begin{bmatrix} 0_{mn} & I_{mn} \\ 0_{mn} & 0_{mn} \end{bmatrix} \begin{Bmatrix} p \\ v \end{Bmatrix} + \begin{bmatrix} 0_{mn} \\ I_{mn} \end{bmatrix} u. \tag{7}$$

Equation (6) can be rewritten as

$$u_i = g_p \left[\left|\mathcal{N}_i\right| (p_i^* - p_i) - \sum_{j\in\mathcal{N}_i} (p_j^* - p_j) \right] + g_v \left[\left|\mathcal{N}_i\right| (v_i^* - v_i) - \sum_{j\in\mathcal{N}_i} (v_j^* - v_j) \right]. \tag{8}$$

Here, $\left|\mathcal{N}_i\right|$ denotes the cardinality of neighbor set $\mathcal{N}_i$ (total number of neighbors of agent i). Equations (7) and (8) can be simplified to

$$u = \left[g_p(\mathcal{L}_n \otimes I_m) \quad g_v(\mathcal{L}_n \otimes I_m) \right] \begin{Bmatrix} p^* - p \\ v^* - v \end{Bmatrix}. \tag{9}$$

Here, $\otimes$ denotes the Kronecker product. Defining the error signals as

$$\begin{cases} e_p = p^* - p \\ e_v = v^* - v \end{cases}, \tag{10}$$

e_p and e_v are the differences between the desired and actual amplitude of agent position and velocity, respectively. With $\dot{p}^* = v^*$ and $\dot{v}^* = 0$, the error dynamics can be evolved from Equations (2) and (10) as

$$\begin{cases} \dot{e}_p = \dot{p}^* - \dot{p} = v^* - v = e_v \\ \dot{e}_v = \dot{v}^* - \dot{v} = -\dot{v} = -u \end{cases}. \tag{11}$$

From Equations (9) and (11),

$$\dot{e}_v = -u = \left[-g_p(\mathcal{L}_n \otimes I_m) \quad -g_v(\mathcal{L}_n \otimes I_m) \right] \begin{Bmatrix} e_p \\ e_v \end{Bmatrix}. \tag{12}$$

From Equations (11) and (12), error dynamics expression [52] is

$$\begin{Bmatrix} \dot{e}_p \\ \dot{e}_v \end{Bmatrix} = \begin{bmatrix} 0_{mn} & I_{mn} \\ -g_p(\mathcal{L}_n \otimes I_m) & -g_v(\mathcal{L}_n \otimes I_m) \end{bmatrix} \begin{Bmatrix} e_p \\ e_v \end{Bmatrix}. \tag{13}$$

The system represented in Equation (13) reaches consensus if and only if $\mathcal{G}_n$ is connected (if there is at least one edge from one node to any other node of the graph, $\mathcal{G}_n$, is said to be connected). In consensus,

$$\begin{cases} \left|\left|e_{p_i}(t) - e_{p_j}(t)\right|\right| \to 0, \\ \left|\left|e_{v_i}(t) - e_{v_j}(t)\right|\right| \to 0, \end{cases} \qquad t \to \infty. \tag{14}$$

The rate at which convergence or consensus is achieved hinges on the values of the constants g_p and g_v, intrinsic properties of the controller. In the context of Structural Health Monitoring (SHM) applications involving carriers, these constants, g_p and g_v, are reliant upon the vehicle controller responsible for steering the system. During convergence or consensus processes, Equation (14) guarantees compliance with Equation (5). With the control law established as per Equation (6), the dynamics of the closed-loop system are characterized by

$$\begin{Bmatrix} \dot{p} \\ \dot{v} \end{Bmatrix} = \begin{bmatrix} 0_{mn} & I_{mn} \\ -g_p(\mathcal{L}_n \otimes I_m) & -g_v(\mathcal{L}_n \otimes I_m) \end{bmatrix} \begin{Bmatrix} p \\ v \end{Bmatrix} + \begin{bmatrix} 0_{mn} & 0_{mn} \\ g_p(\mathcal{L}_n \otimes I_m) & g_v(\mathcal{L}_n \otimes I_m) \end{bmatrix} \begin{Bmatrix} p^* \\ v^* \end{Bmatrix}. \tag{15}$$

In the scenario where agents are limited to sensing only the relative positions and velocities of neighboring agents, they would not fulfill Equation (3), implying an inability to attain predetermined absolute positions within the global coordinate system. To address this, a minimum subset of agents, usually just one, must possess the capability to sense absolute positions. This particular agent plays the role of a leader within the multi-agent system, and by employing a leader–follower methodology, formation consensus can be achieved. Consequently, the control law is modified as follows:

$$\begin{aligned} u_i(t) = g_p &\sum_{j\in\mathcal{N}_i} (p_j(t) - p_i(t) - p_j^*(t) + p_i^*(t)) \\ +g_v &\sum_{j\in\mathcal{N}_i} (v_j(t) - v_i(t) - v_j^*(t) + v_i^*(t)) \\ &+h_i g_p(p_l^* - p_l) + h_i g_v(v_l^* - v_l), \end{aligned} \tag{16}$$

where p_l and v_l denote the position and velocity of the leader, respectively, and

$$h_i = \begin{cases} 1 & \text{if } i = l \\ 0 & \text{otherwise} \end{cases}. \tag{17}$$

Defining the matrix, $H = \mathtt{diag}(h_1, \cdots, h_n)$, the error dynamic is given by

$$\begin{Bmatrix} \dot{e}_p \\ \dot{e}_v \end{Bmatrix} = \begin{bmatrix} 0_{mn} & I_{mn} \\ -g_p\left[(\mathcal{L}_n + H_n) \otimes I_m\right] & -g_v\left[(\mathcal{L}_n + H_n) \otimes I_m\right] \end{bmatrix} \begin{Bmatrix} e_p \\ e_v \end{Bmatrix}, \tag{18}$$

and similarly, the final closed-loop system dynamics is expressed as

$$\begin{aligned} \begin{Bmatrix} \dot{p} \\ \dot{v} \end{Bmatrix} = &\begin{bmatrix} 0_{mn} & I_{mn} \\ -g_p\left[(\mathcal{L}_n + H_n) \otimes I_m\right] & -g_v\left[(\mathcal{L}_n + H_n) \otimes I_m\right] \end{bmatrix} \begin{Bmatrix} p \\ v \end{Bmatrix} \\ + &\begin{bmatrix} 0_{mn} & 0_{mn} \\ g_p\left[(\mathcal{L}_n + H_n) \otimes I_m\right] & g_v\left[(\mathcal{L}_n + H_n) \otimes I_m\right] \end{bmatrix} \begin{Bmatrix} p^* \\ v^* \end{Bmatrix}. \end{aligned} \tag{19}$$

In brief, Figure 3 presents a flowchart illustrating the formation control strategy. This strategy comprises two distinct loops: the inner loop focuses on controlling the individual agent dynamics, as described in Equation (2), while the outer loop manages the overall formation, adhering to the less stringent objective outlined in Equation (5).

The performance evaluation of the proposed formation control algorithm can be based on the disparity between the desired and actual positions of the agents [53]. Therefore, for any given agent i and time instance t, the error is quantified as $\Delta_i(t) = |p_i^*(t) - p_i(t)|$. Over the entire data collection duration denoted as Q, the position error for each agent is computed as $\Delta_i = \frac{1}{Q}\sum_{i=1}^{Q}\Delta_i(t)$. Ultimately, the collective formation error for all n agents is expressed as $\Delta = \frac{1}{n}\sum_{i=1}^{n}\Delta_i$. It is important to note that since the primary goal of the multi-agent system is to populate the elements of the spatiotemporal response matrix, the formation error is exclusively contingent on the agents' positions and not their velocities.

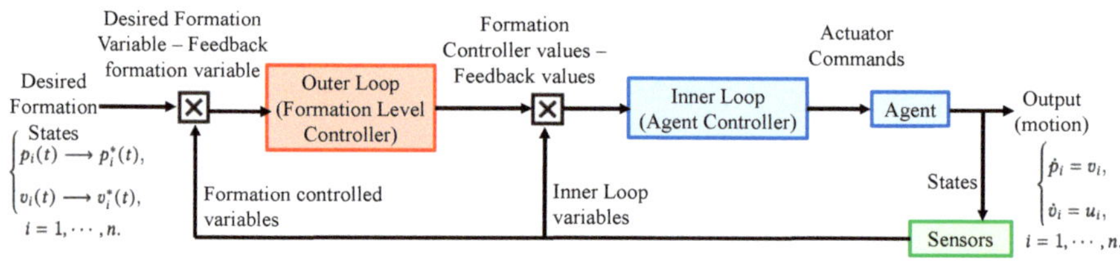

Figure 3. Formation control framework for a multi-agent system.

4. Brief Overview of Full-Field Response Estimation from a Limited Number of Sensors

Using the Formation control algorithm proposed in Section 3, the elements of the spatiotemporal response matrix are sparsely filled as shown in Figure 2b. In order to obtain the full-field response, the compressive sensing framework proposed by the authors [41,42] is used in this paper. Hence, in this section, we briefly discuss the procedure for the sake of completeness.

4.1. Compressive-Sensing-Based Full Signal Reconstruction from Few Measurements

The concept of Compressive Sensing [54] is briefly discussed in this section. A signal $\mathbf{y} \in \mathbb{R}^m$ is sparse, if

$$\mathbf{y} = \mathbf{D}\mathbf{x} = \sum_{j=1}^{n} x_j \mathbf{d}_j = \sum_{j \in \mathcal{S}} x_j \mathbf{d}_j. \tag{20}$$

In this context, $\mathbf{D} \in \mathbb{R}^{m \times n}$ signifies the orthonormal basis matrix, with $\mathbf{d}_j$ representing the jth column of $\mathbf{D}$. Typically, the basis matrix is treated as overcomplete, i.e., $m < n$. The majority of coefficients of x_j are zero in Equation (20). This characteristic results in signal sparsity, which can be expressed as $\mathcal{S} = \{j | x_j \neq 0\}$. The level of sparsity is represented by $s = |\mathcal{S}| = ||\mathbf{x}||_0$, thus indicating that $\mathbf{x} \in \mathbb{R}^n$ represents a sparse vector.

The Compressive Sensing (CS) technique is capable of estimating $\mathbf{y} \in \mathbb{R}^m$ from the noisy measured vector $\mathbf{z} \in \mathbb{R}^p$, where $p << m$.

$$\mathbf{z} = \mathbf{\Theta}\mathbf{y} + \mathbf{e} = \mathbf{\Theta}\mathbf{D}\mathbf{x} + \mathbf{e} = \mathbf{\Phi}\mathbf{x} + \mathbf{e}, \quad \text{where } \mathbf{\Phi} = \mathbf{\Theta}\mathbf{D}, \tag{21}$$

where $\mathbf{\Theta} \in \mathbb{R}^{p \times m}$ constitutes the measurement matrix. The term $\mathbf{e}$ signifies the error or noise constrained within the bound $||\mathbf{e}||_2 \leq \epsilon$. Consequently, the estimation of basis coefficients is attainable by solving the following convex optimization problem:

$$\hat{\mathbf{x}} = \underset{||\mathbf{\Phi}\mathbf{x} - \mathbf{z}||_2 \leq \epsilon}{\arg\min} ||\mathbf{x}||_1, \tag{22}$$

where the ℓ_2 norm is represented by $||\cdot||_2$. The formulation given in Equation (22) can be expressed within an optimization framework known as LASSO [55], as follows:

$$\text{minimize } ||\mathbf{\Phi}\mathbf{x} - \mathbf{z}||_2 + \lambda ||\mathbf{x}||_1. \tag{23}$$

Here, λ represents the regularization parameter. The interior point method [56] is employed to derive the sparse solution $\mathbf{x}$ from Equation (23), subsequently enabling the recovery of the complete signal $\mathbf{y}$ using Equation (20).

As proposed by Amini et al. [57], the determination of the minimum sampling points required for accurate signal reconstruction relies on the basis matrix $\mathbf{D}$. This estimation can be achieved by applying techniques like Singular Value Decomposition (SVD) and Normalized Power Index (NPI), as follows:

$$\mathbf{D} = \mathbf{U}\mathbf{\Sigma}\mathbf{V}; \qquad \mathrm{NPI}_p = \frac{\sum_{i=1}^{p} \sigma_i^2}{\sum_{i=1}^{m} \sigma_i^2}, \tag{24}$$

where $\mathbf{D} \in \mathbb{R}^{m \times n}$, $\mathbf{U} \in \mathbb{R}^{m \times m}$, $\mathbf{V} \in \mathbb{R}^{n \times n}$, and $\mathbf{\Sigma} \in \mathbb{R}^{m \times n}$ with $m < n$. Diagonal values of $\mathbf{\Sigma}$ represent the singular values, and σ_i represents the ith singular value. The minimum sensor number for accurate signal reconstruction is the smallest integer value of p for which NPI$\rightarrow$1.

We use the concept of compressive sensing in the spatial domain for every time instant to obtain the full field response from the response time histories of sparse sensors. However, the knowledge of basis or Dictionary matrix $\mathbf{D}$ in Equation (20) is still required. If no model knowledge is available, then Dictionary learning [58] can be used to obtain $\mathbf{D}$ from the training signals [41]. On the contrary, if the intrinsic physics of the system is known, then physics-informed dictionaries could be easily obtained [42]. Both of these methods are briefly discussed.

4.2. Learning the Basis Functions Using Dictionary Learning

Dictionary learning [58] designs matrix $\mathbf{D} \in \mathbb{R}^{m \times n}$ to attain a good sparse representation $\mathbf{y} \approx \mathbf{D}\mathbf{x}$ for a set of signals $\mathbf{y} \in \mathbb{R}^m$ based on training samples. The sparse vectors, $\mathbf{x} \in \mathbb{R}^n$, consist of few nonzero coefficients. To construct the dictionary, $\mathbf{D}$, matrix $\mathbf{Y} \in \mathbb{R}^{m \times N}$ can be formulated, where columns correspond to training signals and N represents the number of training signals. Assembling this matrix $\mathbf{Y}$ involves arranging individual $\mathbf{y}$ signals in a stack. Consequently, the optimization problem inherent to Dictionary learning can be expressed as follows:

$$\begin{aligned} &\min_{\mathbf{D},\mathbf{X}} ||\mathbf{Y} - \mathbf{D}\mathbf{X}||_F^2 \\ &\text{such that, } ||x_\ell||_0 \le s,\ \ell = 1:N, \\ &||\mathbf{d}_j|| = 1,\ j = 1:n. \end{aligned} \tag{25}$$

Here, $\mathbf{X}$ corresponds to the matrix of sparse representations, while $||\cdot||_F$ denotes the Frobenius norm. Upon resolving the optimization problem defined in Equation (25), each column within the matrix $\mathbf{D}$ serves as a basis function for the signal set $\mathbf{Y}$.

Directly obtaining $\mathbf{D}$ and $\mathbf{X}$ from $\mathbf{Y}$ is difficult as $\mathbf{Y} = \mathbf{D}\mathbf{X}$; hence, it is subdivided into two smaller optimization problems: (a) Sparse Coding and (b) Dictionary Updating. Basically, in Dictionary learning, the objective is to obtain $\mathbf{D}$ and $\mathbf{X}$ from the training signals $\mathbf{Y}$. The typical approach for solving such challenges is alternating minimization, which involves the following steps: (1) During the sparse coding phase, $\mathbf{D}$ remains fixed while $\mathbf{X}$ is estimated, and (2) in the Dictionary updating stage, $\mathbf{X}$ is held constant while $\mathbf{D}$ is estimated. This iterative process continues until a convergence point is reached.

4.3. Obtaining the Basis Functions from Physics-Based Knowledge

One method alternative to Dictionary learning for estimating basis matrix $\mathbf{D}$ is obtained from the closed-form solution of the inherent differential equation of the continuous system. One example of a simple beam is presented in this section. The equation of motion governing an Euler–Bernoulli beam subject to a distributed transverse force can be denoted using the formulation given by Rao [59]:

$$\frac{\partial^2}{\partial x^2}\left[EI(x)\frac{\partial^2 w(x,t)}{\partial x^2}\right] + \rho A(x)\frac{\partial^2 w(x,t)}{\partial t^2} = f(x,t), \tag{26}$$

where $w(x,t)$ signifies the transverse displacement response of the beam, while $f(x,t)$ represents the applied forcing function. Here, E denotes Young's modulus, ρ represents density, and $I(x)$ and $A(x)$ stand for the moment of inertia and cross-sectional area at position x from one end of the beam, respectively. In the case of uniform beams, it is reasonable to assume that the transverse displacement response can be expressed as a linear combination of the beam's normal modes utilizing a separation of variables approach. For a simply supported beam of length L, the deflection equation is expressed as

$$w(x,t) = \sum_{i=1}^{\infty} W_i(x)\eta_i(t) = \sum_{i=1}^{\infty} C_i \sin(\beta_i x)\eta_i(t) = \sum_{i=1}^{\infty} C_i \sin\frac{i\pi x}{L}\eta_i(t) = \sum_{i=1}^{\infty} \tilde{C}_i \sin\frac{i\pi x}{L}. \quad (27)$$

Here, the ith mode is characterized by the mode shape $W_i(x)$ in generalized coordinates. The response time history of the ith mode is denoted as $\eta_i(t)$. The spatial parameter β is connected to the natural frequency ω through relation $\omega = \beta^2\sqrt{\frac{EI}{\rho A}}$. The constant C_i represents the amplitude associated with the ith mode, dependent on the applied forcing function. Consequently, the basis matrix $\mathbf{D}$ can be formulated as

$$\mathbf{D} = \begin{bmatrix} \sin(\beta_1 x_1) & \dots & \sin(\beta_n x_1) \\ \sin(\beta_1 x_2) & \dots & \sin(\beta_n x_2) \\ \vdots & \vdots & \vdots \\ \sin(\beta_1 x_m) & \dots & \sin(\beta_n x_m) \end{bmatrix}. \quad (28)$$

When considering p random measurements across the beam's length at a specific time instance, the representation can be expressed [60] as follows: $z_j = \sum_{q=1}^{n} C_q^* \sin(\beta_q x_j); j = 1, 2, \cdots, p$. This can be compactly represented in matrix form as $\mathbf{z} = \boldsymbol{\Theta}\mathbf{D}\mathbf{x} = \boldsymbol{\Phi}\mathbf{x}$. In this context, $\mathbf{x} = [C_1^*, C_2^*, \cdots, C_n^*]^T$, and it is expected that the sparse solution should exhibit non-zero values for C_q^* if $C_q^* \approx \tilde{C}_i$. It is crucial to note that the sparse solution $\mathbf{x}$ differs from spatial locations x_i.

In summary, Figure 4 illustrates the compressive sensing framework used to estimate the full-field vibration response, employing a network of multi-agent sensors.

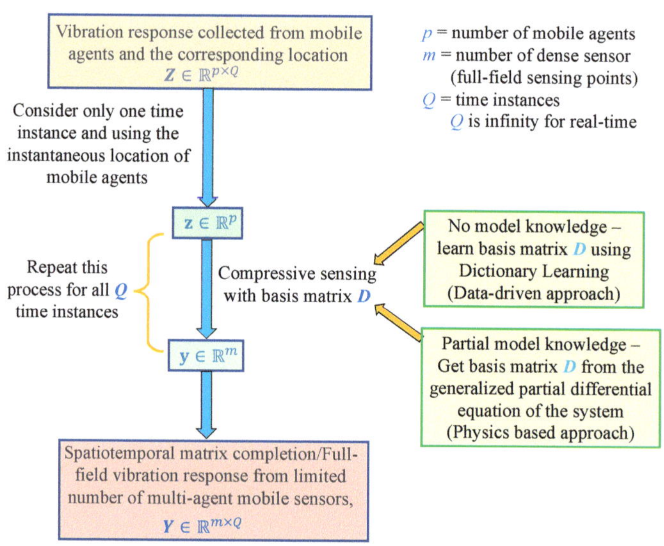

Figure 4. Proposed compressive sensing framework for spatiotemporal response matrix completion.

5. Numerical Analysis and Result

In this section, we showcase the practicality of the proposed formation control framework, as introduced in Section 3, for accomplishing essential formations within multi-agent systems. Initially, we demonstrate the feasibility of achieving user-defined formations from the initial condition of automated mobile sensors. The term "intended formation" refers to the prescribed motion of a collection of mobile sensors, while "initial condition" pertains to the initial position and velocity of this group of mobile sensors. Subsequently, we illustrate the process of estimating the full-field response by employing recorded responses from the array of mobile sensors, using the method outlined in Section 4. In this context, all mobile sensors are treated as point sensors, and their mass is negligibly small in comparison to the total mass of the bridge, which is a realistic assumption. For this numerical analysis, a simply supported bridge is considered as the structure of interest.

5.1. System Description

Bridge decks could be modeled as simply supported beams. In this study, a simply supported steel beam [41] of 50 m in length, 1 m in height, and 0.5 m in width is considered. The beam's Young's modulus is $E = 2.1 \times 10^{11}$ Pa, while its density is $\rho = 7860\ \mathrm{kg/m^3}$. Consequently, the first four natural frequencies of the beam are computed as 0.94, 3.75, 8.44, and 15 Hz. For this scenario, a 1% Rayleigh damping is taken into consideration. The total count of virtual and dense sensing points is set at 4999, resulting in a spatial separation of 0.01 m between the virtual sensing points. The primary aim of this section involves the estimation of vibration time histories for the dense virtual points (4999 in total) using the vibration data collected from a limited number of mobile sensors. In this example, we acquire the system's basis functions through Dictionary learning [41]. In practical scenarios, the training data required for Dictionary Learning can be acquired through various means, including the utilization of cameras [61,62] or alternative full-field sensing techniques such as Digital Image Correlation [63,64]. Utilizing this data-driven basis matrix, compressive sensing is applied across the entire time series to acquire the time histories of all virtual sensing points. Subsequently, the obtained dense time histories need to be compared with against the results of finite element formulation to assess the accuracy of reconstruction. To facilitate this evaluation, a relative error metric ϵ_i [41] is considered as follows:

$$\epsilon_i = \frac{||R_{\text{Exact},i} - R_{\text{Estimated},i}||_2^2}{\frac{1}{m}\sum_{i=1}^{m}||R_{\text{Exact},i}||_2^2} \times 100; \quad i = \text{sensor index}. \tag{29}$$

Here, m is the number of virtual sensing points, and $||\cdot||_2$ indicates the two-norm. $R_{\text{Estimated},i}$ and $R_{\text{Exact},i}$ symbolize the estimated and exact (FEM) responses of the ith virtual sensing point, respectively. Both $R_{\text{Exact},i}$ and $R_{\text{Estimated},i}$ have dimensions of $n_t \times 1$, where n_t corresponds to the number of time samples. The overall average error $\mathcal{E}$ [41] is expressed as follows:

$$\mathcal{E} = \frac{1}{n_s}\sum_{i=1}^{n_s}\epsilon_i, \tag{30}$$

where $\mathcal{E}$ signifies the average error of all the relative errors ϵ_i of independent virtual responses, and hence $\mathcal{E}$ is invariant to the number of virtual sensors.

5.2. Different Types of Formation Control and the Corresponding Reconstruction Result

In this section, we examine two specific configurations termed Formation-1 and Formation-2, with the objective of assessing the capability of the proposed formation control to emulate user-defined formations. In the context of Formation-1, the fleet of mobile sensors traverses the entire bridge back and forth, capturing vibration responses. In contrast, Formation-2 involves the mobile sensors moving back and forth within localized sections of the bridge. Subsequent sections comprehensively delve into the details of these formations. Each individual mobile sensor captures the acceleration response of the bridge's vibration data, a practical choice owing to the convenience of installing accelerometers on vehicles.

5.2.1. Formation-1

This formation involves a total of ten mobile sensors. Half of these sensors (R_1 to R_5) commence from the left segment of the bridge, while the remaining half (R_6 to R_{10}) initiate their movements from the bridge's right segment, as depicted in Figure 5a. Subsequently, R_1 to R_5 advance towards the right extremity of the bridge until the leading vehicle detects proximity to the bridge's end. Likewise, R_6 to R_{10} move towards the bridge's left end. Through synchronized back-and-forth motion, all these mobile sensors record the vibration response data from the bridge. The number of laps conducted entirely depend upon the user's data collection duration. Since the mobile sensors move in opposing directions, creating crossover instances, practical feasibility warrants of two lanes, as depicted in Figure 5.

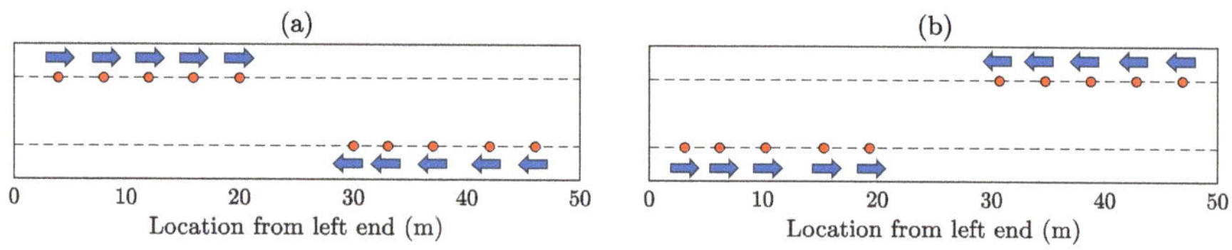

Figure 5. Two instances illustrating Formation 1—(**a**) R_1 to R_5 commencing from the left section of the bridge, while the remaining half (R_6 to R_{10}) initiate movement from the right segment. The "blue" arrows represent the immediate direction of motion for the mobile sensors. (**b**) Once the mobile sensors detect proximity to the bridge's end supports, they alter their movement direction. This sequence persists during the entire sensing duration.

As the direction of movement of agents R_1 to R_5 and R_6 to R_{10} are opposite, crossing occurs between these two sets of agents in the middle region of the bridge. During these occurrences, two agents simultaneously record identical bridge vibration response readings. Consequently, in such scenarios, the elements of the spatiotemporal response matrix are determined as the average measurements derived from these two mobile sensors. The graph connection topology between all the mobile agents is visually depicted in Figure 6. Here, graph $\mathcal{G}$ is considered as an undirected and unweighted graph, where $(j,i) \in \mathcal{E}$ only if $w_{ij} = w_{ji} = 1 \;\; \forall;(i,j) \in \mathcal{E}$ for the sake of simplicity. This study assumes that graph $\mathcal{G}$ has no switching topologies (multi-agent connection topologies remain unchanged over time). The exploration of more intricate agent connection topologies is reserved for future endeavors.

Figure 6. Graph connection topology of Formation-1. Mobile agents are connected only with their neighboring agents. The red line shows the connection between the agents.

Analyzing the graph connection topology in Figure 6, it is evident that the entire graph is disconnected. R_1 to R_5 are interlinked, while R6 to R10 are connected independently. In these scenarios, mobile sensors are connected in a manner where each vehicle can only sense the relative position and velocity of its nearest neighboring mobile sensors. To maintain a global sense of position, at least one mobile sensor, the leader, must measure its position relative to the global coordinates (Equation (16)). For the connected graph of R_1 to R_5, R_1 is designated as the leader; similarly, R_6 serves as the leader for R_6 to R_{10}. Importantly, any vehicle within R_1 to R_5 or R_6 to R_{10} could be assigned as the leader. The reference velocity amplitude v^* is set to 1 m/s for all vehicles, though this choice is user-dependent. However, this can result in the mobile agents leaving the bridge after traversing from one end to the other. The multi-agent system must move back and forth to capture longer vibration data, as depicted in Figure 5. This movement can be achieved by applying the same control strategy discussed in Section 3 at different time windows. The formation error for Formation-1 is calculated as 0.98 m; it arises due to the minimal interconnectivity among the agents. To achieve a more precise formation, it is possible to increase communication among the agents, but this would come at the cost of higher computational demands, particularly concerning wireless communication between the agents. It is important to note, however, that this formation error does not affect the estimation of the full-field response matrix since compressive sensing is employed for this purpose. Compressive sensing is well-suited for reconstructing the entire signal from randomly selected samples. In this context, compressive sensing effectively generates the complete spatial profile of the full field from the randomly positioned multi-agent vehicles at a given time instance. With a time sampling frequency of 100 Hz, this demonstration involves the automated mobile sensors collecting data for 100 s. Consequently, the spatiotemporal matrix assumes dimensions of 10,000 $\times$ 4999 (with 4999 spatial points as discussed in Section 5.1). Data from the automated mobile sensors populate some of the spatiotemporal response matrix elements, while the rest are filled using the compressive sensing algorithm outlined in Section 4. The spatiotemporal response matrix elements filled by the automatic multi-agent mobile sensors are depicted for four-time instances (at t = 5, 10, 15, 20 s) in Figure 7.

Figure 8a showcases all the sparsely populated entries in the spatiotemporal response matrix for the entire 100 s using Formation-1. Given this limited amount of data, matrix completion or full-field sensor data reconstruction is performed for each time instance through the compressive sensing technique outlined in Section 4. The spatial profile of relative errors ϵ_i for the estimated full-field response, as defined in Equation (30), is computed and presented in Figure 8b. The computed average error $\mathcal{E}$ amounts to 1.18%. Notably, relatively large ϵ_i values are observed at locations around 7.02 m and 44.63 m from the left end, corresponding to relative errors of 6.94% and 1.55%, respectively, as depicted in Figure 8b. Worth mentioning is the higher error values near the bridge's ends (0–10 m and 40–50 m from the left end), likely due to fewer sensors being present as vehicles cross each other, particularly during instances like t = 12 s, 37 s, and so on. Given that the vehicles are concentrated in the middle portion nearly half the time, this phenomenon contributes to prominent errors at the ends and negligible errors in the middle. Investigating optimal vehicle movement strategies to minimize reconstruction errors across the entire beam could be a scope of future research.

Reconstructed responses for Location 1 are displayed in Figure 9 for two distinct time segments: 50–55 s and 60–65 s. Figure 8a shows that during the period from 50 to 55 s, the instantaneous location of the automated mobile sensors covers the entirety of the bridge, resulting in the reconstructed time history to be exact to the actual true response. Conversely, in the time span of 60–65 s (Figure 8b), as the automated mobile sensors are intercepting near the middle of the bridge, they become concentrated within that specific region. This concentration leads to discernible disparities between the actual and reconstructed time histories.

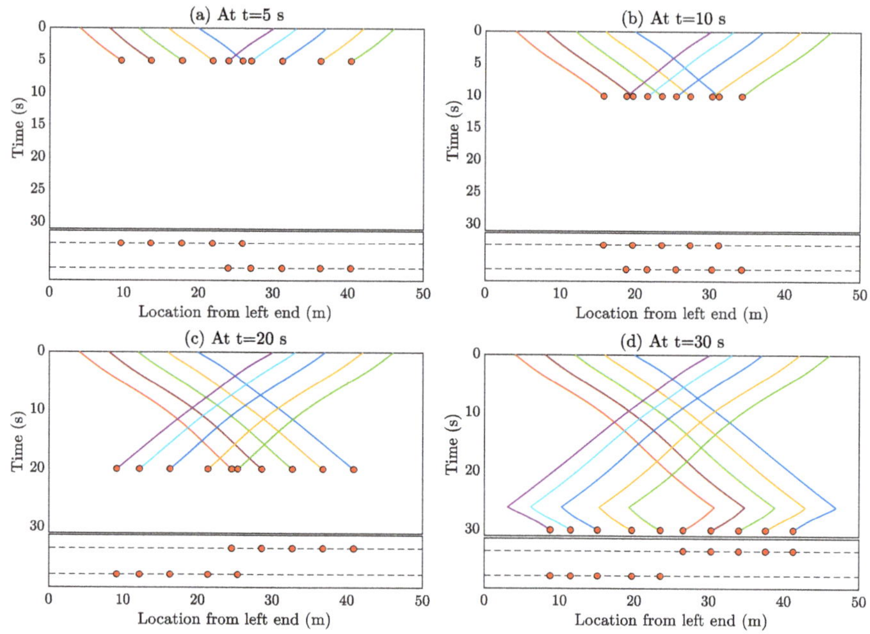

Figure 7. Sparsely filled entries of the spatiotemporal matrix at various time instants due to the data collected by automatic multi-agent mobile sensors in Formation-1. (**a**) at t = 5 s, (**b**) at t = 10 s, (**c**) at t = 15 s, and (**d**) at t = 20 s. The top figure of each subfigure shows the filled entries, and the corresponding bottom figure shows the instantaneous position of the mobile sensor formation.

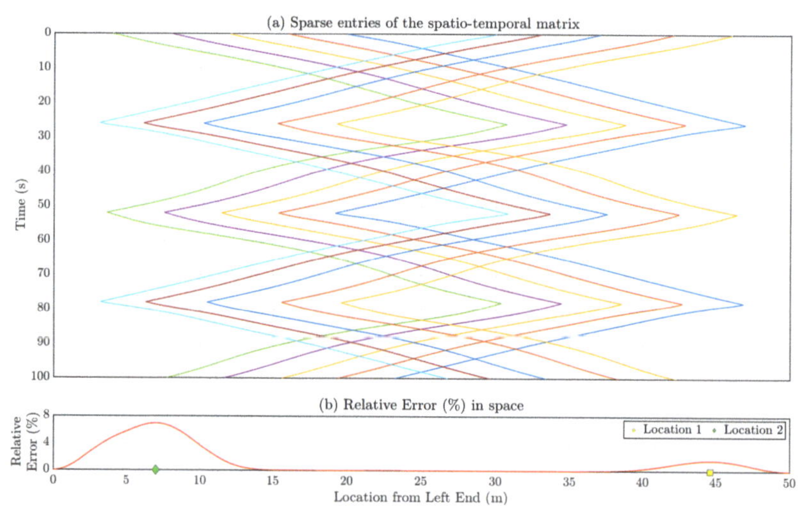

Figure 8. (**a**) Sparse entries in the spatiotemporal response matrix for the entire duration of 100 s using Formation-1 which are used for estimating full-field response. (**b**) Relative reconstruction error (%) is associated with each location along the length of the simply supported bridge. Location 1 corresponds to the highest error observed across the bridge's entire span.

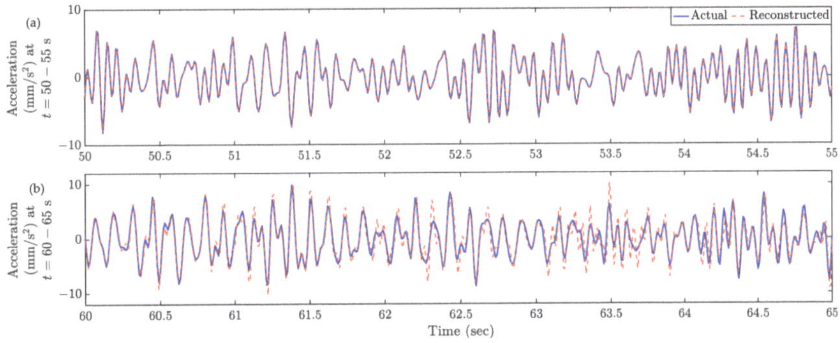

Figure 9. Comparison of reconstructed and actual time history responses at Location 1 (Figure 8) for two time snippets. (**a**) t = 50–55 s, (**b**) t = 60–65 s.

5.2.2. Formation-2

The reconstruction error of Formation-1 is higher near the bridge ends than in the middle section, which is attributed to the multi-agent system's crossing near the bridge midpoint. As the final objective is to achieve highly accurate full-field response estimation, Formation-2 is designed to circumvent situations involving "crossing". In Formation-2, the mobile sensors execute back-and-forth movements within a confined spatial range, as depicted in Figure 10. In this case, we study with only six mobile sensors, as obtained through the formula for the optimal number of sensors required for accurate full-field response estimation as provided in Equation (24). The graph connection topology among all the automated multi-agent mobile sensors is depicted in Figure 11.

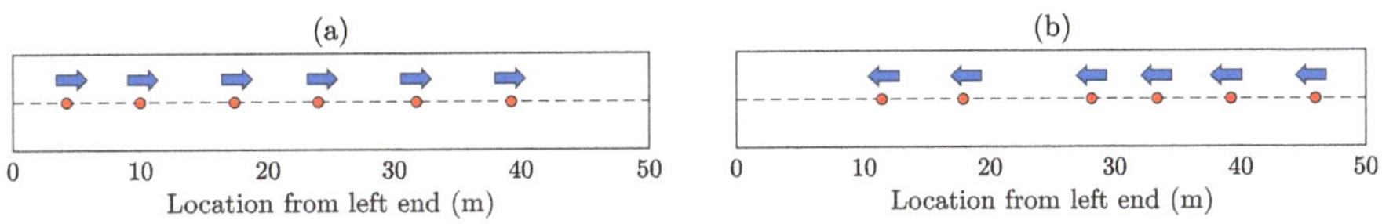

Figure 10. (**a**) Initial positions of R_1 to R_6 in Formation-2. The "blue" arrows indicate the current movement direction of the mobile sensors. (**b**) When the mobile sensors detect their proximity to the bridge end supports, their movement direction is altered. This back-and-forth process continues throughout the data collection phase.

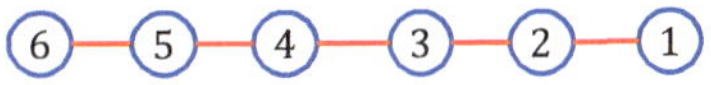

Figure 11. Graph connection topology of Formation-2. Mobile agents are connected only with their neighboring agents. The red line shows the connection between the agents.

Figure 11 shows that the mobile agents are only connected with their neighboring agents, enabling them to sense their relative velocity and relative displacement with respect to the nearest neighboring vehicles. Consequently, the presence of a leader is necessary to determine global position coordination and ensure proper formation maintenance. For Formation-2, the role of the leader is assumed by the agent R_1. The reference velocity amplitude is randomly chosen and kept constant as [1, 1.1, 1.4, 1.6, 1.2, 1.1] m/s for all vehicles throughout the process. Similar to Formation-1, the control strategy is applied in different time windows to achieve the back-and-forth movement of the automatic multi-agent mobile sensors. The formation error of Formation-2 is calculated as 0.54 m and it is important to emphasize that this formation error does not impact the estimation of the full-field response, as explained in Section 5.2.1. The sparsely filled entries contributed by the automated mobile sensors in the spatiotemporal response matrix of dimensions 10,000 $\times$ 4999 are depicted in Figure 12 for four distinct time instances.

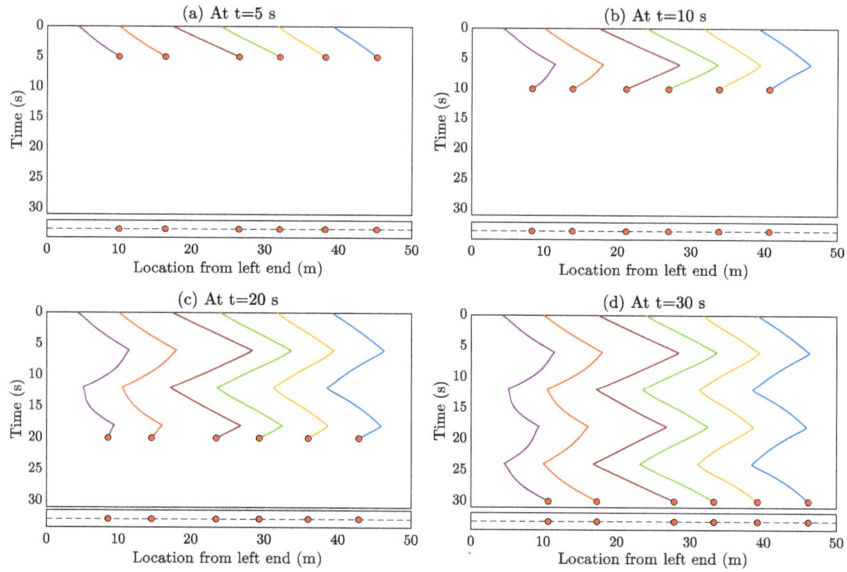

Figure 12. Sparsely filled entries of the spatiotemporal response matrix at different time instants due to Formation-2 motion of multi-agent mobile sensors at (**a**) at t = 5 s, (**b**) at t = 10 s, (**c**) at t = 15 s, and (**d**) at t = 20 s. The top figure of each subfigure shows the filled entries, and the corresponding bottom figure shows the instantaneous position of the mobile sensor formation.

Figure 13a shows the sparsely populated entries of the spatiotemporal response matrix for the total 100 s, when the multi-agent system follows Formation-2. Utilizing the compressive sensing technique, the full-field sensor time history can be derived for each time instance using these sparse entries. The spatial distribution of relative reconstruction error for Formation-2 is depicted in Figure 13b. The average relative error is computed as 0.36%. Notably, Location 1 (18.9 m from the left end) and Location 2 (45.6 m from the left end) exhibit relatively higher relative error values of 0.62% and 1.63%, respectively. The corresponding estimated time histories for these locations are depicted in Figure 14, revealing that the reconstructed response time histories are comparable to the actual response time histories.

We attempted to compare the compressive-sensing-based spatiotemporal matrix completion approach with other state-of-the-art algorithms currently available. Nevertheless, employing matrix completion techniques based on Singular Value Decomposition (SVD) methods [65] and the OptSpace method [66] proved unfeasible due to their failure to converge within an acceptable tolerance limit, even with a large number of iterations. This outcome can be attributed to the dimensions of sparse matrices, which are 10,000 $\times$ 4999, containing only 10,000 $\times$ 6 populated values, resulting in a mere 0.12% of filled entries. Consequently, estimating the remaining unknown values without additional information proved exceedingly difficult. In contrast, our proposed approach applies compressive sensing to each row of the spatiotemporal matrix independently, leveraging knowledge of the underlying basis function obtained either through dictionary learning or a physics-based approach. This approach rendered full-field response estimation feasible, distinguishing it from the other methods.

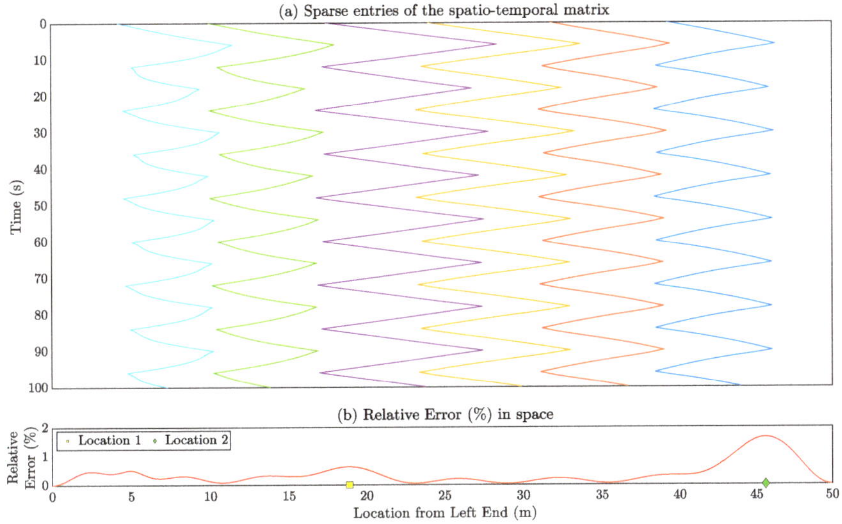

Figure 13. (**a**) Sparsely filled sparse entries of the spatiotemporal matrix for the total time 100 s with Formation 2. (**b**) Relative reconstruction error (%) for each location of the simply supported bridge.

Please note that different types of formations yield varying levels of reconstruction accuracy. As demonstrated in the aforementioned examples, Formation-2 exhibits superior accuracy in estimating full-field vibration responses compared to Formation-1. The primary objective of this paper is to enable user-driven control of the multi-agent system. It is evident from the agent formation control that the proposed algorithm effectively adheres to user-defined formations. Generally, the compressive sensing-based framework yields improved reconstruction accuracy when mobile sensors can continuously span the entire beam. This is in contrast to Formation-1, which yields inferior results due to instances where all mobile sensors cluster in the central portion of the structure. Identifying the optimal formation remains a potential area for future research, which can be motivated from optimal input [67–69] and sensor location [70] for structural system identification literature.

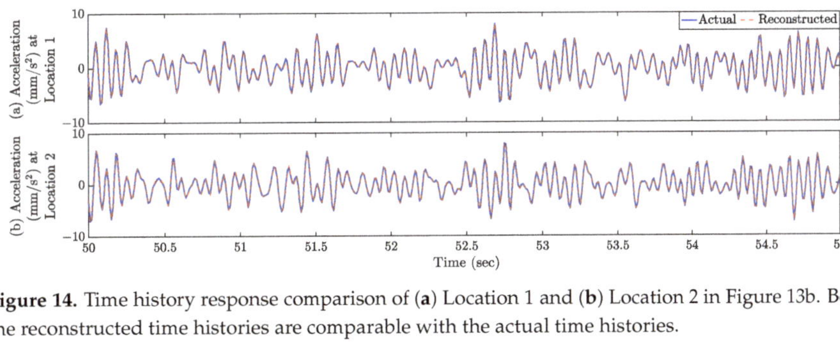

Figure 14. Time history response comparison of (**a**) Location 1 and (**b**) Location 2 in Figure 13b. Both the reconstructed time histories are comparable with the actual time histories.

5.3. Achieving Formation-1 from Any Initial Condition

While Figures 5a and 10a illustrate the initial starting positions of Formation-1 and Formation-2, respectively, a significant advantage of formation control is its capability to achieve any formation from varying initial positions and velocities of the automated multi-agent mobile sensors using the controller outlined in Equation (19). Notably, the multi-agent system requires a certain amount of time to converge to the desired formation from any given position or velocity. Throughout this study, g_p and g_v in Equation (19) are consistently set to one, controlling the convergence rate towards the formation. A practical example is depicted in Figure 15, wherein the initial positions of the mobile sensors are set at 25 m from the left end (bridge midpoint) with zero initial velocity. Utilizing this starting condition, Formation-1 is achieved. Figure 15d illustrates the formation's attainment in approximately 26 s. Consequently, for better accuracy while using the full-field response estimation framework, data collected beyond 26 s are appropriate.

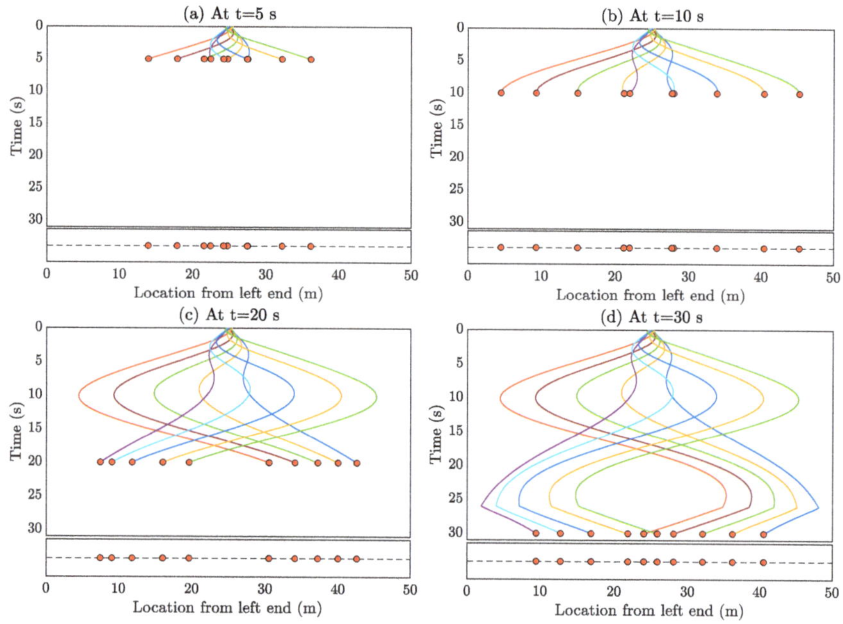

Figure 15. The automated multi-agent mobile sensors achieve Formation-1 where the initial position of all the mobile sensors is in the middle of the bridge. This figure shows the instances of how it achieves the target formation (**a**) t = 5 s, (**b**) t = 10 s, (**c**) t = 15 s, (**d**) t = 20 s.

The requirements for consensus for the undirected graphs are as follows [50]:

(a) Every agent must be connected with at least one other agent; otherwise, achieving consensus becomes unattainable.

(b) The time it takes for all agents to reach a consensus from an arbitrary starting point, known as the convergence time is dependent on g_p, g_v, and the second eigenvalue of the Laplacian matrix ($\mathcal{L}$ in Section 3). Moreover, this convergence time is inversely proportional to the magnitude of the second eigenvalue, which is influenced by the graph connection weights. In essence, increasing the strength of graph connections or weights results in quicker convergence for achieving consensus.

(c) The convergence time of consensus is also influenced by graph connectivity. In this study, the multi-agent system is considered to be connected with only neighboring agents. For instance, if we consider the second eigenvalue of the Laplacian matrix as λ_2, considering a total of n agents, there can be $2^{\binom{n}{2}}$ potential graphs, considering the isomorphic graphs as different graphs. Amidst these diverse graph sets, there are instances where the second eigenvalue of the Laplacian matrix $\hat{\lambda}_2$ exceeds λ_2. Such graphs with a higher second eigenvalue converges faster toward consensus than the neighboring connection graph, as demonstrated in this paper. However, more connections among agents would be attributed to the cost. Therefore, in the pursuit of simplicity and cost effectiveness, we opted to investigate the most straightforward scenarios, such as multi-agent connection with only neighboring agents.

The presented formation control strategy holds potential for modern structural health monitoring through the crowd-sensing of bridge vibration data. It offers the ability to automate the process of collecting vibration data by coordinating the movement of vehicles. The data gathered from smartphones installed in these mobile vehicles can be harnessed to characterize the modal properties of bridge structures under real-world circumstances, which is essential for condition assessment and damage detection frameworks. We discuss the advantages of full-field sensing from a limited number of fixed sensors in detail in our previous studies [41,42]; this paper performs a similar task, i.e., full-field sensing, but with a limited number of automatic mobile sensors.

6. Recommendation for Practical Implementation

Practical implementation recommendations are crucial for the successful execution of full-field vibration response estimation tasks based on multi-agent formation control. Several factors must be taken into account when designing experiments for this purpose.

Sensor arrangement: To successfully achieve the formation control, every agent should be equipped with the IMU (Inertial Measurement Unit) and Wireless Communication Modules. IMUs combine accelerometers and gyroscopes to measure an agent's linear acceleration and angular velocity. They are essential for estimating an agent's orientation and motion dynamics. In this paper, the movement of agent is one-directional; hence, only linear accelerometer is sufficient. For the 2D and 3D formation control problem, IMUs would be necessary. Wireless communication modules (e.g., Wi-Fi, Bluetooth) facilitate communication and coordination among agents as the formation control often requires agents to exchange information with one another. As the proposed formation control requires a "leader" agent, a GPS (Global Positioning System) device should be mounted on the "leader" agent. GPS sensors provide accurate global position information, including latitude, longitude, and altitude, which are used to obtain absolute position estimates of agents.

Data frequency: The data frequency or sampling rate for formation control depends primarily on agent dynamics (faster moving agents needs higher sampling frequency to maintain formation accuracy) and formation precision (precise formation control requires higher data frequency). In this paper, the agents are moving at an approximately 1 m/s velocity; hence, we considered 100 Hz as the data sampling frequency to maintain the preciseness at a cm level.

Noise reduction techniques: Noise reduction techniques play a crucial role in improving the performance and reliability of formation control algorithms, especially in scenarios where sensor data is subject to various sources of noise and uncertainty. The Kalman filter [71] is a widely used technique for estimating the state of a dynamic system while accounting for measurement noise. In formation control, it can be employed to filter noisy sensor measurements, such as GPS positions or IMU data, to obtain more accurate estimates of agent positions and velocities. Additionally, sensor fusion (combining data from multiple sensors) and Predictive Filters (such as the Alpha–Beta filter which can provide a more stable estimate of the current state) can be employed to achieve the desired level of noise reduction and robustness in real-world formation control systems.

Sensor synchronization: Sensor synchronization in formation control is the process of aligning the data from sensors on different agents or vehicles in a formation such that they share a common time reference and are temporally aligned. This synchronization is crucial for achieving accurate and coordinated control of the agents within the formation. However, sensor inaccuracies can lead to errors in position estimation, which can degrade the formation quality. Additionally, delays in communication can disrupt the synchronization of sensors.

Calibration: In many formation control scenarios, agents may have different types of sensors, each with its own calibration requirements and limitations. Coordinating the calibration of heterogeneous sensors can be challenging, as the calibration process for one sensor may not be directly applicable to others.

Scalability: Scalability issues can arise when the formation control system encounters challenges in maintaining performance, coordination, and communication efficiency as the system scales up as the number of agents in the formation grows, communication load and overhead can increase significantly. Also, as the density of agents in a formation increases, the likelihood of collisions or near misses can also rise. To address scalability issues in formation control, decentralization and sparse communication can be adopted, which is scope of future exploration.

Computational demands for real-time data acquisition and processing: As the multi-agent system are connected with each other wirelessly, the computational demand of collecting and transmitting data wirelessly in real time to the server depends on the sampling frequency and data latency. Sometimes, the wireless transmission experience data packet losses which need to recovered, as well in the data server [3]. Efficient wireless protocols and technologies as well as a real-time operating system could be helpful in this regard. Once the data are stored in the server and sparsely populate the spatiotemporal data matrix, the real-time full-field response estimation is very fast. In this paper, each of the rows of the spatiotemporal response matrix requires approximately 0.43 s on average from the sparse data.

7. Discussion

Mobile sensing serves as an alternative to fixed sensing for the acquisition of vibration response data in the field of structural health monitoring. Currently, mobile sensors are operated by assigned drivers, a potentially impractical approach if the array of mobile sensors needs to follow specific patterns to optimize the data collection procedure. To address this issue, we introduce an automated multi-agent mobile sensing framework in this paper. Our proposed method diverges from fully autonomous vehicles which necessitate numerous sensors to maintain the vehicle's position and speed, a potentially economically inefficient arrangement for structural health monitoring objectives. Therefore, the proposed formation control strategy relies solely on vehicles that autonomously manage themselves by gauging the relative velocity and relative position of their nearest neighboring agent/vehicles. In this technique, very few vehicles (often just one) with information about their global position, referred to as the "leader vehicle," are required. This formation control strategy could be useful across various mobile sensing-oriented structural health monitoring technologies. In this work, we utitize the suggested framework for estimating full-field responses using a limited number of mobile sensors. We consider two distinct formations: Formation-1 involves two groups of vehicles crossing each other and traversing back and forth over a bridge during the data collection phase. However, Formation-1 exhibits notable response estimation errors near the bridge's ends due to sensor gaps as the vehicle groups intersect. In contrast, Formation-2 features vehicles moving back and forth locally, resulting in a reduced number of estimation errors. Furthermore, we showcase the capability of mobile sensors to achieve any formation from any initial position and velocity using the proposed formation control framework. This strategy holds the potential to facilitate real-time assessment of changing system parameters or automated localization of damage.

8. Conclusions and Future Work

In this paper, we explored a minimalistic and cost-efficient scenario where the multi-agent system exclusively relies on neighboring connections. Instead, if the number of connections between the multi-agent system increases, then the consensus, as well as the whole framework, i.e., formation control combined with controlling the position of the entire formation over time, is more robust. However, achieving such a dense connection among the multi-agent systems would demand a better and larger number of sensors, potentially leading to higher costs.

This paper primarily investigates mobile sensors with time-invariant graph interaction topology. Exploring directed, weighted, and switching graph topologies could offer insights into the performance of formation control in the domain of vibration sensing and health monitoring, which remains a subject for future investigation. Additionally, in the paper, it is assumed that the vehicles are configured as point mobile sensors with unidirectional movement. In the context of 2D and 3D structures, the adaptation of mobile sensors into multi-dimensional vehicles and the potential consideration of mobile sensors as rigid bodies could be a scope of future study. Furthermore, for 2D and 3D systems, optimal paths for mobile sensors could be identified to maximize sensing information—an aspect not necessary for the current 1D movement scenario. Furthermore, to evaluate the effectiveness of the proposed method, real-life or laboratory experiments can be executed, which could also be explored in future work.

Author Contributions: Conceptualization, D.J. and S.N.; methodology, D.J. and S.N.; software, D.J.; validation, D.J.; formal analysis, D.J.; investigation, D.J.; resources, S.N.; data curation, D.J.; writing—original draft preparation, D.J.; writing—review and editing, S.N.; visualization, D.J.; supervision, S.N.; project administration, S.N.; funding acquisition, S.N. All authors have read and agreed to the published version of the manuscript.

Funding: This research was funded by the Science and Engineering Research Board of India (SERB, grant number: R96701-246000) and Rice University to Debasish Jana and Satish Nagarajaiah. The authors gratefully acknowledge the financial support by SERB-India.

Institutional Review Board Statement: Not applicable.

Informed Consent Statement: Not applicable.

Data Availability Statement: Data will be made available on request.

Conflicts of Interest: The authors declare no conflict of interest. The funders had no role in the design of the study; in the collection, analyses, or interpretation of data; in the writing of the manuscript; or in the decision to publish the results.

References

1. Guan, S.; Bridge, J.A.; Li, C.; DeMello, N.J. Smart radar sensor network for bridge displacement monitoring. *J. Bridge Eng.* **2018**, *23*, 04018102. [CrossRef]
2. Jana, D.; Patil, J.; Herkal, S.; Nagarajaiah, S.; Duenas-Osorio, L. CNN and Convolutional Autoencoder (CAE) based real-time sensor fault detection, localization, and correction. *Mech. Syst. Signal Process.* **2022**, *169*, 108723. [CrossRef]
3. Jana, D.; Nagarajaiah, S.; Yang, Y.; Li, S. Real-time cable tension estimation from acceleration measurements using wireless sensors with packet data losses: Analytics with compressive sensing and sparse component analysis. *J. Civ. Struct. Health Monit.* **2022**, *12*, 797–815. [CrossRef]
4. Malekjafarian, A.; McGetrick, P.J.; OBrien, E.J. A review of indirect bridge monitoring using passing vehicles. *Shock Vib.* **2015**, *2015*, 286139. [CrossRef]
5. Zhu, D.; Yi, X.; Wang, Y.; Lee, K.M.; Guo, J. A mobile sensing system for structural health monitoring: Design and validation. *Smart Mater. Struct.* **2010**, *19*, 055011. [CrossRef]
6. Lynch, J.P.; Loh, K.J. A summary review of wireless sensors and sensor networks for structural health monitoring. *Shock Vib. Dig.* **2006**, *38*, 91–130. [CrossRef]
7. Matarazzo, T.; Vazifeh, M.; Pakzad, S.; Santi, P.; Ratti, C. Smartphone data streams for bridge health monitoring. *Procedia Eng.* **2017**, *199*, 966–971. [CrossRef]
8. Ozer, E.; Feng, M.Q.; Feng, D. Citizen sensors for SHM: Towards a crowdsourcing platform. *Sensors* **2015**, *15*, 14591–14614. [CrossRef]
9. Mei, Q.; Gül, M. A crowdsourcing-based methodology using smartphones for bridge health monitoring. *Struct. Health Monit.* **2019**, *18*, 1602–1619. [CrossRef]
10. Matarazzo, T.J.; Kondor, D.; Milardo, S.; Eshkevari, S.S.; Santi, P.; Pakzad, S.N.; Buehler, M.J.; Ratti, C. Crowdsourcing bridge dynamic monitoring with smartphone vehicle trips. *Commun. Eng.* **2022**, *1*, 29. [CrossRef]
11. Daubechies, I.; Lu, J.; Wu, H.T. Synchrosqueezed wavelet transforms: An empirical mode decomposition-like tool. *Appl. Comput. Harmon. Anal.* **2011**, *30*, 243–261. [CrossRef]
12. Matarazzo, T.J.; Santi, P.; Pakzad, S.N.; Carter, K.; Ratti, C.; Moaveni, B.; Osgood, C.; Jacob, N. Crowdsensing framework for monitoring bridge vibrations using moving smartphones. *Proc. IEEE* **2018**, *106*, 577–593. [CrossRef]
13. Eshkevari, S.S.; Cronin, L.; Matarazzo, T.J.; Pakzad, S.N. Bridge modal property identification based on asynchronous mobile sensing data. *Struct. Health Monit.* **2023**, *22*, 2022–2037. [CrossRef]
14. Yang, Y.B.; Lin, C.; Yau, J. Extracting bridge frequencies from the dynamic response of a passing vehicle. *J. Sound Vib.* **2004**, *272*, 471–493. [CrossRef]

15. Nayek, R.; Mukhopadhyay, S.; Narasimhan, S. Mass normalized mode shape identification of bridge structures using a single actuator-sensor pair. *Struct. Control Health Monit.* **2018**, *25*, e2244. [CrossRef]
16. Hester, D.; González, A. A discussion on the merits and limitations of using drive-by monitoring to detect localised damage in a bridge. *Mech. Syst. Signal Process.* **2017**, *90*, 234–253. [CrossRef]
17. Zhang, B.; Qian, Y.; Wu, Y.; Yang, Y. An effective means for damage detection of bridges using the contact-point response of a moving test vehicle. *J. Sound Vib.* **2018**, *419*, 158–172. [CrossRef]
18. Zhu, X.; Law, S.S.; Huang, L.; Zhu, S. Damage identification of supporting structures with a moving sensory system. *J. Sound Vib.* **2018**, *415*, 111–127. [CrossRef]
19. Sadeghi Eshkevari, S.; Pakzad, S.N.; Takáč, M.; Matarazzo, T.J. Modal identification of bridges using mobile sensors with sparse vibration data. *J. Eng. Mech.* **2020**, *146*, 04020011. [CrossRef]
20. Kim, C.W.; Chang, K.C.; McGetrick, P.J.; Inoue, S.; Hasegawa, S. Utilizing moving vehicles as sensors for bridge condition screening-A laboratory verification. *Sens. Mater.* **2017**, *29*, 153–163.
21. Mei, Q.; Gül, M.; Boay, M. Indirect health monitoring of bridges using Mel-frequency cepstral coefficients and principal component analysis. *Mech. Syst. Signal Process.* **2019**, *119*, 523–546. [CrossRef]
22. Cantero, D.; McGetrick, P.; Kim, C.W.; OBrien, E. Experimental monitoring of bridge frequency evolution during the passage of vehicles with different suspension properties. *Eng. Struct.* **2019**, *187*, 209–219. [CrossRef]
23. Shirzad-Ghaleroudkhani, N.; Mei, Q.; Gül, M. Frequency identification of bridges using smartphones on vehicles with variable features. *J. Bridge Eng.* **2020**, *25*, 04020041. [CrossRef]
24. Yang, Y.; Zhu, Y.; Wang, L.L.; Jia, B.Y.; Jin, R. Structural damage identification of bridges from passing test vehicles. *Sensors* **2018**, *18*, 4035. [CrossRef] [PubMed]
25. Elhattab, A.; Uddin, N.; OBrien, E. Drive-by bridge frequency identification under operational roadway speeds employing frequency independent underdamped pinning stochastic resonance (FI-UPSR). *Sensors* **2018**, *18*, 4207. [CrossRef] [PubMed]
26. Oshima, Y.; Yamamoto, K.; Sugiura, K. Damage assessment of a bridge based on mode shapes estimated by responses of passing vehicles. *Smart Struct. Syst.* **2014**, *13*, 731–753. [CrossRef]
27. Malekjafarian, A.; OBrien, E.J. Identification of bridge mode shapes using short time frequency domain decomposition of the responses measured in a passing vehicle. *Eng. Struct.* **2014**, *81*, 386–397. [CrossRef]
28. OBrien, E.J.; Malekjafarian, A. A mode shape-based damage detection approach using laser measurement from a vehicle crossing a simply supported bridge. *Struct. Control Health Monit.* **2016**, *23*, 1273–1286. [CrossRef]
29. Zhang, Y.; Wang, L.; Xiang, Z. Damage detection by mode shape squares extracted from a passing vehicle. *J. Sound Vib.* **2012**, *331*, 291–307. [CrossRef]
30. OBrien, E.J.; Malekjafarian, A.; González, A. Application of empirical mode decomposition to drive-by bridge damage detection. *Eur. J. Mech.-A/Solids* **2017**, *61*, 151–163. [CrossRef]
31. Yang, Y.; Chang, K. Extraction of bridge frequencies from the dynamic response of a passing vehicle enhanced by the EMD technique. *J. Sound Vib.* **2009**, *322*, 718–739. [CrossRef]
32. Malekjafarian, A.; OBrien, E.J. On the use of a passing vehicle for the estimation of bridge mode shapes. *J. Sound Vib.* **2017**, *397*, 77–91. [CrossRef]
33. Matarazzo, T.J.; Pakzad, S.N. Structural identification for mobile sensing with missing observations. *J. Eng. Mech.* **2016**, *142*, 04016021. [CrossRef]
34. Eshkevari, S.S.; Matarazzo, T.J.; Pakzad, S.N. Bridge modal identification using acceleration measurements within moving vehicles. *Mech. Syst. Signal Process.* **2020**, *141*, 106733. [CrossRef]
35. Eshkevari, S.S.; Pakzad, S. Bridge structural identification using moving vehicle acceleration measurements. In *Dynamics of Civil Structures*; Springer: Berlin/Heidelberg, Germany, 2019; Volume 2, pp. 251–261.
36. Eshkevari, S.S.; Pakzad, S.N. Signal reconstruction from mobile sensors network using matrix completion approach. In *Topics in Modal Analysis & Testing*; Springer: Berlin/Heidelberg, Germany, 2020; Volume 8, pp. 61–75.
37. Eshkevari, S.S.; Takác, M.; Pakzad, S.N.; Eshkevari, S.S. High resolution bridge mode shape identification via matrix completion approach. In Proceedings of the 12th International Workshop on Structural Health Monitoring, Stanford, CA, USA, 10–12 September 2019.
38. Yang, Y.; Nagarajaiah, S. Harnessing data structure for recovery of randomly missing structural vibration responses time history: Sparse representation versus low-rank structure. *Mech. Syst. Signal Process.* **2016**, *74*, 165–182. [CrossRef]
39. Nagarajaiah, S.; Yang, Y. Modeling and harnessing sparse and low-rank data structure: A new paradigm for structural dynamics, identification, damage detection, and health monitoring. *Struct. Control Health Monit.* **2017**, *24*, e1851. [CrossRef]
40. Badue, C.; Guidolini, R.; Carneiro, R.V.; Azevedo, P.; Cardoso, V.B.; Forechi, A.; Jesus, L.; Berriel, R.; Paixao, T.M.; Mutz, F.; et al. Self-driving cars: A survey. *Expert Syst. Appl.* **2021**, *165*, 113816. [CrossRef]
41. Jana, D.; Nagarajaiah, S. Data-driven full-field vibration response estimation from limited measurements in real-time using dictionary learning and compressive sensing. *Eng. Struct.* **2023**, *275*, 115280. [CrossRef]
42. Jana, D.; Nagarajaiah, S. Physics-Guided Real-Time Full-Field Vibration Response Estimation from Sparse Measurements Using Compressive Sensing. *Sensors* **2023**, *23*, 384. [CrossRef]
43. Yang, Y.B.; Yau, J.; Yao, Z.; Wu, Y. *Vehicle-Bridge Interaction Dynamics: With Applications to High-Speed Railways*; World Scientific: Singapore, 2004.

44. Yang, Y.B.; Yau, J.D. Vehicle-bridge interaction element for dynamic analysis. *J. Struct. Eng.* **1997**, *123*, 1512–1518. [CrossRef]
45. Yang, Y.; Lin, C. Vehicle–bridge interaction dynamics and potential applications. *J. Sound Vib.* **2005**, *284*, 205–226. [CrossRef]
46. Ghanem, R.; Shinozuka, M. Structural-system identification. I: Theory. *J. Eng. Mech.* **1995**, *121*, 255–264. [CrossRef]
47. González, A.; OBrien, E.J.; McGetrick, P. Identification of damping in a bridge using a moving instrumented vehicle. *J. Sound Vib.* **2012**, *331*, 4115–4131. [CrossRef]
48. Marulanda, J.; Caicedo, J.M.; Thomson, P. Modal identification using mobile sensors under ambient excitation. *J. Comput. Civ. Eng.* **2017**, *31*, 04016051. [CrossRef]
49. Godsil, C.; Royle, G.F. *Algebraic Graph Theory*; Springer Science & Business Media: Berlin/Heidelberg, Germany, 2001; Volume 207.
50. Ren, W.; Beard, R.W. *Distributed Consensus in Multi-Vehicle Cooperative Control*; Springer: Berlin/Heidelberg, Germany, 2008; Volume 27.
51. Ren, W.; Atkins, E. Distributed multi-vehicle coordinated control via local information exchange. *Int. J. Robust Nonlinear Control IFAC-Affil. J.* **2007**, *17*, 1002–1033. [CrossRef]
52. Oh, K.K.; Park, M.C.; Ahn, H.S. A survey of multi-agent formation control. *Automatica* **2015**, *53*, 424–440. [CrossRef]
53. Wan, S.; Lu, J.; Fan, P.; Letaief, K.B. Information theory in formation control: An error analysis to multi-robot formation. *Entropy* **2018**, *20*, 618. [CrossRef]
54. Baraniuk, R.G. Compressive sensing [lecture notes]. *IEEE Signal Process. Mag.* **2007**, *24*, 118–121. [CrossRef]
55. Tibshirani, R. Regression shrinkage and selection via the lasso. *J. R. Stat. Soc. Ser. B* **1996**, *58*, 267–288. [CrossRef]
56. Kim, S.J.; Koh, K.; Lustig, M.; Boyd, S.; Gorinevsky, D. An interior-point method for large-scale ℓ_1 -regularized least squares. *IEEE J. Sel. Top. Signal Process.* **2007**, *1*, 606–617. [CrossRef]
57. Amini, F.; Hedayati, Y.; Zanddizari, H. Determining the number of measurements for compressive sensing of traffic-induced vibration data. *Measurement* **2020**, *152*, 107259. [CrossRef]
58. Dumitrescu, B.; Irofti, P. *Dictionary Learning Algorithms and Applications*; Springer: Berlin/Heidelberg, Germany, 2018.
59. Rao, S.S. *Vibration of Continuous Systems*; Wiley Online Library: Hoboken, NJ, USA, 2007; Volume 464.
60. Ganesan, V.; Das, T.; Rahnavard, N.; Kauffman, J.L. Vibration-based monitoring and diagnostics using compressive sensing. *J. Sound Vib.* **2017**, *394*, 612–630. [CrossRef]
61. Jana, D.; Nagarajaiah, S. Computer vision-based real-time cable tension estimation in Dubrovnik cable-stayed bridge using moving handheld video camera. *Struct. Control Health Monit.* **2021**, *28*, e2713. [CrossRef]
62. Jana, D.; Nagarajaiah, S.; Yang, Y. Computer vision-based real-time cable tension estimation algorithm using complexity pursuit from video and its application in Fred-Hartman cable-stayed bridge. *Struct. Control Health Monit.* **2022**, *29*, e2985. [CrossRef]
63. Chen, W.; Jana, D.; Singh, A.; Jin, M.; Cenedese, M.; Kosova, G.; Brake, M.R.; Schwingshackl, C.W.; Nagarajaiah, S.; Moore, K.J.; et al. Measurement and identification of the nonlinear dynamics of a jointed structure using full-field data, Part I: Measurement of nonlinear dynamics. *Mech. Syst. Signal Process.* **2022**, *166*, 108401. [CrossRef]
64. Jin, M.; Kosova, G.; Cenedese, M.; Chen, W.; Singh, A.; Jana, D.; Brake, M.R.; Schwingshackl, C.W.; Nagarajaiah, S.; Moore, K.J.; et al. Measurement and identification of the nonlinear dynamics of a jointed structure using full-field data; Part II-Nonlinear system identification. *Mech. Syst. Signal Process.* **2022**, *166*, 108402. [CrossRef]
65. Candes, E.J.; Plan, Y. Matrix completion with noise. *Proc. IEEE* **2010**, *98*, 925–936. [CrossRef]
66. Keshavan, R.H.; Montanari, A.; Oh, S. Matrix completion from a few entries. *IEEE Trans. Inf. Theory* **2010**, *56*, 2980–2998. [CrossRef]
67. Jana, D.; Mukhopadhyay, S.; Ray-Chaudhuri, S. Fisher information-based optimal input locations for modal identification. *J. Sound Vib.* **2019**, *459*, 114833. [CrossRef]
68. Jana, D.; Prajapat, K.; Mukhopadhyay, S.; Chaudhuri, S.R. Optimal input location for modal identification. *Procedia Eng.* **2017**, *199*, 990–995. [CrossRef]
69. Jana, D.; Ghosh, D.; Mukhopadhyay, S.; Ray-Chaudhuri, S. Optimal Input Locations for Stiffness Parameter Identification. In *Model Validation and Uncertainty Quantification*; Springer: Berlin/Heidelberg, Germany, 2020; Volume 3, pp. 399–403.
70. Papadimitriou, C. Optimal sensor placement methodology for parametric identification of structural systems. *J. Sound Vib.* **2004**, *278*, 923–947. [CrossRef]
71. Kalman, R.E. A New Approach to Linear Filtering and Prediction Problems. *J. Basic Eng.* **1960**, *82*, 35–45. [CrossRef]

sensors

MDPI

Article

A Comprehensive Eco-Driving Strategy for CAVs with Microscopic Traffic Simulation Testing Evaluation

Ozgenur Kavas-Torris * and Levent Guvenc

Automated Driving Lab, Department of Mechanical and Aerospace Engineering, The Ohio State University, Columbus, OH 43210, USA; guvenc.1@osu.edu
* Correspondence: kavastorris.1@buckeyemail.osu.edu

Abstract: In this paper, a comprehensive deterministic Eco-Driving strategy for Connected and Autonomous Vehicles (CAVs) is presented. In this setup, multiple driving modes calculate speed profiles that are ideal for their own set of constraints simultaneously to save fuel as much as possible, while a High-Level (HL) controller ensures smooth and safe transitions between the driving modes for Eco-Driving. This Eco-Driving deterministic controller for an ego CAV was equipped with Vehicle-to-Infrastructure (V2I) and Vehicle-to-Vehicle (V2V) algorithms. This comprehensive Eco-Driving strategy and its individual components were tested by using simulations to quantify the fuel economy performance. Simulation results are used to show that the HL controller ensures significant fuel economy improvement as compared to baseline driving modes with no collisions between the ego CAV and traffic vehicles, while the driving mode of the ego CAV was set correctly under changing constraints. For the microscopic traffic simulations, a 6.41% fuel economy improvement was observed for the CAV that was controlled by this comprehensive Eco-Driving strategy.

Keywords: eco-driving; ecological cooperative adaptive cruise control; velocity trajectory; dynamic programming; traffic simulation

Citation: Kavas-Torris, O.; Guvenc, L. A Comprehensive Eco-Driving Strategy for CAVs with Microscopic Traffic Simulation Testing Evaluation. *Sensors* **2023**, *23*, 8416. https://doi.org/10.3390/s23208416

Academic Editors: Yafei Wang, Chao Huang, Hailong Huang, Zhiqiang Zuo, Bo Leng and Peng Hang

Received: 20 September 2023
Revised: 6 October 2023
Accepted: 10 October 2023
Published: 12 October 2023

1. Introduction

Fuel economy enhancement in road vehicles is a problem that researchers around the world have been working to improve for decades. Eco-Driving is a term used to describe the energy-efficient use of road vehicles. Some researchers have focused on improving the powertrain efficiency to improve the fuel economy in vehicles [1,2], whereas others have worked on utilizing Connected and Autonomous Vehicle (CAV) technologies for the same purpose [3–5]. Longitudinal autonomy and connectivity have also been utilized to achieve fuel savings for individual and platooning vehicles [6]. Robust control and model regulation were also used for vehicle control [7–9]. A parameter space with robustness was also utilized as another method for vehicle control [10–13], (p. 20 [14]), [15,16]. The problem being addressed here is how to improve the fuel efficiency of a vehicle (Eco-Driving) by using connectivity with the infrastructure and nearby vehicles. Existing solutions are first presented in the literature review below, followed by the contributions made in this paper. The aim of this paper is to improve the fuel economy, which is shown in the simulation experiment results parts of this paper.

Developments in Vehicle-to-Infrastructure (V2I) communication technology have enhanced the capabilities of CAVs. Researchers are able to study and enhance the fuel economy in vehicles by using vehicle connectivity technology in CAVs. CAVs can use roadway infrastructure information through V2I, where they receive information about traffic lights and STOP signs in order to reduce fuel consumption for conventional vehicles and battery power for electric vehicles. In V2I, CAVs receive traffic light and STOP sign locations, as well as the Signal Phase and Timing (SPaT) from traffic lights. Using this information, longitudinal control algorithms can be developed to modify the speed of the ego CAV in order to save fuel.

There is ample work in academia on algorithms utilizing V2I technology. Altan et al. developed a V2I algorithm and tested it at one signalized intersection to quantify how much fuel it saved for one connected vehicle [17]. Cantas et al. [18] and Kavas-Torris et al. [19] studied the fuel saving performance of the Pass-at-Green (PaG) V2I application in a microscopic traffic simulator through Monte Carlo simulations, as well as Hardware-in-the-loop (HIL) tests. Kavas-Torris et al. [20] analyzed the PaG V2I algorithm through microscopic traffic simulations, where varying but realistic traffic flows were present around the ego CAV. Sun et al. used a data-driven approach, where the optimal speed profile for a CAV thorough an intersection showed 40% fuel savings [21]. Asadi and Vahidi utilized V2I and a radar for a control algorithm that reduced the fuel consumption and idling time at traffic lights [22]. Li et al. also utilized SPaT to improve fuel savings [23]. Li et al. used V2I for Eco-Driving through Eco-Departure from a signalized intersection for CAVs with internal combustion engines [24].

Drivers interact with other drivers during daily driving activities and are bound by the speed of the slower preceding vehicle that they are following. To consider the Eco-Driving of a CAV in traffic, the preceding vehicle information also has to be taken into account by control algorithms in the ego CAV, such as the lead vehicle position and speed. Vehicles can also communicate with each other through V2V to obtain acceleration information and use it for fuel economy, emissions and safety benefits.

Cruise Control (CC) systems aim to keep the vehicle speed constant to aid the drivers on roadways and are particularly helpful for freeway driving [25]. CC designs usually employ classical control methods while approaches like fuzzy logic control have also been used [26]. They help in safety and are useful as Driver Assist Systems (DAS); however, CC models do not adjust the ego vehicle speed with respect to the outside input, such as the preceding vehicle's position and speed.

Adaptive Cruise Control (ACC) has been widely used for saving fuel and improving safety for vehicles [27,28]. ACC is a valuable part of the Advanced Driver Assistance Systems (ADAS) and SAE Level 2 automated vehicles are equipped with ACC for car-following scenarios [29]. An ego vehicle equipped with a classical ACC uses cameras and radars to detect and track the preceding vehicle and actuators to control the ego vehicle speed [28]. Kural and Aksun Güvenç designed an ACC model by using Model Predictive Control (MPC) [30]. By reducing the unnecessary accelerations and decelerations as much as possible in the ego vehicle, ACC systems help to improve performance and indirectly save fuel. However, V2V technology is not utilized in ACC systems.

Cooperative Adaptive Cruise Control (CACC) enables V2V to be used for car-following scenarios [31]. In CACC, the ego vehicle receives information about the preceding vehicle from the preceding vehicle itself via V2V communication. Hu et al. utilized V2V technology for car following with an optimal look-ahead control framework for fuel savings [32]. Cantas et al. implemented a CACC algorithm, where the ego CAV received the acceleration of the preceding vehicle through V2V [33]. Kianfar et al. designed a CACC architecture that is capable of driving within a vehicle platoon while minimizing inter-vehicular spacing, attenuating shock waves and ensuring safety [34]. Rasool et al. used Pontryagin's Minimum Principle (PMP) to improve fuel efficiency during car following with CACC [35]. Güvenç et al. designed and tested a CACC system for the Grand Cooperative Driving Challenge (GCDC) [36]. Naus et al. used the frequency-domain approach to design and experimentally validate a string-stable CACC system [37].

Ecological Cooperative Adaptive Cruise Control (Eco-CACC) is an improvement over the CACC system and aims to improve fuel efficiency by using road information while utilizing CACC in car-following scenarios or vehicle platoons. Zhai et al. designed an Eco-CACC model for a heterogeneous platoon with a time delay between the platoon agents [38]. Yang et al. modeled an Eco-CACC algorithm to compute the fuel-optimum vehicle trajectory through a signalized intersection that also handles queue effects [39]. Almannaa et al. designed an Eco-CACC model to reduce fuel consumption and achieve travel time savings around signalized intersections and also tested the system through field implementation [40].

There is also research on the energy management of vehicles using V2I and V2V in the recent literature. Zhang et al. focused on using a chaining neural network and an improved equivalent consumption minimization strategy (ECMS) with an equivalent factor (EF) to minimize energy consumption in a hybrid electric vehicle and showed a benefit ranging from 0.2% to 5% over the ECMS with a traditional adaptation law [41]. He et al. proposed an improved MPC-based strategy for energy management utilizing V2V and V2I for a hybrid vehicle [42]. Ma et al. used V2V for platooning and V2I for passing at intersections for a homogeneous platoon of connected electric vehicles [43].

In this study, a comprehensive Eco-Driving strategy was developed for a CAV equipped with V2I and V2V algorithms. The validation of the proposed strategy was carried out by using realistic simulations with other traffic generated by a microscopic traffic simulator. This study shows the relative fuel savings that each component provides to CAVs, how each component can be improved and what constitutes the largest effect on fuel savings. It has been shown that the complete Eco-Driving architecture presented in this paper is applicable to be used in real life in actual vehicles. The main contribution of this paper is the development and simulation validation of an integrated Eco-Driving system that uses V2I to handle realistic situations with infrastructure (STOP signs and traffic lights) and V2V to handle interactions with other vehicles. The other contributions that help this main contribution are summarized as follows:

1. V2I and V2V algorithms were developed to control the longitudinal motion of a CAV for Eco-Driving.
2. The High-Level (HL) controller was also tested in a traffic simulator with realistic traffic flow. The traffic vehicles were controlled by the traffic simulator and had default car-following models, which enabled them to change lanes when they were behind slower vehicles. Thus, the traffic vehicles created dynamically changing constraints on the HL controller. It was observed that the HL controller ensured that no collisions were observed between the ego CAV and traffic vehicles, and the driving mode of the ego CAV was set correctly under changing constraints.
3. The High-Level (HL) controller designed for the comprehensive Eco-Driving of a CAV enabled fuel savings.

The rest of this paper is organized as follows: Section 2 describes the comprehensive Eco-Driving strategy for CAVs that was developed in this work. Section 3 details the deterministic High-Level controller. The microscopic traffic simulation environment is introduced in Section 4. Section 5 discusses the simulation results and comparative analysis based on various performance measures, followed by conclusions summarized in Section 6.

2. Complete Eco-Driving Strategy for a Connected and Automated Vehicle (CAV)

The schematic in Figure 1 displays a complete picture of the comprehensive Eco-Driving strategy for CAVs proposed in this paper. Firstly, the CAV needs to have a speed profile, which is called Eco-Cruise, that is route-dependent and fuel-optimal. This Eco-Cruise speed profile would assume normal operating conditions, meaning it would assume no surrounding traffic and infrastructure around the CAV. Additionally, the speed limit of the route and ride comfort with desired and safe acceleration and deceleration limits need to be enforced as constraints during the calculation of this fuel-optimal speed profile. This speed profile takes the route elevation into account, as well as the constraints of the vehicle, and can be calculated offline by using Dynamic Programming (DP). The Eco-Cruise mode shown in Figure 1 is the default driving mode, meaning that when the ego CAV does not interact with other vehicles or is not in the vicinity of traffic signs, Eco-Cruise is active to consume as little fuel as possible.

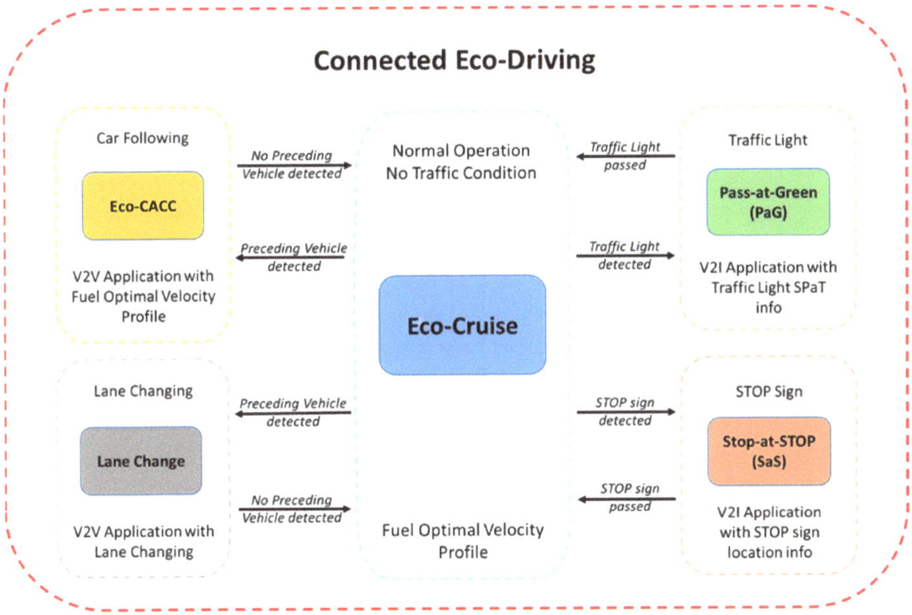

Figure 1. Comprehensive Eco-Driving architecture of CAVs.

CAVs interact with roadway infrastructure, such as traffic lights and STOP signs. For the Eco-Driving of a CAV, when there is an upcoming traffic light and the traffic light Signal Phase and Timing (SPaT) information is broadcast from a Roadside Unit (RSU), then the ego CAV goes into Pass-at-Green (PaG) mode (the green-colored block in Figure 1). In this mode, the ego CAV picks up the traffic light state and duration, as well as the location, from the upcoming traffic light. Then, the V2I longitudinal control algorithm on the ego CAV makes a decision about either passing the traffic light or stopping for a red light. In order to pass the traffic light, the ego CAV can accelerate to a higher speed, keep its speed constant or decelerate to a lower speed. If one of these three states is possible, then the PaG calculates a smooth speed profile for the ego CAV to follow so that the fuel economy and ride comfort are maximized. For the state where the vehicle is not able to pass, the PaG calculates a smooth Eco-Approach to the traffic light so that the vehicle decelerates smoothly and spends as little time as possible while idling during the red light. Once the light turns green, the PaG calculates a smooth Eco-Departure speed profile from the traffic light.

For the Eco-Driving of a CAV, the ego CAV also interacts with STOP signs on roadways. STOP signs are usually not equipped with any type of V2I equipment; therefore, another tool needs to be used to obtain the STOP sign location information. In this architecture, the ego CAV is equipped with eHorizon (Autoliv Inc., Ogden, UT, USA, 2020), an electronic horizon that has a detailed map in it. Once the ego CAV gets close to the STOP sign location, it goes into Eco-Stop mode (the red-colored block in Figure 1). In Eco-Stop mode, using the STOP sign location information, an Eco-Approach speed profile is calculated that enables the ego CAV to decelerate smoothly in a fuel-optimal manner and stop at the STOP sign. After the ego CAV waits for 5 s at the STOP sign during the simulations, the Eco-Departure is subsequently activated to get the vehicle to speed up to the speed limit. A perception sensor like a camera and image processing should be used in conjunction with the electronic horizon map in practice to be certain of the STOP sign location and presence. While a 5 s wait period is fine for the fuel economy computations in this paper, the CAV should use perception and communication sensors to assess the safety of operation before proceeding to depart the STOP sign.

Other than the roadway infrastructure, CAVs also interact with other surrounding traffic agents. CAVs are equipped with perception sensors; hence, they can detect nearby objects or vehicles. For the Eco-Driving of a CAV, once the ego CAV detects a preceding vehicle, it needs to go into Eco-Cooperative Adaptive Cruise Control (Eco-CACC) mode (the light-orange-colored block in Figure 1). This mode uses V2V communication so that the ego CAV obtains the preceding vehicle's information and uses that information to follow the preceding vehicle in a fuel-efficient manner.

When the preceding vehicle's movement is too erratic or the preceding vehicle is moving too slowly, the ego CAV goes into Lane-Change mode (the gray-colored block in Figure 1). In Lane-Change mode, the ego CAV obtains the surrounding vehicles' information, such as the vehicles' speed and acceleration, as well as the vehicles' position. Then, the model determines if it is safe to change lanes and executes lane changing. The main goal of a Lane Change in the Eco-Driving of a CAV is to make sure the ego CAV can maintain the optimal Eco-Cruise speed to obtain maximum fuel savings while also ensuring the safety of the ego vehicle and other nearby vehicles in adjacent lanes. If the leader vehicle changes lanes, it is not a leader vehicle anymore and the ego vehicle will revert back to Eco-Cruise until a new leader vehicle is encountered. If the Lane-Change mode commanded a Lane Change for the ego vehicle, but a new vehicle from adjacent traffic lanes joined the target lane, then the ego vehicle would either go back to the Eco-Cruise or car-following modes, depending on the speed of this new vehicle in front.

2.1. Fuel Optimization with Eco-Cruise

Dynamic Programming is a well-known solution that is used to find optimal benchmark solutions to various optimal control problems. Dynamic Programming (DP) [9] was used in the calculation of the fuel-optimal Eco-Cruise speed profile for a conventional vehicle. For Eco-Cruise, the problem was to minimize the road load acting on the vehicle (Figure 2) so that the fuel consumed by the vehicle would also be minimized.

$$Road\ Load = F_{rolling} + F_{aero} + F_{grade} \tag{1}$$

$$Road\ Load = mgr_0\cos(\alpha(s)) + \left(\frac{1}{2}\rho_{air}A_fC_Dv^2\right) + mg\sin(\alpha(s)) \tag{2}$$

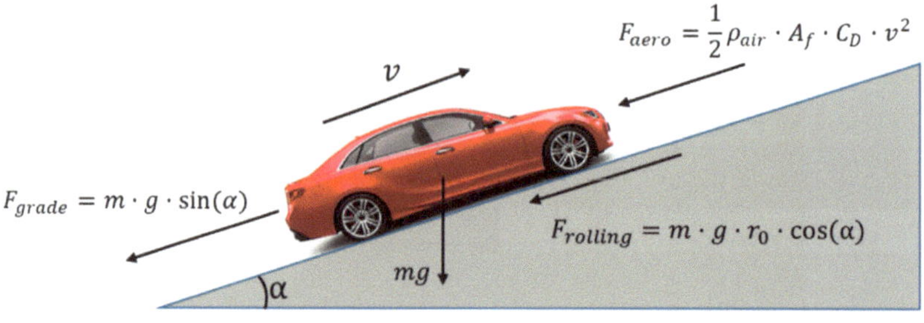

Figure 2. Road forces acting on a vehicle.

The road load (Figure 2) equation given in (1) has three parts. The first part is the rolling resistance $F_{rolling}$, and this term depends on the tire properties, vehicle speed and road conditions. In Equation (2), m is the vehicle mass with the rotating inertia factor, r_0 is a parameter of the rolling resistance equation and α is the road grade. The second term is F_{aero} and refers to the aerodynamic drag term. F_{aero} depends on the vehicle speed and frontal cross-section area of the vehicle. In Equation (2), ρ_{air} is the density of air, A_f is the front cross-sectional area, C_D is the drag coefficient and v is the vehicle speed. F_{grade} is the road grade term, and it depends on the vehicle mass and the road grade.

The power that needs to be provided from the engine in a vehicle to beat road load and enable acceleration can be expressed as follows:

$$P = F_x v = \left(m_e \frac{dv}{dt} + \frac{1}{2} \rho_{air} \cdot A_f \cdot C_D \cdot v^2 + m \cdot g \cdot r_0 \cdot \cos(\alpha) + m \cdot g \cdot \sin(\alpha) \right) v \quad (3)$$

where P is the power, F_x is the force required at the tires and $m_e \frac{dv}{dt}$ is the force required to accelerate. The rest of the terms in Equation (3) come from the road load acting on the vehicle, which was given in Equation (2). Using the power expression given in Equation (3), the fuel rate that is consumed by the vehicle when it is traveling can be expressed as follows:

$$\dot{m}_f = \frac{P/\eta_t + P_{accessories}}{\eta_e} \quad (4)$$

where $\dot{m}_f$ is the fuel rate, $P_{accessories}$ is the power required to keep the accessories running, η_t is the transmission efficiency and η_e is the engine efficiency. This expression for the fuel rate given in Equation (4) can be used as the cost function that needs to be minimized for this analysis. Further details for this optimal control formulation can be found in [44].

In this paper, the fuel-optimal DP solution presented here was used for different driving modes. Firstly, the driving mode called Eco-Cruise, where the fuel-optimal speed profile is calculated offline by using road information, was found by using DP. Additionally, the Eco-Stop mode, where the ego vehicle approaches a STOP sign fuel-economically, also utilized DP. Finally, the Eco-Departure mode, where the ego vehicle departs from a traffic light or STOP sign, also used DP. These solutions were all distance-based solutions, as presented earlier.

In the DP solution, the whole trip horizon is divided into segments. Additionally, the solution space is also divided into nodes. The solution starts from the end point, where the desired vehicle speed and vehicle location are known. The cost in terms of the fuel rate seen in Equation (4) is assigned to each link to move from the current node to each previous neighboring node in backward propagation. Then, the feasibility constraint of going from one node to the next is checked, where the acceleration and deceleration, as well as the jerk-rate limits, are enforced. Details about this approach can be found in [44].

2.2. Vehicle-to-Infrastructure (V2I) Interactions of a CAV

A vehicle traveling from a starting location to a traffic light (or a STOP sign) can be seen in Figure 3. In Figure 3, x_{ego} is the position, v_{ego} is the speed and a_{ego} is the acceleration of the ego vehicle. $TL_{location}$ is the traffic light location, TL_{SPaT} is the traffic light state and duration and $STOP_{location}$ is the location of the STOP sign.

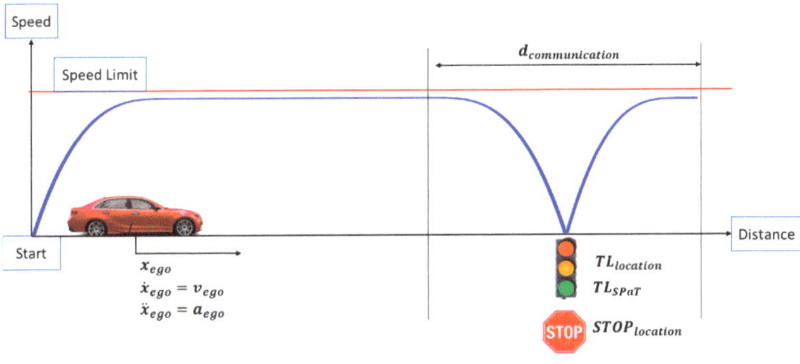

Figure 3. V2I interaction as an optimal control problem.

In this paper, when it comes to V2I communication, the aim is to design control algorithms that minimize the fuel consumption in a vehicle. Fuel consumption can be reduced through the utilization of V2I so that the vehicle control algorithms can obtain roadway infrastructure information and use it to consume less fuel. The optimal control problem can be defined with the objective function (5):

$$\underbrace{minimize}_{T_e(t),\, F_b(t)} \; J(u(t)) = L_N\left(s\left(t_f\right), v\left(t_f\right)\right) + \int_0^{t_f} L_k(s(t), v(t), T_e(t), F_b(t), t)dt \tag{5}$$

where L_k is the running cost and L_N is the terminal cost. Additionally, $u(t)$ is the input, $s(t)$ is the distance, t_f is the final time, $v(t)$ is the vehicle velocity, $T_e(t)$ is the engine torque and $F_b(t)$ is the brake force. The states are subject to

$$\frac{ds(t)}{dt} = v(t) \tag{6}$$

$$\frac{dv(t)}{dt} = K_{T_e}T_e - K_{F_b}F_b - gr_0\cos(\alpha(t)) - \frac{1}{2m}\rho_{air}A_fC_Dv(t)^2 - g\sin(\alpha(t)) \tag{7}$$

where Equation (6) expresses that the derivative of the position is equal to the speed. In Equation (7), m is the vehicle mass, r_0 is a parameter of the rolling resistance equation, $\alpha(t)$ is the road grade, ρ_{air} is the density of air, A_f is the front cross-sectional area, C_D is the drag coefficient, $v(t)$ is the vehicle speed and g is the gravitational acceleration. The vehicle model expression given in Equation (7) shows that road load and brake force subtracted from the total powertrain force to the wheels is equal to the vehicle acceleration. There are initial and final algebraic constraints on the states of the position and speed, and they are as follows:

$$s(0) = s_{initial} = 0 \tag{8}$$

$$s\left(t_f\right) = s_{final} = s_f \tag{9}$$

$$v(0) = v_{initial} = v_i = 0 \tag{10}$$

$$v\left(t_f\right) = v_{final} = v_f = 0 \tag{11}$$

$$v_{min}(t, s(t)) \leq v(t) \leq v_{max}(t, s(t)) \tag{12}$$

where $s_{initial}$ (8) is the initial vehicle position and s_f (9) is the final vehicle position. Additionally, v_i (10) is the initial vehicle speed, v_f (11) is the final vehicle speed and v_{min} (12) is the minimum allowable speed for the vehicle. The speed limit of the roadway is enforced as the v_{max} (12) constraint, which is the maximum allowable speed of the vehicle. There are also algebraic constraints on the input's engine torque T_e (13) and brake force F_b (14):

$$T_{e,min}(v(t), t) \leq T_e(t) \leq T_{e,max}(v(t), t) \tag{13}$$

$$0 \leq F_b(t) \leq F_{b,max}(v(t), t) \tag{14}$$

The optimal control problem posed here was solved by using Dynamic Programming for the case where the ego CAV approaches a STOP sign. Different Eco-Approach profiles were calculated for approaching the STOP sign, and depending on the instantaneous speed of the ego CAV when it was within 300 m of the STOP sign, the appropriate profile was chosen during the simulations.

For the interactions between the ego CAV and traffic lights, Pass-at-Green (PaG) was used. PaG is a V2I application that uses roadway infrastructure information to eliminate or decrease idling at red lights to decrease the fuel consumption for the ego vehicle. PaG operates under deterministic control by using the input, which includes the distance to the upcoming traffic light, Signal Phase and Timing (SPaT) information received from the upcoming traffic light, instantaneous actual vehicle speed, maximum acceleration and maximum deceleration limits and jerk limit for ride comfort. Using these inputs, the PaG calculates a smooth and fuel-economic speed profile so that the vehicle can pass the upcoming traffic light.

Depending on the distance to the traffic light and the SPaT information from the upcoming traffic light, the PaG chooses one of four options for the recommended vehicle-speed trajectory. These PaG states are as follows:

- *Cruise State:* the vehicle keeps its speed constant and passes the traffic light when the light is green.
- *Increase Speed State:* The vehicle accelerates to a higher speed, travels at a constant speed when it is passing the green traffic light and then decelerates to the initial lower speed. The vehicle obeys speed limits, as well as acceleration, deceleration and jerk limits for ride comfort.
- *Eco-Approach State:* The vehicle cannot catch the current green light; therefore, it decelerates to a stop at the traffic light. Then, after the traffic light turns green, the vehicle smoothly accelerates to a higher speed and passes the traffic light. The vehicle obeys speed limits, as well as acceleration and deceleration limits for ride comfort.
- *Decrease Speed State:* The vehicle decelerates to a lower speed, travels at a constant speed when it is passing the traffic light and then accelerates to the initial higher speed. The vehicle obeys speed limits as well as acceleration and deceleration limits for ride comfort.

More information on the PaG can be found in [17–19,44].

2.3. Vehicle-to-Vehicle (V2V) Interactions of a CAV

An ego CAV following a lead connected vehicle can be seen in Figure 4. x_{ego} and x_{lead} are the positions of the ego and lead vehicles, respectively. $\dot{x}_{ego}$ and $\dot{x}_{lead}$ are the speeds of the ego and lead vehicle, respectively. $\ddot{x}_{ego}$ and $\ddot{x}_{lead}$ are the accelerations of the ego and lead vehicle, respectively. It should be noted that the sinusoidal-looking perturbation in the speed profile of Figure 4 is for illustration purposes only and represents a perturbation (not necessarily sinusoidal) that the ego vehicle does not want to follow.

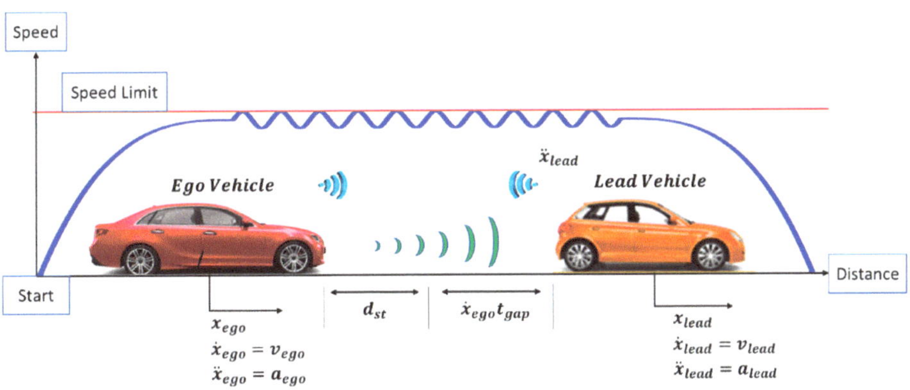

Figure 4. Car following a CAV as an optimal control problem.

Fuel consumption in CAVs can be reduced by the utilization of V2V so that the vehicle control algorithms can obtain the lead vehicle's information and use it to consume less fuel.

In order to prevent a collision from happening between the lead vehicle and the ego CAV, the following algebraic constraint also needs to be enforced. These constraints are as follows:

$$x_{actual} = x_{lead} - x_{ego},\ x_{actual} > 0 \quad (15)$$

where x_{actual} (15) is the actual distance between the lead and the ego vehicle, and it needs to always be larger than zero to ensure that the vehicles do not collide.

ACC, CACC and Eco-CACC models with Proportional-Derivative (PD) feedback control and a constant time-gap spacing policy were designed in order for the ego CAV to safely follow the lead vehicle. Eco-CACC used a preceding acceleration feedforward compensator that filtered high-frequency acceleration disturbances of the preceding vehicle. More information about the V2V models can be found in [45,46].

3. The High-Level Controllers for V2I, V2V and V2I + V2V

In this section, the deterministic control algorithms that were developed for the Eco-Driving of a CAV are explored further.

3.1. High-Level (HL) Controller for V2I with No Traffic

The High-Level (HL) controller for V2I with no traffic handles how the ego CAV behaves when it is traveling on a roadway with no other vehicle around it and is implemented as a state-flow chart. The aim is to determine when the CAV has to switch between the different driving modes of the Eco-Driving of the CAV architecture presented in Figure 1. This controller ensures the seamless transition from one driving mode to the next.

Depending on deterministic conditions, such as the current upcoming traffic light state and duration, the distance between the infrastructure elements (the traffic lights and STOP signs) and the ego vehicle, as well as the instantaneous vehicle speed, the controller is tasked to make a decision to switch between driving modes. The flow chart for the deterministic control algorithm for the fuel-economic Eco-Driving of a single CAV with no traffic can be seen in Figure 5.

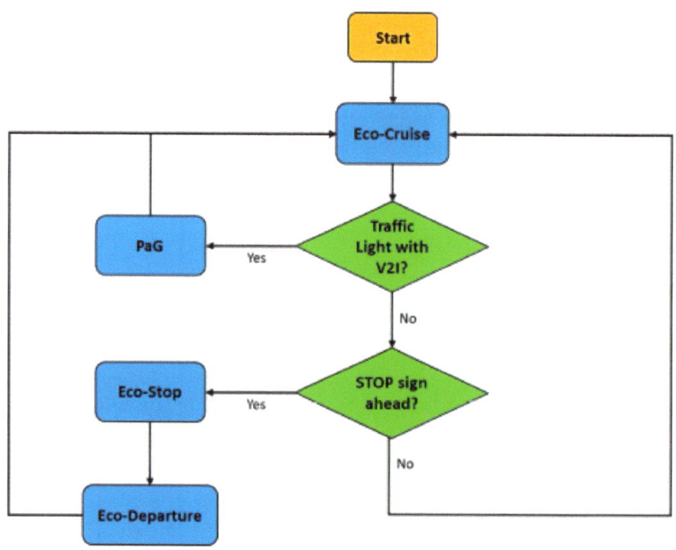

Figure 5. Flowchart for the HL controller with no traffic for V2I.

As seen in Figure 5, the ego CAV aims to maintain its speed as close to the Eco-Cruise speed as possible. The Eco-Cruise speed is the fuel-economic speed profile that is route-dependent and is calculated offline prior to the trip. In case there is an upcoming traffic light, the Pass-at-Green (PaG) V2I algorithm takes over control of the ego vehicle. If there is a STOP sign, then Eco-Stop mode is activated to make the vehicle stop smoothly at the sign. After stopping at the STOP sign for a few seconds, Eco-Departure takes over and makes the ego vehicle accelerate smoothly. The HL controller makes sure the correct driving mode is active and mode transitions are smooth to save as much fuel as possible.

3.2. High-Level (HL) Controller for V2V with Traffic

This High-Level (HL) controller for V2V with traffic aims to make transitions between driving modes correctly and smoothly so that the ego vehicle speed does not jump abruptly when the driving mode changes. The flowchart for the controller is seen in Figure 6, where the Eco-Cruise speed is the fuel-economic and road-dependent speed profile for the ego vehicle to follow to consume less fuel. When there is a preceding vehicle with no V2V communication, the ACC model is activated and the ego CAV safely follows the lead vehicle. If the preceding vehicle is equipped with V2V and does not have an erratic driver, then the CACC takes over and follows the lead vehicle smoothly while keeping a safe distance between vehicles to prevent a collision. If the preceding vehicle with V2V has an erratic driver, then Ecological Cooperative Adaptive Cruise Control (Eco-CACC) takes over control to follow the erratic leader without responding to its high-frequency accelerations in order to maintain fuel savings and safety. If the leader is erratic and lane changing is possible for the ego vehicle, then the ego vehicle changes its lane and maintains the Eco-Cruise speed.

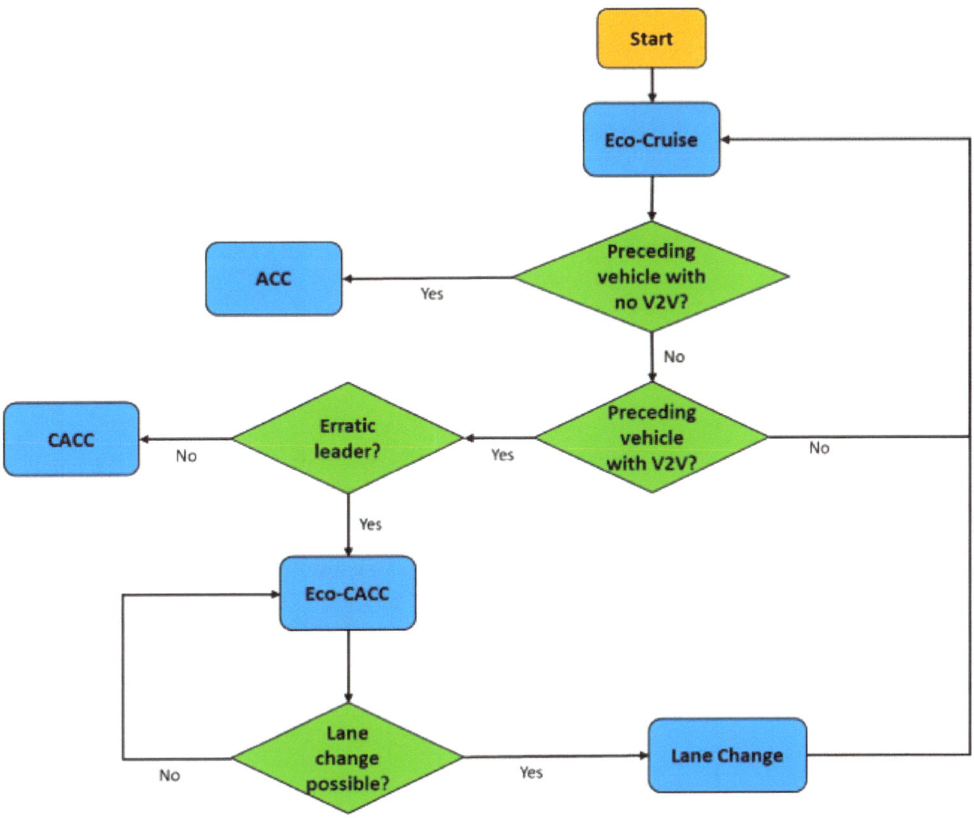

Figure 6. Flowchart for the HL controller with traffic for V2V.

The driving modes shown in Figure 6 have different controllers, and when they all run simultaneously during testing, the recommended vehicle speeds from each driving mode are usually different. If driving modes were to switch immediately with no transition, then the recommended speeds would not be continuous and cause the actual ego vehicle speed to jump abruptly. To overcome this problem, a Transition State is added to smoothly transition between driving modes. The algebraic equation for the Transition State to smoothly increase the vehicle speed is as follows:

$$v_{trans} = v_{trig} + v_{chg,acc} \tag{16}$$

$$v_{chg,acc} = \left(v_{lim} - v_{trig}\right)\left(\left(\frac{t_{act} - t_{trig}}{4\left(v_{lim} - v_{trig}\right)} - 1\right)^3 + 1\right) \tag{17}$$

where v_{trans} (16) is the recommended transition speed for the vehicle, v_{trig} is the vehicle speed when the driving-mode transition started and $v_{chg,acc}$ is the speed change needed for the ego vehicle to travel at the higher speed limit. In Equation (17), v_{lim} is the actual speed limit of the road, t_{act} is the actual simulation time and t_{trig} is the time instant when the driving-mode transition starts. The third-order power equation that comprises the variables seen in Equation (17) ensures that the recommended speed is smooth when driving modes are switched and the ego CAV accelerates.

When the Eco-Cruise speed is smaller than the instantaneous vehicle speed, the following Equation (18) ensures that the vehicle decelerates slowly. In Equation (18), $v_{chg,dec}$ is the speed change needed for the ego vehicle to travel at the lower speed limit. In Equation (19), $v_{lim,low}$ is the user-set lower speed limit, t_{act} is the actual simulation time and t_{trig} is the time instant when the driving-mode transition starts. A third-order power equation that comprises the variables seen in Equation (19) ensures that the recommended speed is smooth when driving modes are switched and the ego CAV decelerates. When the Eco-Cruise speed catches up to the vehicle speed, then the recommended speed for the CAV to follow switches back to the Eco-Cruise speed:

$$v_{trans} = v_{trig} - v_{chg,\ dec} \tag{18}$$

$$v_{chg,dcc} = \left(v_{lim,low} - v_{trig}\right)\left(\left(\frac{t_{act} - t_{trig}}{4\left(v_{lim,low} - v_{trig}\right)} - 1\right)^3 + 1\right) \tag{19}$$

3.3. High-Level (HL) Controller for V2V and V2I with Traffic

The HL controller for V2V and V2I with traffic was designed as a state-flow diagram in Simulink, and the flowchart for the HL controller decision-making process can be seen in Figure 7. The default mode is the Eco-Cruise mode, where the precalculated fuel-economic DP profile is the desired speed profile for the vehicle. The Eco-Cruise speed profile also makes sure the ego vehicle drives in a fuel-economic manner around STOP signs. When there is a lead vehicle in close proximity to the ego vehicle, car-following models are activated to safely and closely follow the preceding vehicle. When there is a traffic light ahead, the mode is switched to the PaG V2I algorithm. After the ego vehicle passes the traffic light, depending on the instantaneous speed of the vehicle, the transition modes are activated (speed up or speed down).

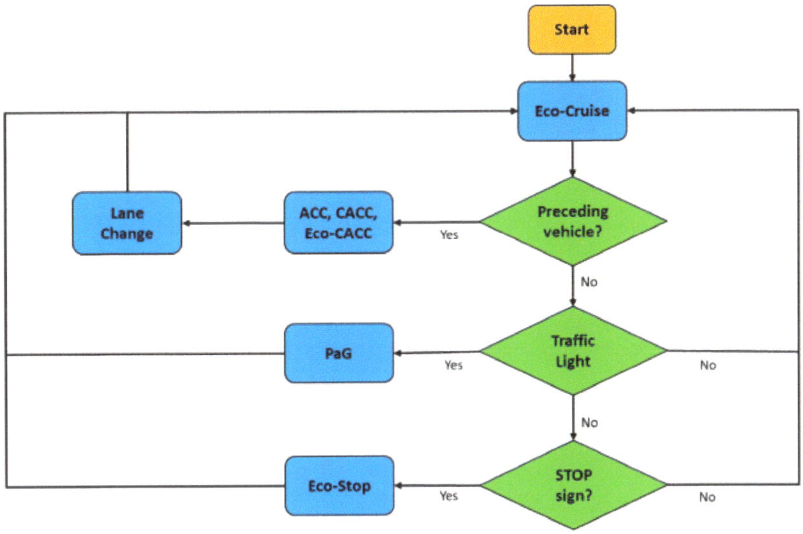

Figure 7. Flowchart for the HL controller with traffic for V2V and V2I.

4. Microscopic Traffic Simulation Environment

A simulation environment was set up by using Simulink and the Vissim commercial traffic simulator to run co-simulations by using the COM interface capability of Vissim [19,47]. Details about setting up a COM connection between Simulink and Vissim can be found in [47]. Other than the COM interface between Simulink and Vissim, there was no specific Matlab Simulink package that was installed for the simulation experiments. During the co-simulations, realistic traffic information was being sent from Vissim to Simulink. The ego vehicle with a mid-sized vehicle powertrain was being controlled by the High-Level (HL) controller in Simulink. The fuel consumption model was also in Simulink, and the realistic fuel consumption values were achieved by using multi-dimensional tables that replicated the behavior of a real vehicle engine. The HL controller determined which action to take and which driving mode to activate in response to the realistic traffic and infrastructure information received from the traffic simulator.

The simulation environment designed in Vissim is called the Arlington Route and it has one STOP sign, five traffic lights and is 6873 m long. The Arlington Route can be seen below in Figure 8.

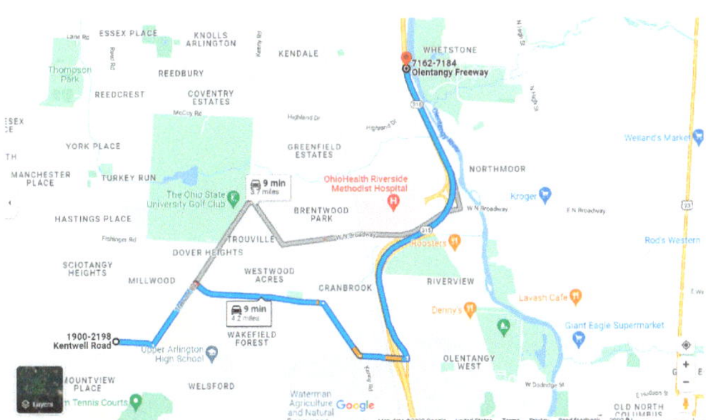

Figure 8. Arlington Route from "Google Maps" (2022).

The speed limit, traffic sign locations and route-dependent fuel-economic DP solution for the Eco-Cruise driving mode for the Arlington Route can be seen below in Figure 9.

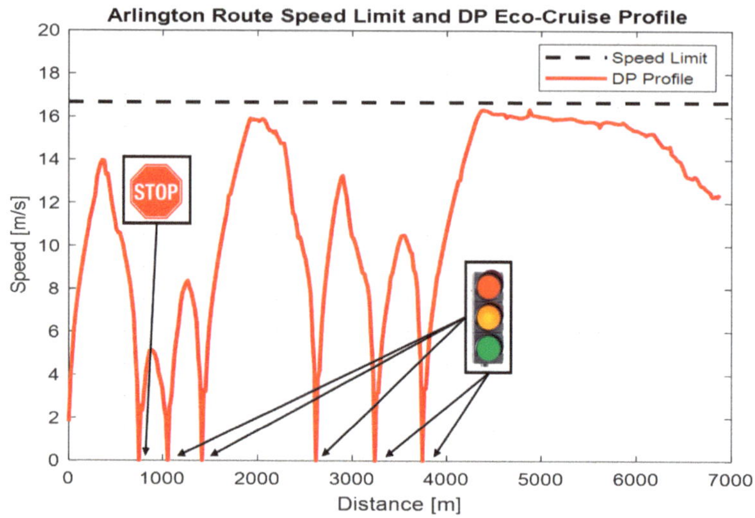

Figure 9. Characteristics of the Arlington Route.

The pink ego vehicle approaching a traffic light at an intersection with other traffic vehicles around it during the traffic simulation can be seen in Figure 10. During the simulation, the ego vehicle was controlled by the HL controller to save fuel by smoothly approaching traffic lights and STOP signs. At the same time, whenever there was a vehicle in front of the ego vehicle and the distance between the ego and lead vehicles was less than 50 m, ACC, CACC or Eco-CACC were activated to prevent collisions between the vehicles during car following.

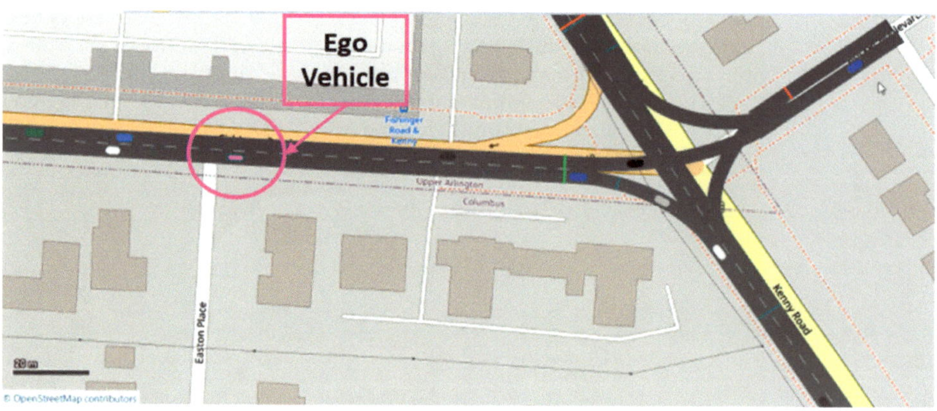

Figure 10. Ego vehicle approaching an intersection in the traffic simulation.

The traffic-vehicle compositions were the same at each simulation. Additionally, the traffic simulator spawned vehicles at a common start time for each simulation, meaning that the vehicle with a specific ID entered the roadway at the same timestamp across all simulation cases. This unity ensures that the simulation results can be compared with each other since the traffic vehicles that interact with the ego vehicle appear in the simulator at the same time. Additionally, the traffic light periods for each traffic light were the same across all simulations. When it comes to experimental parameters, the distance traveled by vehicles, the inter-vehicular distance between the ego and leader vehicle, vehicle speed, simulation time, distance to traffic lights and STOP signs, SPAT for traffic lights and HL controller state were recorded and analyzed for system performance.

Depending on the test case and whether there were other traffic vehicles around the ego vehicle for V2V, or V2I communication with the road infrastructure, one of the three HL controllers presented in Section 3 was used.

5. Results and Discussion

To assess the fuel economy performance of the V2I and V2V algorithms in a traffic network, five different simulations were run. For case 1, the ego vehicle was commanded to follow the fuel-economic DP profile in Eco-Cruise mode with no other traffic vehicles around in the simulation.

For the second case, the ego vehicle was commanded to follow the same Eco-Cruise speed as the first case while also interacting with STOP signs by using Eco-Stop and traffic lights by using PaG. For case 2, the HL controller for V2I with no traffic presented in Section 3.1 was utilized during the simulations.

The third simulation case built on top of the second simulation case, where Eco-Cruise, Eco-Stop and PaG were all working in tandem, and there were also traffic vehicles around the ego vehicle. Whenever the ego vehicle was in the vicinity of a lead vehicle, the ACC mode was activated. The fourth simulation case used the same V2I models, and when there was a lead vehicle ahead, the CACC mode was activated. The fifth and final simulation case used the same V2I models as the fourth case, except the car-following model that was used when there was a lead vehicle in front of the ego vehicle for this case was Eco-CACC. For cases 3, 4 and 5, the HL controller for V2V and V2I with traffic, which was presented in Section 3.3, was used.

The speed profile for the ego vehicle when there were no other traffic vehicles around can be seen in Figure 11. The light blue line represents the ego vehicle speed when it was commanded to follow the DP offline-calculated Eco-Cruise profile in Figure 9. The light red line represents the vehicle speed when the vehicle was around a traffic light, and the SPaT information was used to modify the speed profile for case 2.

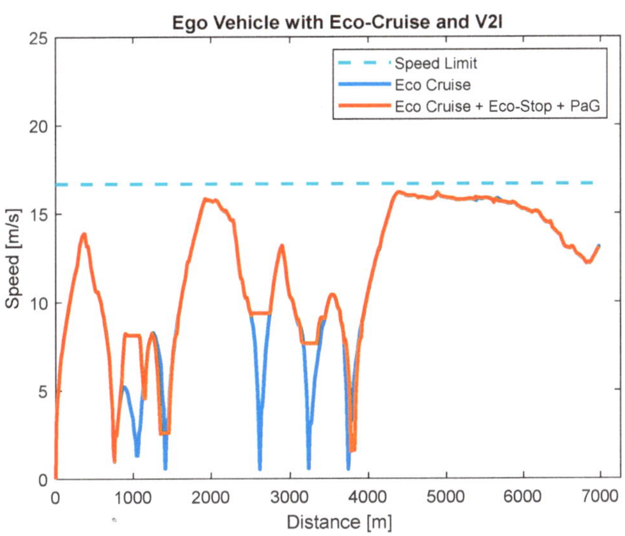

Figure 11. Ego vehicle with Eco-Cruise only (case 1) and Eco-Cruise + Eco-Stop + PaG (case 2).

The results of the third simulation case, where there were other traffic vehicles in the traffic simulator and the ego vehicle was equipped with the V2I algorithms and ACC, can be seen in Figure 12. Whenever the distance between the ego vehicle and the lead vehicle was below 50 m, the ACC took over control to make sure no collision could occur. If the distance between the ego and the lead vehicle was larger than 50 m, the HL controller commanded the ego vehicle to either follow the Eco-Cruise trajectory or the PaG trajectory to save fuel. During this simulation, around 500 s, the PaG commanded the vehicle to accelerate to pass the traffic light, which was not observed in cases 4 and 5. This acceleration-to-pass behavior observed in case 3 resulted in the ego vehicle having the highest fuel consumption among cases 3, 4 and 5.

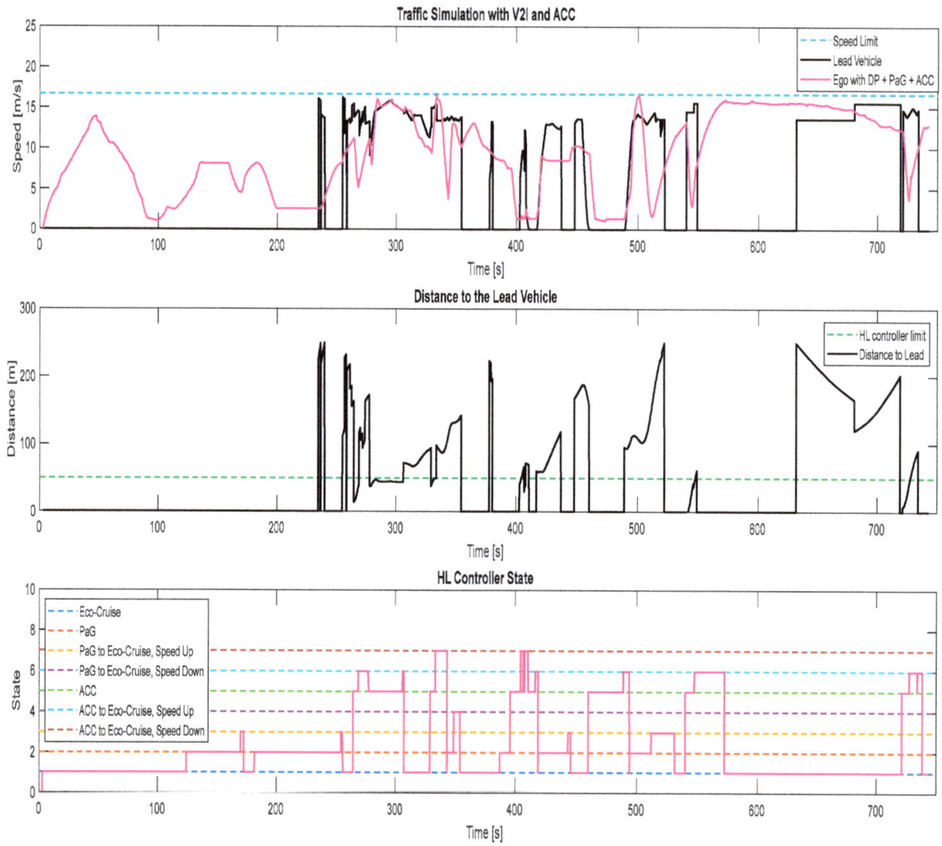

Figure 12. Traffic simulation for ego vehicle with V2I and ACC, case 3.

The results of the fourth simulation case, where there were other traffic vehicles in the traffic simulator and the ego vehicle was equipped with the V2I algorithms and CACC, can be seen in Figure 13. The HL controller handled having a preceding vehicle ahead of the ego vehicle the same as the ACC case. Towards the end of the simulation in case 4, the ego vehicle switched into car-following mode with CACC. In CACC mode, the ego vehicle tried to follow the lead vehicle at a safe distance. During the simulation, the lead vehicle was driving faster than the ego vehicle, which resulted in the ego vehicle accelerating to a higher speed to keep up with the lead vehicle around 730 s. This resulted in the ego vehicle having a higher fuel consumption in case 4 (Figure 13), where the ego vehicle used CACC compared to case 5 (Figure 14), where the ego vehicle used Eco-CACC for car following.

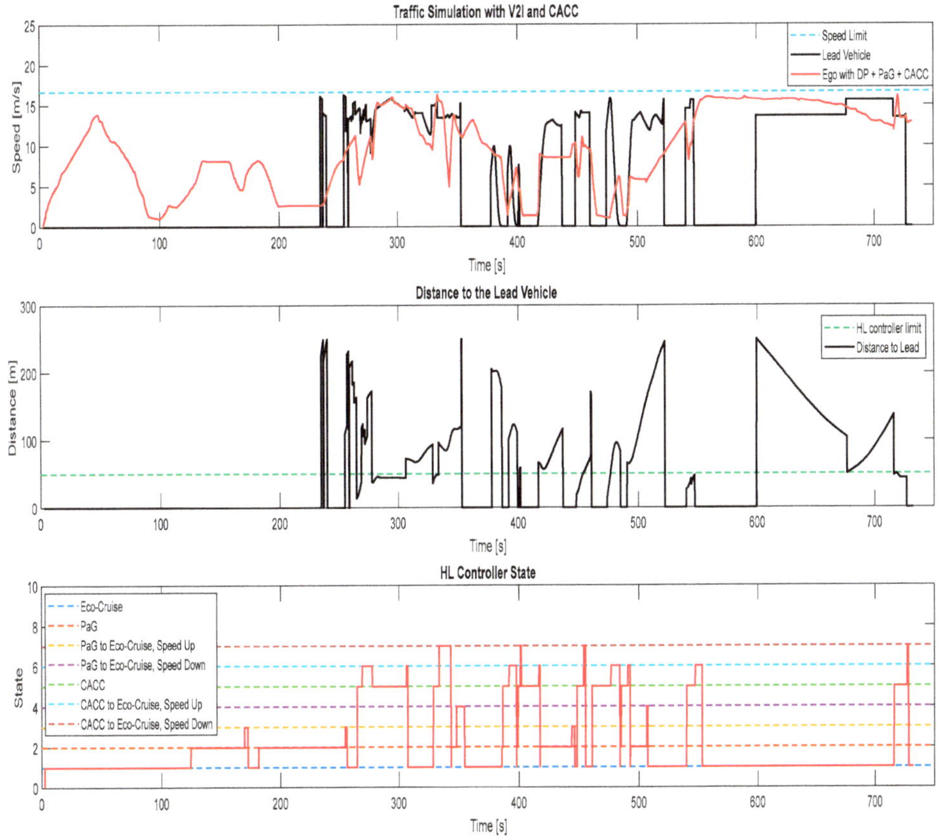

Figure 13. Traffic simulation for ego vehicle with V2I and CACC, case 4.

The results of the fifth and the final simulation case, where there were other traffic vehicles in the traffic simulator and the ego vehicle was equipped with the V2I algorithms and Eco-CACC, can be seen in Figure 14. Similar to the previous cases with ACC and CACC, the HL controller handled the state transitions.

The fuel consumed by the ego vehicle in each of the two simulation cases, where there was no other traffic flow around the ego vehicle, was recorded, and the percentage of the fuel consumption reduction in the models was calculated with respect to the simulation case 1 (Table 1). During case 1 and case 2, there were no other vehicles around the ego vehicle to interact with by using V2V. When the ego vehicle could use V2I in case 2, the fuel consumed by the ego vehicle decreased compared to using the Eco-Cruise-only simulation in case 1, where the ego vehicle stops at all traffic lights and STOP signs.

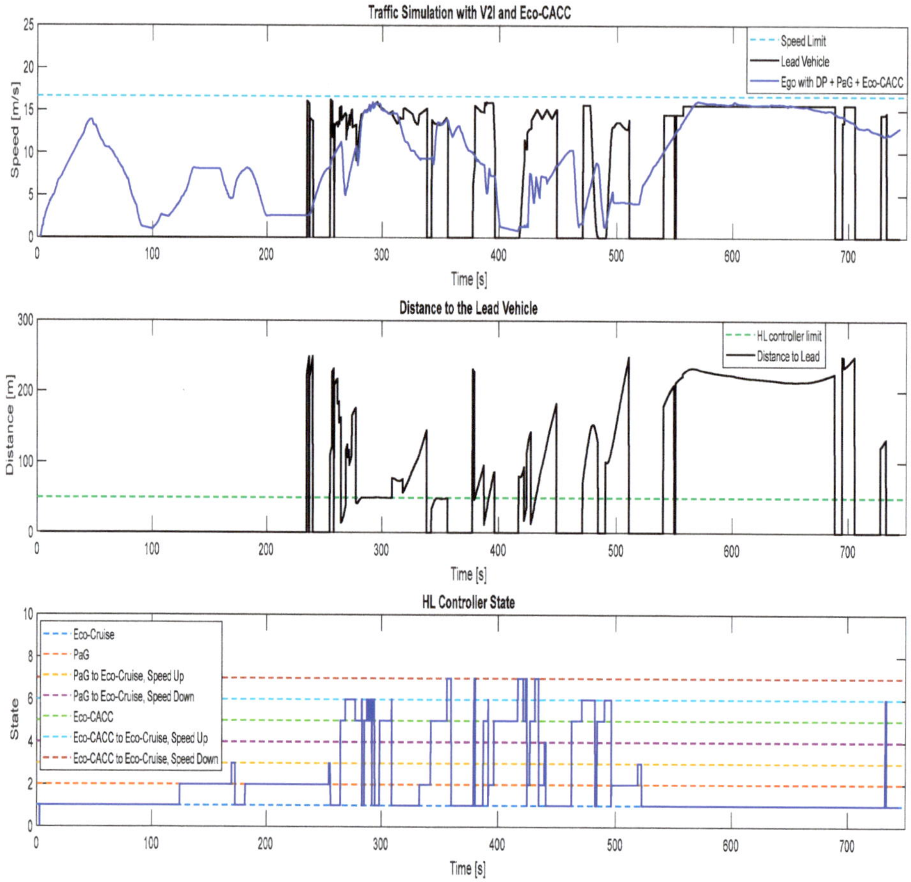

Figure 14. Traffic simulation for ego vehicle with V2I and Eco-CACC, case 5.

Table 1. Results for the fuel economy of the ego vehicle in no-traffic network.

Simulation Case Number	Simulation Scenario Name	Total Fuel Consumption (g)	% Fuel Consumption Reduction with Respect to Case #1
1	Eco-Cruise only (no traffic, vehicle stops at all traffic lights)	395.85	-
2	Eco-Cruise with Eco-Stop and PaG (no traffic, V2I only)	382.17	3.46%

Three simulations were run, where there was another vehicle around the ego vehicle, and the results are summarized in Table 2. Traffic vehicles that constrained the motion of the ego vehicle were present for cases 3, 4 and 5. Compared to ACC for car following in case 3, using CACC in case 4 resulted in a 1.51% fuel economy improvement. Moreover, using the Eco-CACC in case 5 was even more beneficial in reducing the fuel consumed by the ego vehicle. The fuel consumption decreased by 6.41% when using Eco-CACC in case 5 compared to using ACC in case 3.

Table 2. Results for the fuel economy of the ego vehicle in a traffic network.

Simulation Case Number	Simulation Scenario Name	Total Fuel Consumption (g)	% Fuel Consumption Reduction with Respect to Case #3
3	Eco-Cruise with Eco-Stop and PaG and ACC (V2I + no V2V)	454.20	-
4	Eco-Cruise with Eco-Stop and PaG and CACC (V2I + V2V)	447.37	1.51%
5	Eco-Cruise with Eco-Stop and PaG and Eco-CACC (V2I + V2V)	425.12	6.41%

6. Conclusions and Future Work

In this paper, a comprehensive Eco-Driving strategy with V2I and V2V algorithms was tested in a realistic microscopic traffic simulation environment, where a real-life route in Columbus, Ohio, USA, was modeled in a traffic simulator with the same number of lanes, speed limits, traffic lights and STOP signs. When PaG was active and used traffic infrastructure information in case 2, 3.46% less fuel was consumed compared to only using the Eco-Cruise speed profile case 1. For the simulation cases that required car following, it was shown that using CACC and Eco-CACC with V2V was more beneficial than using only ACC. The ego vehicle consumed 1.51% and 6.41% less fuel as compared to ACC only (case 3) for car following when CACC (case 4) and Eco-CACC (case 5) were used, respectively. Moreover, it was seen that Eco-CACC, which was modeled with a filter to attenuate the acceleration of the lead vehicle, consumed less fuel than CACC, which used the lead vehicle acceleration without filtering it.

For future work, the different driving modes that were presented here can be combined as part of an MPC with varying constraints under different driving conditions to improve the complete Eco-Driving strategy of the CAV presented in this paper.

There is also potential for improvement for the High-Level (HL) controller. In the simulation results, it was seen that for some cases during car following, the HL controller switched between different driving modes very rapidly. In real-life implementations, this rapid switching between driving modes would diminish the ride comfort for the passengers. To eliminate this rapid switching issue in the HL controller, a dead zone can be included in the controller. When controllers have dead zones, they do not respond to the change in the input within the dead zone region [48]. By exploring the addition of a dead zone to the HL controller, the rapid switching issue might be eliminated.

Within the scope of this paper, it was assumed that the functional safety of the ego CAV was satisfied and there were no malicious agents for the V2I, V2V and V2X communication. However, in real life, there could be cyber-security threats to the functional safety of a CAV due to malicious road agents. For example, malicious agents could broadcast inaccurate acceleration information to other CAVs on the roadway, or they could intentionally drive in an erratic manner. For safe and reliable real-life implementation and VIL testing, the cyber-security and functional safety aspects of CAVs need to be explored further.

To obtain real-world behavior when the CAVs are deployed, datasets dedicated to CAVs are needed. These were not created in the current paper, but there are such papers in the literature where such data are collected. For example, the paper in [49] presented a dedicated dataset for analyzing CAVs' behavior.

Author Contributions: Conceptualization, O.K.-T.; methodology, O.K.-T.; software, O.K.-T.; validation, O.K.-T.; formal analysis, O.K.-T.; investigation, O.K.-T.; resources, O.K.-T.; data curation, O.K.-T.; writing—original draft preparation, O.K.-T. and L.G.; writing—review and editing, O.K.-T. and L.G.; visualization, O.K.-T.; supervision, L.G.; funding acquisition, L.G. All authors have read and agreed to the published version of the manuscript.

Funding: This research received no external funding.

Acknowledgments: This work is supported by the Automated Driving Laboratory (ADL) at Ohio State University (OSU).

Conflicts of Interest: The authors declare no conflict of interest.

References

1. Saboohi, Y.; Farzaneh, H. Model for developing an eco-driving strategy of a passenger vehicle based on the least fuel consumption. *Appl. Energy* **2009**, *86*, 1925–1932. [CrossRef]
2. Emekli, M.E.; Güvenç, B.A. Explicit MIMO Model Predictive Boost Pressure Control of a Two-Stage Turbocharged Diesel Engine. *IEEE Trans. Control Syst. Technol.* **2017**, *25*, 521–534. [CrossRef]
3. Vahidi, A.; Sciarretta, A. Energy saving potentials of connected and automated vehicles. *Transp. Res. Part C Emerg. Technol.* **2018**, *95*, 822–843. [CrossRef]
4. Typaldos, P.; Papamichail, I.; Papageorgiou, M. Minimization of Fuel Consumption for Vehicle Trajectories. *IEEE Trans. Intell. Transp. Syst.* **2020**, *21*, 1716–1727. [CrossRef]
5. Fredette, D.; Ozguner, U. Dynamic Eco-Driving's Fuel Saving Potential in Traffic: Multi-Vehicle Simulation Study Comparing Three Representative Methods. *IEEE Trans. Intell. Transport. Syst.* **2018**, *19*, 2871–2879. [CrossRef]
6. Yang, Y.; Ma, F.; Wang, J.; Zhu, S.; Gelbal, S.Y.; Kavas-Torris, O.; Aksun-Guvenc, B.; Guvenc, L. Cooperative ecological cruising using hierarchical control strategy with optimal sustainable performance for connected automated vehicles on varying road conditions. *J. Clean. Prod.* **2020**, *275*, 123056. [CrossRef]
7. Aksun-Guvenc, B.; Guvenc, L.; Ozturk, E.S.; Yigit, T. Model regulator based individual wheel braking control. In Proceedings of the 2003 IEEE Conference on Control Applications, CCA 2003, Istanbul, Turkey, 25 June 2003; Volume 1, pp. 31–36. [CrossRef]
8. Aksun-Guvenc, B.; Guvenc, L. Robust steer-by-wire control based on the model regulator. In Proceedings of the International Conference on Control Applications, Glasgow, UK, 18–20 September 2002; Volume 1, pp. 435–440. [CrossRef]
9. Güvenç, L.; Srinivasan, K. Friction compensation and evaluation for a force control application. *Mech. Syst. Signal Process.* **1994**, *8*, 623–638. [CrossRef]
10. Güvenç, L.; Güvenç, B.A.; Demirel, B.; Emirler, M.T. *Control of Mechatronic Systems*; IET: London, UK, 2017. Available online: https://search.ebscohost.com/login.aspx?direct=true&scope=site&db=nlebk&db=nlabk&AN=1531301 (accessed on 26 October 2020).
11. Aksun-Guvenc, B.; Guvenc, L. Robustness of disturbance observers in the presence of structured real parametric uncertainty. In Proceedings of the 2001 American Control Conference, (Cat. No. 01CH37148), Arlington, VA, USA, 25–27 June 2001; Volume 6, pp. 4222–4227. [CrossRef]
12. Aksun-Güvenç, B.; Guvenc, L. The Limited Integrator Model Regulator and Its Use in Vehicle Steering Control. *Turk. J. Eng. Environ. Sci.* **2002**, *26*, 473–482.
13. Emirler, M.T.; Wang, H.; Güvenç, B.A.; Güvenç, L. Automated Robust Path Following Control Based on Calculation of Lateral Deviation and Yaw Angle Error. In Proceedings of the ASME 2015 Dynamic Systems and Control Conference, American Society of Mechanical Engineers Digital Collection, Columbus, OH, USA, 28–30 October 2015. [CrossRef]
14. Orun, B.; Necipoglu, S.; Basdogan, C.; Guvenc, L. State feedback control for adjusting the dynamic behavior of a piezoactuated bimorph atomic force microscopy probe. *Rev. Sci. Instrum.* **2009**, *80*, 063701. [CrossRef] [PubMed]
15. Demirel, B.; Güvenç, L. Parameter Space Design of Repetitive Controllers for Satisfying a Robust Performance Requirement. *IEEE Trans. Autom. Control* **2010**, *55*, 1893–1899. [CrossRef]
16. Oncu, S.; Karaman, S.; Guvenc, L.; Ersolmaz, S.S.; Ozturk, E.S.; Cetin, E.; Sinal, M. Robust Yaw Stability Controller Design for a Light Commercial Vehicle Using a Hardware in the Loop Steering Test Rig. In Proceedings of the 2007 IEEE Intelligent Vehicles Symposium, Istanbul, Turkey, 13–15 June 2007; pp. 852–859. [CrossRef]
17. Altan, O.D.; Wu, G.; Barth, M.J.; Boriboonsomsin, K.; Stark, J.A. GlidePath: Eco-Friendly Automated Approach and Departure at Signalized Intersections. *IEEE Trans. Intell. Veh.* **2017**, *2*, 266–277. [CrossRef]
18. Cantas, M.R.; Kavas, O.; Tamilarasan, S.; Gelbal, S.Y.; Guvenc, L. Use of Hardware in the Loop (HIL) Simulation for Developing Connected Autonomous Vehicle (CAV) Applications. In Proceedings of the WCX SAE World Congress Experience, Detroit, MI, USA, 9–11 April 2019; SAE International: Warrendale, PA, USA, 2019. [CrossRef]
19. Kavas-Torris, O.; Cantas, M.R.; Gelbal, S.Y.; Guvenc, L. Performance Evaluation of the Pass-at-Green (PaG) Connected Vehicle V2I Application. In Proceedings of the WCX SAE World Congress Experience, Detroit, MI, USA, 21–23 April 2020; p. 2020-01-1380. [CrossRef]
20. Kavas-Torris, O.; Cantas, M.R.; Cime, K.M.; Guvenc, B.A.; Guvenc, L. The Effects of Varying Penetration Rates of L4–L5 Autonomous Vehicles on Fuel Efficiency and Mobility of Traffic Networks. In Proceedings of the WCX SAE World Congress Experience, Detroit, MI, USA, 21–23 April 2020; p. 2020-01-0137. [CrossRef]
21. Sun, C.; Shen, X.; Moura, S. Robust Optimal ECO-driving Control with Uncertain Traffic Signal Timing. In Proceedings of the 2018 Annual American Control Conference (ACC), Milwaukee, WI, USA, 27–29 June 2018; pp. 5548–5553. [CrossRef]
22. Asadi, B.; Vahidi, A. Predictive Cruise Control: Utilizing Upcoming Traffic Signal Information for Improving Fuel Economy and Reducing Trip Time. *IEEE Trans. Control Syst. Technol.* **2011**, *19*, 707–714. [CrossRef]

23. Lin, Q.; Li, S.E.; Du, X.; Zhang, X.; Peng, H.; Luo, Y.; Li, K. Minimize the Fuel Consumption of Connected Vehicles Between Two Red-Signalized Intersections in Urban Traffic. *IEEE Trans. Veh. Technol.* **2018**, *67*, 9060–9072. [CrossRef]
24. Li, S.E.; Xu, S.; Huang, X.; Cheng, B.; Peng, H. Eco-Departure of Connected Vehicles With V2X Communication at Signalized Intersections. *IEEE Trans. Veh. Technol.* **2015**, *64*, 5439–5449. [CrossRef]
25. Bjornberg, A. Autonomous intelligent cruise control. In Proceedings of the IEEE Vehicular Technology Conference (VTC), Stockholm, Sweden, 8–10 June 1994; Volume 1, pp. 429–433. [CrossRef]
26. Shaout, A.; Jarrah, M.A. Cruise control technology review. *Comput. Electr. Eng.* **1997**, *23*, 259–271. [CrossRef]
27. Xiao, L.; Gao, F. A comprehensive review of the development of adaptive cruise control systems. *Veh. Syst. Dyn.* **2010**, *48*, 1167–1192. [CrossRef]
28. Pan, C.; Huang, A.; Chen, L.; Cai, Y.; Chen, L.; Liang, J.; Zhou, W. A review of the development trend of adaptive cruise control for ecological driving. *Proc. IMechE Part D J. Automob. Eng.* **2021**, *1*, 18. [CrossRef]
29. J3016B: Taxonomy and Definitions for Terms Related to Driving Automation Systems for on-Road Motor Vehicles—SAE International. Available online: https://www.sae.org/standards/content/j3016_201806/ (accessed on 8 December 2020).
30. Kural, E.; Güvenç, B.A. Model Predictive Adaptive Cruise Control. In Proceedings of the 2010 IEEE International Conference on Systems, Man and Cybernetics, Istanbul, Turkey, 10–13 October 2010; pp. 1455–1461. [CrossRef]
31. Wang, Z.; Wu, G.; Barth, M.J. A Review on Cooperative Adaptive Cruise Control (CACC) Systems: Architectures, Controls, and Applications. In Proceedings of the 2018 21st International Conference on Intelligent Transportation Systems (ITSC), Maui, HI, USA, 4–7 November 2018; pp. 2884–2891. [CrossRef]
32. Hu, X.; Wang, H.; Tang, X. Cyber-Physical Control for Energy-Saving Vehicle Following With Connectivity. *IEEE Trans. Ind. Electron.* **2017**, *64*, 8578–8587. [CrossRef]
33. Cantas, M.R.; Gelbal, S.Y.; Guvenc, L.; Guvenc, B.A. *Cooperative Adaptive Cruise Control Design and Implementation*; SAE Technical Paper 2019-01-0496; SAE International: Warrendale, PA, USA, 2019. [CrossRef]
34. Kianfar, R.; Augusto, B.; Ebadighajari, A.; Hakeem, U.; Nilsson, J.; Raza, A.; Tabar, R.S.; Irukulapati, N.V.; Englund, C.; Falcone, P.; et al. Design and Experimental Validation of a Cooperative Driving System in the Grand Cooperative Driving Challenge. *IEEE Trans. Intell. Transp. Syst.* **2012**, *13*, 994–1007. [CrossRef]
35. Rasool, E.; Liaquat, M.; Bhatti, A.I.; Mehmood, A. Cooperative Adaptive Cruise Control with Fuel Efficiency Using PMP Technique. In Proceedings of the 2019 5th International Conference on Control, Automation and Robotics (ICCAR), Beijing, China, 19–22 April 2019; pp. 396–400. [CrossRef]
36. Güvenç, L.; Uygan, I.M.C.; Kahraman, K.; Karaahmetoglu, R.; Altay, I.; Senturk, M.; Emirler, M.T.; Karci, A.E.H.; Guvenc, B.A.; Altug, E.; et al. Cooperative Adaptive Cruise Control Implementation of Team Mekar at the Grand Cooperative Driving Challenge. *IEEE Trans. Intell. Transp. Syst.* **2012**, *13*, 1062–1074. [CrossRef]
37. Naus, G.J.L.; Vugts, R.P.A.; Ploeg, J.; van de Molengraft, M.J.G.; Steinbuch, M. String-Stable CACC Design and Experimental Validation: A Frequency-Domain Approach. *IEEE Trans. Veh. Technol.* **2010**, *59*, 4268–4279. [CrossRef]
38. Zhai, C.; Chen, X.; Yan, C.; Liu, Y.; Li, H. Ecological Cooperative Adaptive Cruise Control for a Heterogeneous Platoon of Heavy-Duty Vehicles With Time Delays. *IEEE Access* **2020**, *8*, 146208–146219. [CrossRef]
39. Yang, H.; Rakha, H.; Ala, M.V. Eco-Cooperative Adaptive Cruise Control at Signalized Intersections Considering Queue Effects. *IEEE Trans. Intell. Transp. Syst.* **2017**, *18*, 1575–1585. [CrossRef]
40. Almannaa, M.H.; Chen, H.; Rakha, H.A.; Loulizi, A.; El-Shawarby, I. Field implementation and testing of an automated eco-cooperative adaptive cruise control system in the vicinity of signalized intersections. *Transp. Res. Part D Transp. Environ.* **2019**, *67*, 244–262. [CrossRef]
41. Zhang, F.; Xi, J.; Langari, R. Real-Time Energy Management Strategy Based on Velocity Forecasts Using V2V and V2I Communications. *IEEE Trans. Intell. Transp. Syst.* **2017**, *18*, 416–430. [CrossRef]
42. He, H.; Wang, Y.; Han, R.; Han, M.; Bai, Y.; Liu, Q. An improved MPC-based energy management strategy for hybrid vehicles using V2V and V2I communications. *Energy* **2021**, *225*, 120273. [CrossRef]
43. Ma, F.; Yang, Y.; Wang, J.; Li, X.; Wu, G.; Zhao, Y.; Wu, L.; Aksun-Guvenc, B.; Guvenc, L. Eco-driving-based cooperative adaptive cruise control of connected vehicles platoon at signalized intersections. *Transp. Res. Part D Transp. Environ.* **2021**, *92*, 102746. [CrossRef]
44. Kavas-Torris, O.; Cantas, M.R.; Gelbal, S.Y.; Aksun-Guvenc, B.; Guvenc, L. Fuel Economy Benefit Analysis of Pass-at-Green (PaG) V2I Application on Urban Routes with STOP Signs. *Int. J. Veh. Des.* **2021**, *83*, 258–279. [CrossRef]
45. Torris, O.K. Eco-Driving of Connected and Automated Vehicles (CAVs). Ph.D. Thesis, The Ohio State University, OhioLINK Electronic Theses and Dissertations Center, Columbus, OH, USA, 2022.
46. Kavas-Torris, O.; Guvenc, L. Modelling and Analysis of Car Following Algorithms for Fuel Economy Improvement in Connected and Autonomous Vehicles (CAVs). *arXiv* **2022**, arXiv:2203.12078.
47. Cantas, M.R.; Fan, S.; Kavas, O.; Tamilarasan, S.; Guvenc, L.; Yoo, S.; Lee, J.H.; Lee, B.; Ha, J. Development of Virtual Fuel Economy Trend Evaluation Process. In *WCX SAE World Congress Experience*; SAE International: Warrendale, PA, USA, 2019. [CrossRef]

48. Kim, J.-H.; Park, J.-H.; Lee, S.-W.; Chong, E.K.P. Control of systems with deadzones using PD controllers with fuzzy precompensation. In Proceedings of the 8th IEEE International Symposium on Intelligent Control, Chicago, IL, USA, 25–27 August 1993; pp. 451–456. [CrossRef]
49. Xia, X.; Meng, Z.; Han, X.; Li, H.; Tsukiji, T.; Xu, R.; Zheng, Z.; Ma, J. An automated driving systems data acquisition and analytics platform. *Transp. Res. Part C Emerg. Technol.* **2023**, *151*, 104120. [CrossRef]

sensors

MDPI

Article

Generating Edge Cases for Testing Autonomous Vehicles Using Real-World Data

Dhanoop Karunakaran [1], Julie Stephany Berrio Perez [2] and Stewart Worrall [2,*]

[1] Emerging Technologies, IAG, Sydney, NSW 2000, Australia; dhanoop.karunakaran@iag.com.au
[2] Australian Centre for Robotics, University of Sydney, Sydney, NSW 2008, Australia; stephany.berrioperez@sydney.edu.au
* Correspondence: stewart.worrall@sydney.edu.au

Abstract: In the past decade, automotive companies have invested significantly in autonomous vehicles (AV), but achieving widespread deployment remains a challenge in part due to the complexities of safety evaluation. Traditional distance-based testing has been shown to be expensive and time-consuming. To address this, experts have proposed scenario-based testing (SBT), which simulates detailed real-world driving scenarios to assess vehicle responses efficiently. This paper introduces a method that builds a parametric representation of a driving scenario using collected driving data. By adopting a data-driven approach, we are then able to generate realistic, concrete scenarios that correspond to high-risk situations. A reinforcement learning technique is used to identify the combination of parameter values that result in the failure of a system under test (SUT). The proposed method generates novel, simulated high-risk scenarios, thereby offering a meaningful and focused assessment of AV systems.

Keywords: autonomous vehicles; testing; edge case generation; scenario-based testing; parametric representation; data-driven method

Citation: Karunakaran, D.; Berrio Perez, J.S.; Worrall, S. Generating Edge Cases for Testing Autonomous Vehicles Using Real-World Data. *Sensors* **2024**, *24*, 108. https://doi.org/10.3390/s24010108

Academic Editors: Yafei Wang, Enrico Meli, Chao Huang, Hailong Huang, Zhiqiang Zuo, Bo Leng and Peng Hang

Received: 10 October 2023
Revised: 19 December 2023
Accepted: 21 December 2023
Published: 25 December 2023

1. Introduction

The Society of Automotive Engineers (SAE) outlines six levels of driving automation [1]. As automation levels increase, autonomous systems must ensure safety without human intervention. Therefore, comprehensive evaluations are crucial before deploying these systems on public roads. Unlike deterministic approaches like distance-based testing, scenario-based testing (SBT) offers a promising alternative by assessing systems against meaningful driving scenarios and reducing test efforts for safety assurance [2–4]. SBT concentrates on creating high-risk traffic situation test cases for evaluating a system's performance. In SBT, there are three levels of representations [5]:

- *Functional scenarios*, which comprise the highest abstraction layer and contain a linguistic description of a scenario.
- *Logical scenarios*, which detail the range and distribution of parameters that describe a specific event.
- *Concrete scenarios*, which arise from specifying a value for each parameter, and are sampled from the distribution defined in the logical scenario.

Moreover, concrete scenarios can be classified into the following three types based on their occurrence probability: *typical*, *critical*, and *edge case* scenarios.

- *Typical scenarios* represent common, real-world operating behaviors with a low likelihood of leading to high-risk situations.
- *Critical scenarios* entail higher-risk situations with safety concerns, occurring less frequently than *typical scenarios*.
- *Edge cases* refer to statistical outliers, presenting challenging scenarios that are rarely encountered in normal driving activities.

Identifying a combination of parameter values that describe realistic scenarios is challenging due to the complex, non-linear relationships between parameters. A considerable effort is required to pinpoint the optimal combination of parameters that accurately capture all aspects of a given scenario.

Highly automated vehicles (HAVs) will eliminate the need for human drivers, making it crucial to evaluate their operation in edge case scenarios to ensure safe deployment [6]. A representative test case, for example, is a cut-in scenario where the HAV might collide during a lane change.

In the literature, generating concrete scenarios for critical events is a burgeoning research area in SBT. Prior studies have centered on manually constructing simulated scenarios utilizing expert knowledge [7–9]. However, these methods may produce scenarios that do not accurately reflect real-world traffic participant behavior. Moreover, many existing techniques focus on scenario generation without giving adequate attention to edge cases [2,10,11]. A recent study by De Gelder et al. applied kernel density estimation to fit a distribution to three parameters from a given scenario [2]. This work generates test cases using Monte Carlo simulation but does not explicitly direct the search toward edge cases.

Generating high-risk scenarios for testing HAVs is important in ensuring their safety and reliability. These scenarios, particularly edge cases, are crucial for assessing how these vehicles perform in rare or unexpected situations, thereby establishing public trust and meeting regulatory safety standards. By simulating realistic driving conditions, including rare and unusual situations, these tests provide a comprehensive evaluation of autonomous systems, thus uncovering potential weaknesses that might not be evident in normal driving conditions. This approach is more efficient than traditional distance-based testing methods and crucial for the overall advancement of autonomous vehicle technology. It allows for focused and effective testing, thus ensuring the vehicles are well equipped to handle a wide range of driving scenarios.

This paper presents a method for generating realistic and challenging concrete scenarios to evaluate HAV subsystems using real-world data. We exemplify our approach using lane change events on urban roads. However, the proposed methodology can be extended to accommodate various road scenarios, thereby highlighting its versatility and applicability. In this work, we used the CARLA simulator, a tool that is engineered to facilitate autonomous urban driving system development, training, and validation [12]. This versatile software allows us to control all digital assets, which encompasses static and dynamic actors, thus enabling the creation or playback of diverse scenarios.

We transformed raw data into meaningful scenarios through a three-step process. First, we extracted lane change maneuvers from the collected data and converted them into a parameterized representation [13], thereby validating the parameterization's effectiveness using metrics to compare the parameterized scenario representation with the original real-world trajectory. Next, we constructed a search space, or *parameter space*, using extracted parameter values, and we then compared the following three different representations: independent univariate normal, multivariate normal, and a multimodal–multivariate distribution. Finally, we treated CARLA's open-source simulation tool's collision avoidance system (CAS) as the system under testing (SUT). We employed an RL method to optimize the parameter values for realistic and challenging scenarios, with the reward function biasing learning toward edge cases.

We summarize our paper's contributions as follows:

- A novel method for identifying and parameterizing real-world lane change scenarios, thereby demonstrating a strong resemblance between reproduced and real-world trajectories.
- A unique urban lane change dataset (81 cut-in and 53 cut-out) with raw parameter values and OpenSCENARIO representations for each event (available online).
- An extended data-driven methodology [14] for generating problematic concrete scenarios for a specific CAS, which was evaluated through quantitative and qualitative analysis.

2. Related Work

This section provides an overview of the previous works on scenario extraction methods, scenario datasets, and concrete scenario generation methods.

De Gelder et al. in [15] implemented a scenario extraction method that relies heavily on real-time tagging. Although the technique enabled automated tagging in real-time and used a combination of tags to mine scenarios, extracting scenarios using other published datasets is not easy. Moreover, the approach was highly dependent on the accuracy and consistency of the real-time tagging process, which may not be reliable. In contrast, Krajewski et al. in [16] demonstrated a more robust scenario extraction process for lane change maneuvers by measuring when the vehicles crossed a lane in the data collected from cameras fitted to drones. This method did not depend on real-time tagging, which made it more reliable and easier to apply to other datasets. Xinxin et al. in [17] presented a scenario extraction framework that relied on various computer vision techniques to extract scenarios from video data. While this approach showed promise, it also had limitations, particularly regarding the accuracy of the extracted scenarios. Similarly, the study of [18] proposed a methodology to generate concrete scenarios by extracting scenario parameters from the HighD dataset for assessing an active lane-keeping system (ALKS). While their approach was promising it also had limitations, particularly regarding the complexity of the generated scenarios and the need for accurate data to ensure their validity. Despite their limitations, both of the studies of [17,18] proposed scenario extraction frameworks and generated scenarios in a standard format, which may be helpful for researchers and practitioners in the field.

Several literature contributions have focused on the publication of verification and validation datasets. For instance, the Safety Pilot Model Deployment (SPMD) [19] program, launched by the University of Michigan Transportation Research Institute (UMTRI) with the support of various US departments, collected 73 miles of data and stored it in a text format. However, extracting scenarios from this dataset requires extensive post-processing, which can be time-consuming and inconvenient for end users [20]. Krajewski et al. [16] published the HighD dataset, one of the most extensive highway vehicle trajectory datasets. While it consists of over 16 h of measurements from six locations with around 100,000 vehicles, the trajectories were recorded from a bird's eye perspective. In the study of [18], a method to extract scenario parameters from the HighD dataset to create concrete scenarios was proposed. While this method is promising, it focuses primarily on highway scenarios, which limits its usefulness for researchers and practitioners interested in urban driving scenarios.

The generation of concrete scenarios has been explored in the literature, but these methods have limitations that must be considered. For instance, Barbier et al. proposed in [10] a technique to evaluate a system's behavior by formulating it as a statistical model-checking problem. While they computed the statistical characteristics of the system under test (SUT) by identifying the system failures in randomly generated scenarios in a simulation, they did not bias their search toward edge cases. In contrast, in [2], real-world scenarios were parameterized and stored in a database. Monte Carlo simulations were then employed to generate test cases from the parameterized representation. However, the Monte Carlo simulation is a random search. Depending on the scenario, it may be less likely to find a falsification result than a directed search, as described in this paper. Zhao et al. in [11] extracted a statistical model of vehicle behavior from real-world data in a lane following scenario, and they then skewed the statistical model to create more critical scenarios. The performance of this model-based approach depends on having an adequate model representation of real-world behavior, which may not always be the case.

Recent research has focused on generating concrete scenarios using a manually built combination of parameters. For instance, Gangopadhyay et al. proposed using Bayesian optimization (BO) to create challenging scenarios [7]. BO utilizes the Bayes rule to learn the model, and it then finds challenging scenarios by using the learned model. Similarly, in [8], an evolutionary algorithm (EA) was employed to search for critical scenarios from parameter space. Also, Zhou et al. [21] introduced a challenging scenario generation approach for automated driving systems through using genetic-based algorithms. Other researchers, such as Koren et al., Liu et al., and Lu et al. [9,22,23], have employed an RL-based approach to search for collision scenarios from a manually built search space. These methods do not account for real-world traffic participant behavior in the scenario generation process, as their parameter space does not consider the correlations between the different variables.

The authors in [24] presented a novel scenario generation method that utilizes kernel density estimation (KDE) to approximate the probability density function (PDF) of scenario parameters, thus enabling the generation of realistic scenarios. In addition to that, they introduced a novel metric to quantify the extent to which the generated scenarios reflect real-world scenarios. This work was limited as it did not explicitly demonstrate the generation of more critical scenarios.

In our previous work, ref. [14], we proposed a method to generate concrete scenarios for assessing the performance of CAS at pedestrian crossings. Our approach had the advantage of biasing the learning toward high-risk events, which enabled the creation of many concrete scenarios through using a parameter space built from expert knowledge. However, it did not fully encode realistic traffic participant behavior, as expected in a data-driven approach. In this paper, we address this limitation by building the parameter space from real-world data, which enables the generation of more realistic and challenging scenarios in the context of a lane change maneuver.

3. Background

This section introduces several fundamental concepts used in our approach to generate concrete scenarios for assessing the performance of collision avoidance systems. These concepts include REINFORCE RL, responsibility sensitive safety (RSS), OpenX formats, the data collection vehicle, coordinate frames, and the lane change scenario types used in our method.

- Reinforce RL

Reinforcement learning (RL) algorithms aim to find an optimal policy that maximizes reward by interacting with the environment modeled as a Markov decision process (MDP) [25]. RL is typically implemented in three ways: dynamic programming, Monte Carlo methods, and temporal difference learning. Our approach employs the Monte Carlo method REINFORCE, a policy gradient algorithm that directly manipulates the policy to find the optimal one that maximizes expected return [26]. In this algorithm, the policy is defined by the weights of the neural networks [27]. The learning process updates the weights to find the optimal policy that predicts the desired action given a state.

- Responsibility Sensitive Safety (RSS)

Responsibility Sensitive Safety (RSS) is a formal method proposed by Intel's Mobileye that computes the minimum distance required to keep a vehicle safe [28]. RSS aims to guarantee that an agent will not cause an accident rather than to ensure that an agent will not be involved in an accident [28]. Our work focuses on the safe longitudinal distance, as well as on the minimum distance required for the ego vehicle to stop in time if a vehicle or object in front brakes abruptly.

- OpenX Formats: OpenSCENARIO and OpenDRIVE

The OpenX formats, including OpenSCENARIO and OpenDRIVE, enable the construction of simulations based on real-world scenarios using programs that support the format [29–31]. These formats facilitate the sharing of test scenarios that have the potential to influence safety profoundly. OpenSCENARIO describes the dynamic contents, such as the behavior of the traffic participants and weather conditions, while OpenDRIVE can represent a road network and the surrounding environment. Our approach uses version 1.1 for both formats, stored as XML files.

- Data Collection Vehicle

Our data collection vehicle is a Volkswagen Passat station wagon fitted with Ibeo and SICK lidars. The Ibeo HAD feature fusion detection and tracking system provides tracking data for capturing the trajectory of dynamic traffic participants. The algorithms and methodology presented in this paper are not specific to this exact sensor arrangement. It is possible to employ this framework on a different platform with minor modifications to the code.

- Coordinate Frames

Our scenario extraction framework runs on the robot operating system (ROS). Two coordinate frames, *base_link* and *odom*, reference the vehicle's position in the environment. The Frenet reference frame is used to represent the positions of the traffic participants, thus enabling the trajectories and interactions to be described with fewer parameters.

Lane Change Scenario Types

Our approach considers two lane change scenarios: cut-in and cut-out. A cut-in scenario is when a vehicle moves into the ego vehicle's lane, while a cut-out scenario is when a front vehicle moves out of the ego vehicle's lane. These scenarios are essential for assessing the performance of collision avoidance systems.

4. Edge Case Focused Concrete Scenario Generation

This section describes our methodology for generating concrete scenarios, which involves extracting lane change scenarios from real-world data and generating edge case concrete scenarios. The process, as shown in Figure 1, consists of two stages.

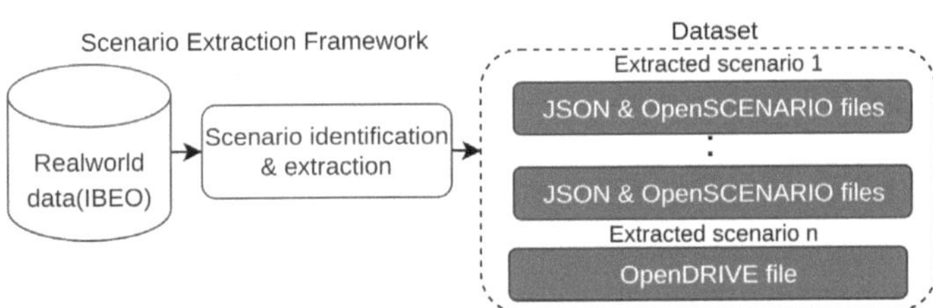

Figure 1. Architecture of the proposed method. The concrete scenario generation method is an end-to-end approach that involves identifying and converting real-world scenarios into a parametric representation to build a dataset and generate concrete scenarios from these parameters.

4.1. Lane Change Scenario Extraction

In the initial stage, we propose a novel approach for extracting lane change scenarios from real-world data and represent them in a parameterized form. The set of parameters, previously introduced in our work [13], characterizes the scenarios. We transform the extracted parameterized collections of lane change interactions into OpenSCENARIO files, thus allowing us to replay the trajectories in a simulator. The proposed scenario extraction technique utilizes the point clouds obtained from the rear, downwards-facing SICK lidar that are configured in a push-broom layout in combination with odometry and object tracking data. This module generates individual scenarios in the OpenSCENARIO format and their corresponding road structure in the OpenDRIVE format. The values of each parameter are stored in JSON files. The framework of the scenario extraction is depicted in Figure 2.

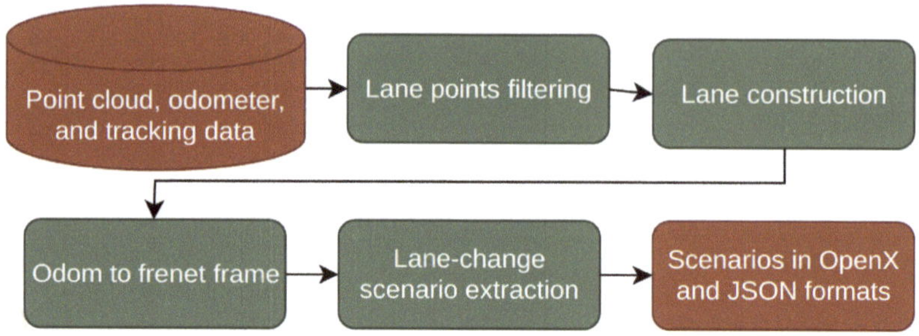

Figure 2. The scenario extraction approach pipeline involves the identification and parameterization of lane change scenarios in an open format using the data logged by the sensor system.

4.1.1. Tracking and Point Cloud Processing

The data collection vehicle is equipped with an object-tracking system to capture the road participants' trajectories in real-time. We used the *TF* transformation module in the robot operating system (ROS) to convert the tracking relative observations from the *base_link* (ego vehicle) frame to the *odom* (map) frame.

We used the lidar readings to find the location of the road lane markings, which helped us to find where and when a lane change occurs. To achieve this, we filtered the point cloud data from the push-broom lidar. Initially, the lidar points were sorted from the road center toward the lane boundaries. Then, we evaluated the first and second derivatives of the angle between the adjacent points to obtain the points hitting the road [32]. This allowed us to identify the curbs, as well as the separate road and non-road points. We grouped the points belonging to the lane markings by utilizing the intensity information in the point cloud. The reflective paint used to draw lane markings typically produces higher intensity readings than other road points.

After identifying and grouping the lidar points belonging to the lane markings, we converted them to the global *odom* frame and merged them to form the lanes. A more detailed explanation of this process can be found in [33]. We assigned a numerical label to each lane, and its location and tracking information were converted into a Frenet frame, where the data is represented in longitudinal s and lateral displacement t. This lane representation allows us to detect lane change scenarios based on specific criteria such as a lateral displacement to the lane center, which is not directly available in the Cartesian coordinate system.

4.1.2. Scenario Extraction Logic and Parameters

To detect the lane change scenarios, we compared the lane number and lateral distance of the tracked vehicles to the ego path at each time step. A cut-in scenario was detected when the lateral displacement between any front and side vehicle and the ego vehicle lane approached zero, while a cut-out scenario was detected when a tracked vehicle moved away from the ego vehicle's lane and the lateral displacement increased. The scenarios' duration was set from eight seconds prior to the lane crossing to four seconds ahead, thereby capturing a total of 12 s of trajectory for each scenario, which is longer than the average lane change duration estimated by Toledo et al. [34].

To parameterize the real-world lane change scenarios, we used the list of parameters from our previous work [13]. These parameters came from four control points: *scenario start*, *cut start*, *cut end*, and *scenario end*. The scenario set the initial parameters for the ego and adversary vehicles, and the remaining parameters configured the adversary vehicle's dynamic properties over time.

The output of this stage was the parametric representation of the extracted scenarios, which we used to create OpenSCENARIO files that described the trajectories in simulation. We also used OpenDRIVE files to describe the road network where the trajectories were executed. To replay the parameterized scenarios, we used the OpenSCENARIO player

Esmini [35]. Additionally, we stored the representations of each captured scenario in the JSON files to facilitate the construction of the parameter space in the subsequent stage. The collection of OpenSCENARIO, OpenDRIVE, and JSON files have been made publicly available [33].

4.2. Edge Case Scenario Generation

The proposed method for concrete scenario generation involves building a parameter search space based on the built scenario dataset and employing a reinforcement learning (RL)-based approach, as illustrated in Figure 3. Specifically, this work focused on cut-in scenarios, although the same process can be used for other road events.

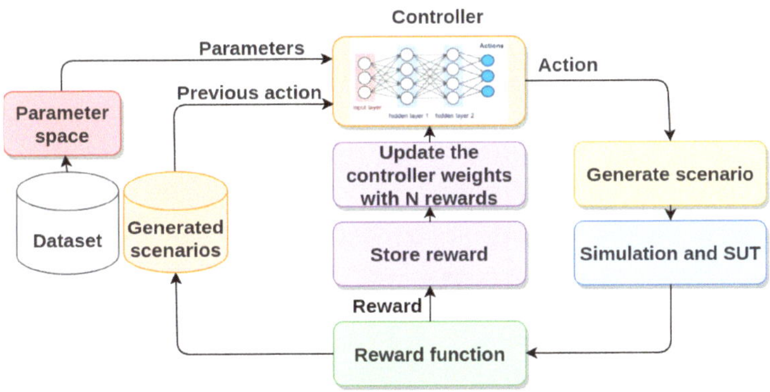

Figure 3. Architecture of the concrete scenario generation technique.

The parameter space comprises the set of possible values for each parameter to generate new scenarios. In our previous work, independent distributions were introduced for each of the five parameters, and these were randomly sampled to learn the combination of the values that led to problematic scenarios. However, this approach was unable to account for parameter correlations, thus potentially resulting in unrealistic scenarios. To overcome this issue, this paper employed a multivariate–multimodal distribution to model the values of seven parameters that describe the extracted scenarios, thereby providing a more realistic range of values that account for the correlations observed in real-world vehicle interactions. The parameters used to recreate lane-changing vehicle trajectories were adversary vehicle trigger distance, velocity at the cut start, duration from start to cut start, velocity at the cut end, time from cut start to cut end, final velocity, and the duration from cut end to scenario end.

Algorithm 1 outlines the RL-based technique for generating concrete scenarios. In earlier stages, for each episode, the controller predicts parameter values as actions to generate a new concrete scenario in simulation by sampling from the parameter space. In later stages, the controller indicates actions based on the learned policy.

The state in the RL context encapsulates the current conditions of the environment. For the concrete scenario generation focused on lane changing, the state can be represented as a vector of the current values of the following seven parameters: adversary vehicle trigger distance, velocity at the cut start, duration from start to cut start, velocity at the cut end, time from cut start to cut end, final velocity, and the duration from cut end to scenario end. Formally, the state at time t can be represented as

$$S_t = [d_{trigger}, v_{start}, t_{start}, v_{end}, t_{end}, v_{final}, t_{scenario}]. \tag{1}$$

The action taken by the controller is to generate a new set of parameter values for the next scenario. Thus, an action A_t at time t can be represented as a vector of the new parameter values:

$$A_t = [d'_{trigger}, v'_{start}, t'_{start}, v'_{end}, t'_{end}, v'_{final}, t'_{scenario}] \tag{2}$$

Algorithm 1: Pseudocode for the Proposed Concrete Scenario Generation Method

Result: Optimal policy π_θ in which controller can generate a challenging scenario
1. Build a multivariate, multimodal distribution as parameter space;
while *until the threshold* **do**
  if *exploration_decay > threshold* **then**
    2. The controller generates action to create a scenario in simulation by sampling from the distribution;
  end
  else
    3. The controller generates action to create a scenario in simulation using policy π_θ;
  end
  4. Run the scenario in simulation and get the reward;
  5. Store the action and reward for 50 episodes;
  if *episode%50 == 0* **then**
    6. Evaluate the gradient of the objective function J using the below expression:

$$\nabla J(\theta) \approx \frac{1}{N} \sum_{\tau \in N} \sum_{t=0}^{T-1} \nabla_\theta log\pi_\theta(A_t, S_t) R(\tau) \tag{3}$$

    Where N is the number of trajectories and R is the total return of a trajectory;
    7. Update the weights of the controller using the above function;
  end
end

The reward function guides the learning process by considering the minimum safe distance using RSS and the occurrence of collisions. The reward function quantifies the risk in the lane change maneuver. A high reward signifies a dangerous lane change, while a low reward means the vehicles are driving safely. At the end of each episode, the total number of high-risk timesteps is normalized between -0.1 and 0.1, which is denoted as *n_highrisk* in the reward function.

$$R = \begin{cases} n\_highrisk & non\ collision \\ 0.25 & collision \end{cases} \tag{4}$$

Through experimentation, we found that a smaller reward value improves exploration, resulting in a better chance of finding challenging scenarios, hence the values in the reward function. The controller is updated after storing action–reward pairs for 50 episodes, and the NN's weights are updated to lead toward an optimal policy. The controller operates for 2500 episodes in this configuration.

The objective function $J(\theta)$ was used to evaluate and improve the policy. It was approximated using the total return R of trajectories over a batch of episodes and $log\pi_\theta$. The $log\pi_\theta$ in the objective function is equivalent to categorical cross entropy, so we can use categorical cross entropy as a loss function. The loss function is multiplied with return R to enable the learning in the direction of the maximum return. The gradient of J with respect to θ guides the update of the policy parameters, and it can be represented as follows:

$$\nabla J(\theta) \approx \frac{1}{N} \sum_{\tau \in N} \sum_{t=0}^{T-1} \nabla_\theta log\pi_\theta(A_t, S_t) R(\tau). \tag{5}$$

The controller (policy) iteratively samples actions (sets of parameter values) from the parameter space, and the environment provides feedback in the form of rewards based on the riskiness of the generated lane-changing scenarios. The goal is to learn an optimal policy that maximizes the likelihood of generating challenging and realistic concrete scenarios.

5. Results

The proposed method uses a data-driven parameter space constructed from the parametric representations of real-world events to generate problematic scenarios. This section presents the results of the scenario extraction process from the collected data and the RL learning of high-risk scenarios.

5.1. Scenario Parameterization

We developed a novel technique for extracting lane change scenarios from real-world data. The parameterized cut-in scenarios were replayed using the Esmini player. The positions of the ego and challenging vehicles were recorded at six timesteps during a cut-in maneuver.

To illustrate the comparison between the real-world scenario and its corresponding simulated version, Figure 4a shows the trajectories of a cut-in scenario. The blue trajectory represents the real-world data, and the red represents the parameterized scenario. The numbers within the circles indicate the seconds that have elapsed since the beginning of the lane change maneuver. The initial longitudinal positions of both trajectories were similar. Toward the end, at the 9th second, we can see a slight variation in the longitudinal position. Figure 4b shows a comparison of a parameterized cut-out scenario and its corresponding real-world scenario. Our scenario extraction method allowed us to build a dataset of lane change events from real-world data, which consists of 81 cut-in scenarios and 53 cut-out scenarios.

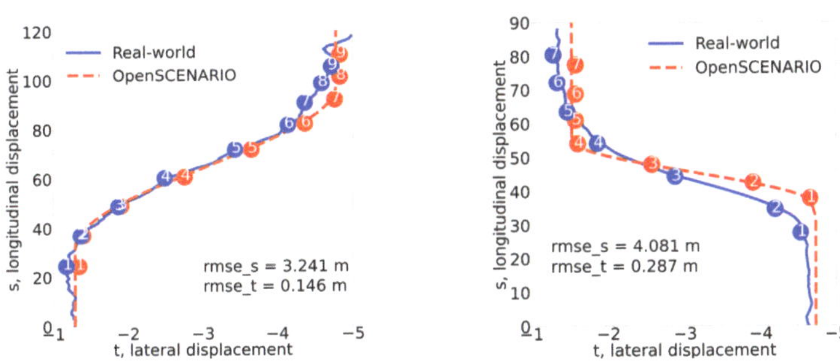

(a) (b)

Figure 4. Alignment of the temporal (seconds from the cut start indicated within circles) and spatial parameters between the real-world scenarios and their corresponding parameterized cut-in and cut-out scenarios. y and x axis correspond to longitudinal and lateral displacement in meters. (**a**) Cut in. (**b**) Cut out.

5.2. Training

The proposed method aims to maximize the reward function using RSS and a binary collision metric. The average reward was computed over 50 episodes. As shown in Figure 5, the reward tended to increase with additional iterations, thus indicating that the generated scenarios were of an increasingly higher risk. This suggests that the controller had learned to predict the action that generates a challenging scenario with maximum return, which, in this example, involved a collision.

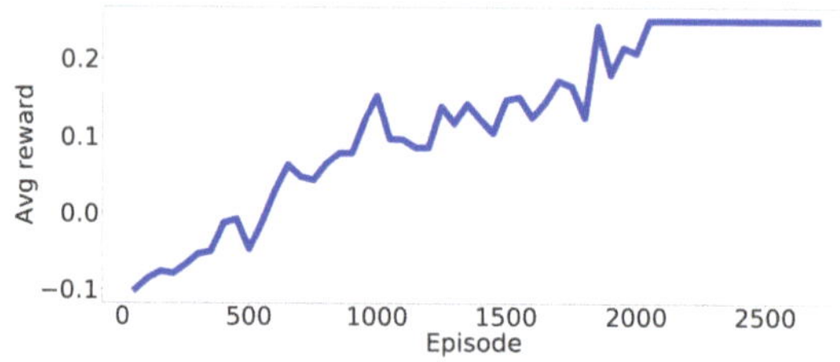

Figure 5. Average reward per episode showing the overall learning of the method.

Figure 6a,c show the highest-risk scenario learned by the method. As shown in Figure 6a, the adversary vehicle initiated a lane change maneuver around the 4th timestep and ended up in a collision with the ego vehicle on the 21st. The minimum safe distance computed using longitudinal RSS is always higher than the actual relative distance between the vehicles, thus indicating a higher risk profile. The parameter values for this scenario are shown in Table 1. Upon analysis, we found that the collision occurred due to the restricted field of view of the SUT, which failed to detect and respond to the adversary vehicle in the blind spot.

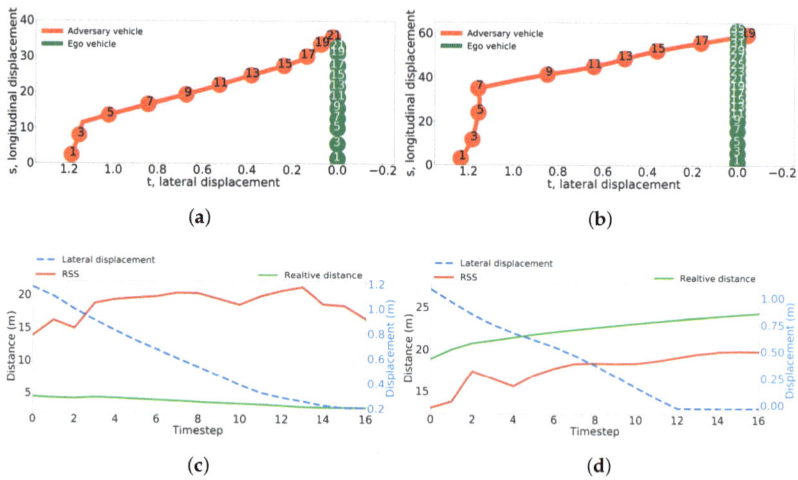

Figure 6. The method learned challenging and non-challenging cut-in scenarios from the multivariate, multimodal distribution, as illustrated in (**a**,**b**), where the y axis represents longitudinal displacement in meters, the x axis indicates lateral displacement in meters, and the numbers inside the circles denote the timesteps following the commencement of the cut-in. The safe threshold for RSS is also shown along with the actual relative distance. In cases where the RSS distance was lower than the actual distance, it indicated a dangerous scenario. (**a**) Challenging scenario representation. (**b**) Non-challenging scenario representation. (**c**) RSS performance in a challenging scenario. (**d**) RSS performance in a non-challenging scenario.

Table 1. Parameter values for a challenging scenario.

Parameter	Challenging
trigger_dist	4.0 m
cutin_vel	9.5 m/s
start_to_cutin_dist	8.0 m
cutin_end_vel	7.5 m/s
cutin_start_to_cutin_end_time	5.5 s
final_vel	8.0 m/s
cutin_end_to_final_time	3.0 s

In contrast, an example of a low-risk scenario is illustrated in Figure 6b,d. In Figure 6b, the adversary vehicle initiated a lane change maneuver at the 7th timestep and reached the ego vehicle's lane around the 18th timestep. The comparison of both vehicles' longitudinal positions at the 19th timestep indicated that the longitudinal position of the ego vehicle was safely behind the adversary vehicle. Figure 6d illustrates the risk profile of a non-challenging scenario, where the minimum safe distance computed by RSS is always lower than the relative distance between the ego vehicle and adversary vehicle, suggesting lower levels of risk.

As the seven parameters defined each scenario, the correlation between these parameters was encoded in a seven-dimensional space, which cannot be visualized directly. To show a subset of this distribution, we used two-dimensional plots, as illustrated in Figure 7. The green dots indicate the samples from the distribution that represent different collected scenarios, while the red dot shows the collision scenarios generated during the learning phase. The subfigures of Figure 7 illustrate the generated challenging scenarios and the distribution samples related to the following three pairs of scenario parameters: *cutin_vel* over *trigger_dist* (Figure 7a), *start_to_cutin_time* over *cutin_vel* (Figure 7b), and *cutin_start_to_end_time* over *cutin_end_vel* (Figure 7c). Our proposed method utilizes correlated data during the learning phase, thus leading to generated challenging scenarios that are consistent with the parameter space. In contrast, random sampling without considering the correlation can lead to unrealistic scenarios that are unlikely to occur in the real world.

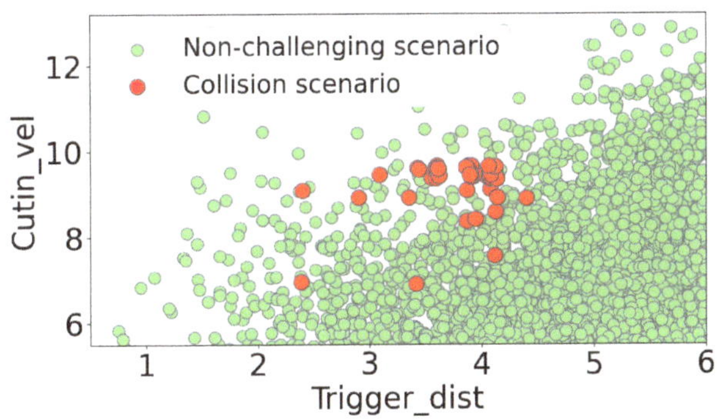

(**a**)

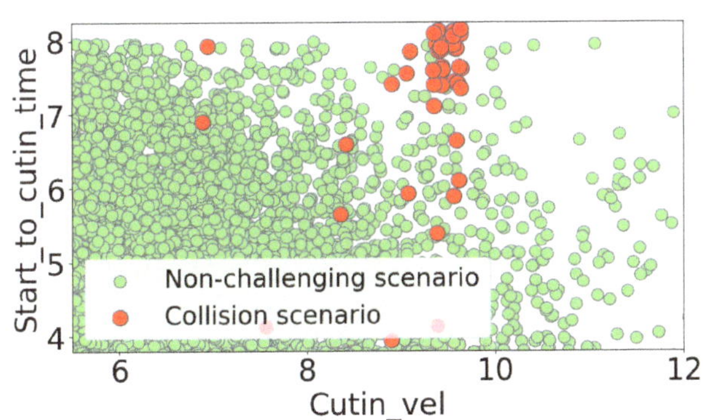

(**b**)

Figure 7. *Cont.*

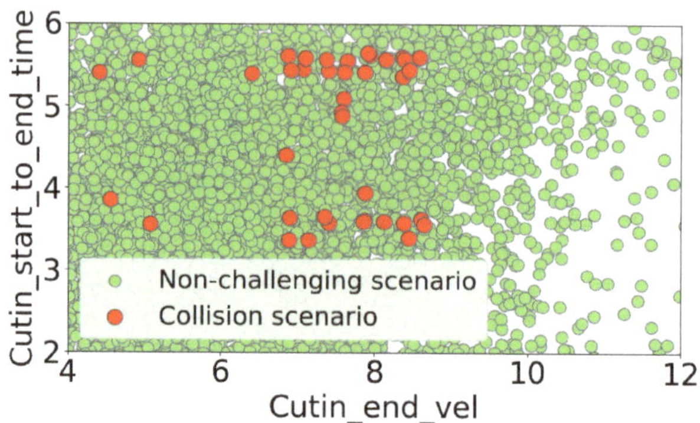

(c)

Figure 7. These graphs illustrate the multivariate, multi-modal distribution that represents the seven-dimensional parametric representation of the dataset. Green dots indicate normal scenarios, while red dots indicate collisions. The distribution is not uniform, and the correlations between the parameters are captured. The generated scenarios from this distribution are more realistic than those from uniform random sampling as they are more similar to the captured dataset. The parameter pairs shown in the graphs highlight the different correlations in the dataset. (**a**) Cutin_vel vs. trigger_dist; (**b**) start_to_cutin_time vs cutin_vel; and (**c**) cutin_start_to_end vs. cutin_end_vel.

5.3. Experiments

We compared three parameter spaces using independent univariate, multivariate normal, and multivariate multimodal distributions. In the first experiment, we fitted an independent multimodal distribution to each parameter. We used kernel density estimation to fit a multivariate normal distribution for the second experiment, while the third experiment used a multimodal, multivariate distribution to build the parameter space and to sample new high-risk scenarios.

For the first experiment, we created a scenario by combining random samples from each univariate distribution. The resulting learned high-risk scenario, as shown in Figure 8a,b, depicted an adversary vehicle initiating a lane change maneuver when it was ahead of the ego vehicle but ending up in a collision by hitting the back of the ego vehicle. However, this type of interaction was not close to any of the scenarios from the real-world dataset, and it was not representative of a common scenario.

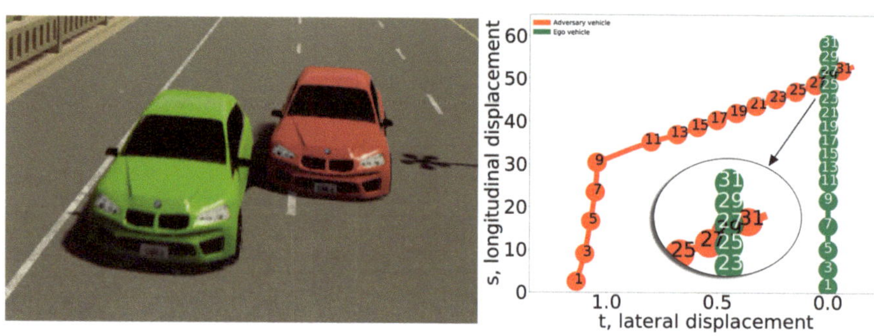

(a) (b)

Figure 8. *Cont.*

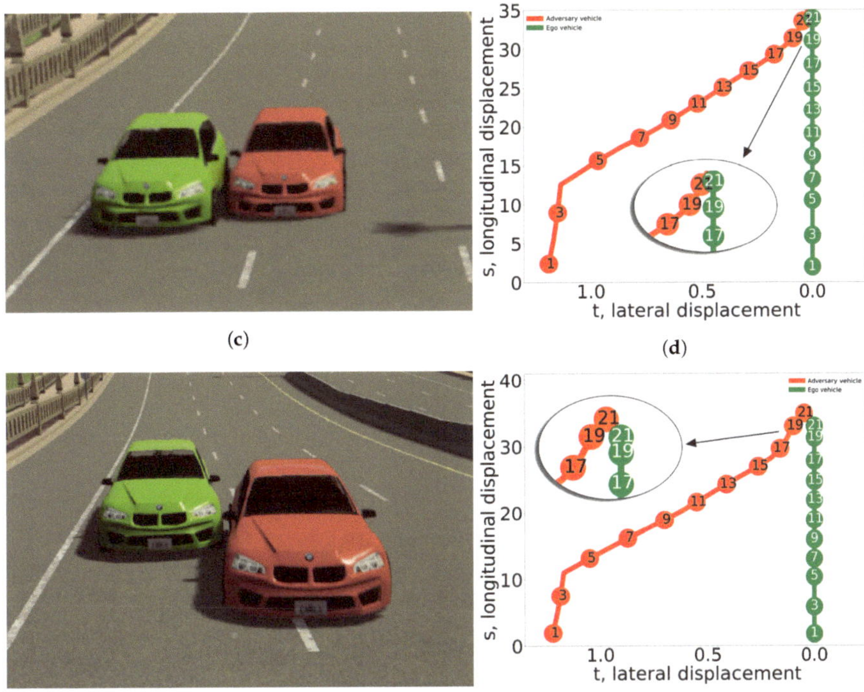

(**c**) (**d**)

(**e**) (**f**)

Figure 8. A qualitative analysis was conducted on the three different parameter spaces constructed using different representations of the dataset. The RL process output was used for each parameter space and a comparison was made. Figures (**a**,**b**) represent the learned scenario based on the independent univariate distribution, (**c**,**d**) represent the multivariate normal distribution, and (**e**,**f**) represent the multivariate multimodal distribution. The third parameter space, which is multivariate and multimodal, generated a collision interaction that closely matched the trajectory of the real-world data. The y axis represents longitudinal displacement in meters and the x axis represents lateral displacement in meters, with the numbers inside the circles indicating the timesteps following the commencement of the maneuver. (**a**) Independent. (**b**) Trajectory. (**c**) Multivariate. (**d**) Trajectory. (**e**) Multimodal. (**f**) Trajectory.

For the second experiment, we used kernel density estimation (KDE) to fit a multivariate normal distribution that incorporated the correlations between the different variables. The resulting collision scenario, as shown in Figure 8c,d, was closer to the trajectories collected in the real-world dataset. However, the vehicle collided with the side of the ego vehicle, and the adversary vehicle was slightly behind the ego vehicle at the 21st timestep, thereby resulting in a side collision. The sampling from the multivariate, unimodal normal distribution to the data was not likely to accurately represent the original data as most of the parameters were multimodal.

For the third and final experiment, we used a multimodal, multivariate distribution to describe the parameters in the dataset. We created these distributions to build the parameter space and sampled from them to generate and extract new high-risk scenarios. The resulting trajectory from our RL-based algorithm, as shown in Figure 8e, was much closer to the original real-world dataset. The learned parameters were close to the original data, and the algorithm was able to find a collision that was a cut-in scenario where the adversary vehicle misjudged the merge and was slightly too close to the ego vehicle. Figure 8f shows that the adversary vehicle began the lane change maneuver around the 4th timestep and collided with the ego vehicle around the 21st.

We conducted a quantitative analysis to compare the three different parameter spaces based on their likelihood estimates. We employed the RL-based algorithm to generate high-risk scenarios and used the Python library *scipy* to compute the likelihoods. We compared the likelihoods of the scenarios generated using independent univariate, multivariate normal and multivariate multimodal distributions. We found that the multivariate, multimodal distribution had the highest likelihood value of existing within the real-world data, which we believe better represents the collected data. Therefore, we used it as the ground truth model to compare the likelihood of a learned high-risk scenario. We fitted the distributions using the *gaussian_kde* function, and we computed the likelihood of the selected scenarios in the distribution using the pdf function. The likelihood for the converged RL scenario using the independent univariate distribution was extremely low (e^{-95}). This result indicated that the learned scenarios based on this approach significantly deviated from the original real-world dataset. The experiment based on the multivariate normal parameter space gave a significantly higher likelihood of existing within the real-world data (e^{-17}). However, the likelihood was still far lower than the multivariate, multimodal distribution, which had the highest likelihood value (e^{-12}), thus indicating that the scenarios generated using this distribution were more realistic and representative of the original dataset.

6. Conclusions and Future Work

We propose a method that generates realistic and challenging scenarios in simulation by using a data-driven parameter space and RL-based technique. Parameterized scenarios from real-world data enable the reproduction of events and the building of data-driven parameter spaces that encode realistic traffic behaviors.

We compared the following three parameter spaces for generating high-risk scenarios in autonomous vehicle testing: independent univariate, multivariate normal, and multivariate multimodal distributions. The first experiment, using univariate distributions, produced unrealistic scenarios that did not match real-world data. The second experiment, employing a multivariate normal distribution, yielded more realistic side collision scenarios but still lacked accuracy in representing complex real-world situations. The third experiment was most successful using a multimodal, multivariate distribution, which closely mirrored real-world driving scenarios (particularly in simulating realistic lane change maneuvers).

Our quantitative analysis reinforced these findings, with the multimodal, multivariate distribution showing the highest likelihood of resembling real-world scenarios. This contrasted with the low likelihood values of the scenarios generated from univariate distributions. The multivariate normal distribution was better but less effective than the multimodal approach. Our research demonstrates that a multivariate, multimodal distribution is the most effective in creating realistic and challenging scenarios for autonomous vehicle testing, which is crucial for ensuring the safety and reliability of these systems in real-world conditions.

In future work, we plan to integrate more complex traffic situations to enhance the robustness of our models. We also see potential in developing a more interactive simulation environment, where the autonomous vehicle's responses can dynamically alter the scenario in real-timedh, thus providing a more comprehensive assessment of its decision-making capabilities.

Author Contributions: Conceptualization, D.K., J.S.B.P. and S.W.; Methodology, D.K., J.S.B.P. and S.W.; Software, D.K.; Validation, J.S.B.P. and S.W.; Formal analysis, D.K.; Writing—original draft, D.K.; Writing—review & editing, J.S.B.P. and S.W.; Supervision, J.S.B.P. and S.W.; Project administration, S.W. All authors have read and agreed to the published version of the manuscript.

Funding: This work has been funded by the Australian Centre for Robotics (ACFR), University of Sydney and Insurance Australia Group (IAG) and iMOVE CRC and supported by the Cooperative Research Centres program, an Australian Government initiative.

Data Availability Statement: The code, dataset, and videos are available online at https://github.com/dkarunakaran/concrete_scenario_generation_real_world.

Conflicts of Interest: Author Dhanoop Karunakaran was employed by the company IAG. The remaining authors declare that the research was conducted in the absence of any commercial or financial relationships that could be construed as a potential conflict of interest.

References

1. Standard, SAE. J3016-Taxonomy and Definitions for Terms Related to Driving Automation Systems for on-Road Motor Vehicles. Available online: https://www.sae.org/standards/content/j3016_202104/preview/ (accessed on 22 December 2023).
2. de Gelder, E.; Paardekooper, J.P. Assessment of automated driving systems using real-life scenarios. In Proceedings of the 2017 IEEE Intelligent Vehicles Symposium (IV), Los Angeles, CA, USA, 11–14 June 2017; pp. 589–594.
3. Elrofai, H.; Worm, D.; den Camp, O.O. Scenario identification for validation of automated driving functions. In *Advanced Microsystems for Automotive Applications*; Springer: Berlin/Heidelberg, Germany, 2016; pp. 153–163.
4. Ponn, T.; Breitfuß, M.; Yu, X.; Diermeyer, F. Identification of Challenging Highway-Scenarios for the Safety Validation of Automated Vehicles Based on Real Driving Data. In Proceedings of the 2020 Fifteenth International Conference on Ecological Vehicles and Renewable Energies (EVER), Monte-Carlo, Monaco, 10–12 September 2020; pp. 1–10.
5. Menzel, T.; Bagschik, G.; Maurer, M. Scenarios for development, test and validation of automated vehicles. In Proceedings of the 2018 IEEE Intelligent Vehicles Symposium (IV), Changshu, China, 26–30 June 2018; pp. 1821–1827.
6. Koopman, P.; Kane, A.; Black, J. Credible autonomy safety argumentation. In Proceedings of the 27th Safety-Critical Systems Symposium, Bristol, UK, 5–7 February 2019.
7. Gangopadhyay, B.; Khastgir, S.; Dey, S.; Dasgupta, P.; Montana, G.; Jennings, P. Identification of test cases for automated driving systems using Bayesian optimization. In Proceedings of the 2019 IEEE Intelligent Transportation Systems Conference (ITSC), Auckland, New Zealand, 27–30 October 2019; pp. 1961–1967.
8. Althoff, M.; Lutz, S. Automatic generation of safety-critical test scenarios for collision avoidance of road vehicles. In Proceedings of the 2018 IEEE Intelligent Vehicles Symposium (IV), Changshu, China, 26–30 June 2018; pp. 1326–1333.
9. Koren, M.; Alsaif, S.; Lee, R.; Kochenderfer, M.J. Adaptive stress testing for autonomous vehicles. In Proceedings of the 2018 IEEE Intelligent Vehicles Symposium (IV), Changshu, China, 26–30 June 2018; pp. 1–7.
10. Barbier, M.; Renzaglia, A.; Quilbeuf, J.; Rummelhard, L.; Paigwar, A.; Laugier, C.; Legay, A.; Ibañez-Guzmán, J.; Simonin, O. Validation of Perception and Decision-Making Systems for Autonomous Driving via Statistical Model Checking. In Proceedings of the 2019 IEEE Intelligent Vehicles Symposium (IV), Paris, France, 9–12 June 2019; pp. 252–259. [CrossRef]
11. Zhao, D.; Huang, X.; Peng, H.; Lam, H.; LeBlanc, D.J. Accelerated evaluation of automated vehicles in car-following maneuvers. *IEEE Trans. Intell. Transp. Syst.* **2017**, *19*, 733–744. [CrossRef]
12. Dosovitskiy, A.; Ros, G.; Codevilla, F.; Lopez, A.; Koltun, V. CARLA: An Open Urban Driving Simulator. In Proceedings of the 1st Annual Conference on Robot Learning, Mountain View, CA, USA, 13–15 November 2017; pp. 1–16.
13. Karunakaran, D.; Berrio, J.S.; Worrall, S.; Nebot, E. Parameterisation of lane-change scenarios from real-world data. In Proceedings of the 2022 IEEE 25th International Conference on Intelligent Transportation Systems (ITSC), Macau, China, 8–12 October 2022.
14. Karunakaran, D.; Berrio, J.S.; Worrall, S.; Nebot, E.M. Critical concrete scenario generation using scenario-based falsification. *arXiv* **2022**, arXiv:1708.06374.
15. De Gelder, E.; Manders, J.; Grappiolo, C.; Paardekooper, J.P.; Den Camp, O.O.; De Schutter, B. Real-world scenario mining for the assessment of automated vehicles. In Proceedings of the 2020 IEEE 23rd International Conference on Intelligent Transportation Systems (ITSC), Rhodes, Greece, 20–23 September 2020; pp. 1–8.
16. Krajewski, R.; Bock, J.; Kloeker, L.; Eckstein, L. The highd dataset: A drone dataset of naturalistic vehicle trajectories on german highways for validation of highly automated driving systems. In Proceedings of the 2018 21st International Conference on Intelligent Transportation Systems (ITSC), Maui, HI, USA, 4–7 November 2018; pp. 2118–2125.
17. Xinxin, Z.; Fei, L.; Xiangbin, W. CSG: Critical Scenario Generation from Real Traffic Accidents. In Proceedings of the 2020 IEEE Intelligent Vehicles Symposium (IV), Las Vegas, NV, USA, 19 October–13 November 2020; pp. 1330–1336.
18. Tenbrock, A.; König, A.; Keutgens, T.; Weber, H. The ConScenD Dataset: Concrete Scenarios from the highD Dataset According to ALKS Regulation UNECE R157 in OpenX. In Proceedings of the 2021 IEEE Intelligent Vehicles Symposium Workshops (IV Workshops), Nagoya, Japan, 11–17 July 2021; pp. 174–181.
19. Bezzina, D.; Sayer, J. *Safety Pilot Model Deployment: Test Conductor Team Report*; Report No. DOT HS; National Highway Traffic Safety Administration: Washington, DC, USA, 2014; Volume 812, p. 18.
20. Zhao, D.; Guo, Y.; Jia, Y.J. Trafficnet: An open naturalistic driving scenario library. In Proceedings of the 2017 IEEE 20th International Conference on Intelligent Transportation Systems (ITSC), Yokohama, Japan, 16–19 October 2017; pp. 1–8.
21. Zhou, R.; Liu, Y.; Zhang, K.; Yang, O. Genetic algorithm-based challenging scenarios generation for autonomous vehicle testing. *IEEE J. Radio Freq. Identif.* **2022**, *6*, 928–933. [CrossRef]
22. Liu, Y.; Zhang, Q.; Zhao, D. A Reinforcement Learning Benchmark for Autonomous Driving in Intersection Scenarios. In Proceedings of the 2021 IEEE Symposium Series on Computational Intelligence (SSCI), Orlando, FL, USA, 5–7 December 2021; pp. 1–8.

23. Lu, C. Test Scenario Generation for Autonomous Driving Systems with Reinforcement Learning. In Proceedings of the 2023 IEEE/ACM 45th International Conference on Software Engineering: Companion Proceedings (ICSE-Companion), Melbourne, Australia, 14–20 May 2023; pp. 317–319.
24. de Gelder, E.; Hof, J.; Cator, E.; Paardekooper, J.P.; den Camp, O.O.; Ploeg, J.; De Schutter, B. Scenario parameter generation method and scenario representativeness metric for scenario-based assessment of automated vehicles. *IEEE Trans. Intell. Transp. Syst.* **2022**, *23*, 18794–18807. [CrossRef]
25. Van Otterlo, M.; Wiering, M. Reinforcement learning and markov decision processes. In *Reinforcement Learning*; Springer: Berlin/Heidelberg, Germany, 2012; pp. 3–42.
26. Levine, S. Policy Gradients. In Proceedings of the CS 294-112: Deep Reinforcement Learning. Available online: http://rail.eecs.berkeley.edu/deeprlcourse-fa18/ (accessed on 22 December 2023).
27. Matlab. Reinforcement Learning Using Deep Neural Networks. 2023. Available online: https://www.mathworks.com/help/deeplearning/ug/reinforcement-learning-using-deep-neural-networks.html# (accessed on 1 December 2023).
28. Shalev-Shwartz, S.; Shammah, S.; Shashua, A. On a Formal Model of Safe and Scalable Self-driving Cars. *CoRR* **2017**, abs/1708.06374.
29. OpenSCENARIO—Association for Standardization of Automation and Measuring Systems. Available online: https://www.asam.net/standards/detail/openscenario/ (accessed on 6 June 2023).
30. OpenDRIVE—Association for Standardization of Automation and Measuring Systems. Available online: https://www.asam.net/standards/detail/opendrive/ (accessed on 6 June 2023).
31. MSC. Open Standards—Essential for Self-Driving. Available online: https://d10tcz9jtwksbg.cloudfront.net/wp-content/uploads/2020/02/Open-Standards-Essential-for-Self-Driving.pdf (accessed on 5 June 2022).
32. Zhou, W.; Berrio, J.S.; Worrall, S.; Nebot, E. Automated Evaluation of Semantic Segmentation Robustness for Autonomous Driving. *IEEE Trans. Intell. Transp. Syst.* **2020**, *21*, 1951–1963. [CrossRef]
33. Karunakaran, D. Techniques to Generate Lanes and OpeSCENARIO & OpenDRIVE Files. Available online: https://github.com/dkarunakaran/concrete_scenario_generation_real_world/blob/master/README.md (accessed on 4 December 2022).
34. Toledo, T.; Zohar, D. Modeling duration of lane changes. *Transp. Res. Rec.* **2007**, *1999*, 71–78. [CrossRef]
35. esmini OpenSCENARIO Player. Available online: https://github.com/esmini/esmini (accessed on 4 June 2023).

sensors

MDPI

Article

Post-Takeover Proficiency in Conditionally Automated Driving: Understanding Stabilization Time with Driving and Physiological Signals

Timotej Gruden *, Sašo Tomažič and Grega Jakus

Faculty of Electrical Engineering, University of Ljubljana, Tržaška cesta 25, 1000 Ljubljana, Slovenia; saso.tomazic@fe.uni-lj.si (S.T.); grega.jakus@fe.uni-lj.si (G.J.)
* Correspondence: timotej.gruden@fe.uni-lj.si; Tel.: +386-1-4768-951

Abstract: In the realm of conditionally automated driving, understanding the crucial transition phase after a takeover is paramount. This study delves into the concept of post-takeover stabilization by analyzing data recorded in two driving simulator experiments. By analyzing both driving and physiological signals, we investigate the time required for the driver to regain full control and adapt to the dynamic driving task following automation. Our findings show that the stabilization time varies between measured parameters. While the drivers achieved driving-related stabilization (winding, speed) in eight to ten seconds, physiological parameters (heart rate, phasic skin conductance) exhibited a prolonged response. By elucidating the temporal and cognitive dynamics underlying the stabilization process, our results pave the way for the development of more effective and user-friendly automated driving systems, ultimately enhancing safety and driving experience on the roads.

Keywords: takeover; stabilization; conditionally automated driving; driving simulator; user study; physiology

Citation: Gruden, T.; Tomažič, S.; Jakus, G. Post-Takeover Proficiency in Conditionally Automated Driving: Understanding Stabilization Time with Driving and Physiological Signals. *Sensors* **2024**, *24*, 3193. https://doi.org/10.3390/s24103193

Academic Editors: Chao Huang, Yafei Wang, Peng Hang, Zhiqiang Zuo, Bo Leng and Hailong Huang

Received: 27 March 2024
Revised: 15 May 2024
Accepted: 16 May 2024
Published: 17 May 2024

1. Introduction

In recent years, the advent of conditionally automated vehicles (Society of Automotive Engineers—SAE Level 3 [1]) has promised transformative shifts in transportation, offering the potential to enhance road safety, efficiency, and convenience [2–5]. Conditionally automated driving systems, which delegate all driving tasks to the vehicle only when the appropriate conditions are met but still require human supervision, promise significant potential to alleviate driver fatigue, reduce human error, and mitigate traffic congestions. However, realizing these benefits hinges upon a critical factor: the seamless transition of control between the automated driving mode and human operation, commonly referred to as the "takeover" process.

While conditionally automated vehicles offer the allure of a more relaxed driving experience, they also introduce unique challenges, particularly during the handover of control from the automation to the human driver. Maggi et al. define takeover as "the process with which one agent takes back control of part or all of the dynamic driving task" and handover as "the parallel process with which one agent relinquishes part or all of the dynamic driving task" [6]. This transition phase represents a period of increased risk and uncertainty, where drivers must swiftly re-engage in the driving task to ensure safety. Research has highlighted the potential pitfalls associated with takeover situations, including delayed reaction times, reduced situational awareness, and increased likelihood of accidents [7–13]. Multitasking, for example, is a problem already present in conventional vehicles and is only likely to increase with the introduction of conditionally automated vehicles [14,15].

A crucial but understudied aspect of takeover scenarios is the concept of stabilization time: the time it takes a driver to regain full control and exhibit consistent, safe driving behavior following a transition from automated to manual driving. Understanding stabilization time is paramount for optimizing the design and implementation of conditionally automated systems, as it directly impacts the safety and user experience of such technologies.

In this article, we aim to address the pressing need for a comprehensive investigation of stabilization time in conditionally automated driving environments. With driving simulator user studies and advanced physiological signals analysis, we seek to determine and elucidate the factors influencing stabilization time. By shedding light on the temporal and cognitive dynamics during the takeover process, our research seeks to enhance the efficacy and acceptance of conditionally automated vehicles, paving the way for safer and more seamless integration into our transportation systems.

1.1. Related Work

The integration of automated vehicles into existing transportation systems necessitates a comprehensive regulatory framework to ensure their safe and responsible use. The planned amendment to the Vienna Convention on Road Traffic [16] provides that a takeover can be requested at any time either by the vehicle or by the driver. According to the SAE J3016 standard [1], conditionally automated vehicles must provide "the sustained and operational design domain-specific performance of the entire dynamic driving task under normal operation with the expectation that the user is receptive to automated driving-system requests to intervene, as well as to relevant system failures in other vehicle systems, and will respond appropriately." This does not require the driver to monitor the journey while the system is in operation. Working party no. 29 of the World Forum for Harmonization of Rules on Vehicles and their Systems and Units proposes amendments to the United Nations' regulations on automated lane keeping systems (UN Regulation No. 157 and UN Regulation No. 79) [17]. In this context, it was noted that the rules for fully automated systems which can decide on the appropriate maneuver to take in relation to the traffic situation have not yet been finalized [18]. The International Organization for Standardization (ISO) offers two standards on this topic: ISO 26262-1:2018 (Road vehicles—Functional safety) [19] and ISO 21448:2022 (Road vehicles—Safety of the intended functionality) [20]. These two standards complement each other and together form the basis for safety and advanced support systems in first- and second-level automated vehicles, but do not interfere with further levels of automation. Therefore, the transition to further levels of automation and the associated problem of the takeover process are still under research.

1.1.1. On Takeover Process

If proper conditions are met, e.g., a dedicated road infrastructure is in place, normal weather and visibility conditions prevail, advanced driver assistance systems (ADAS) are working, etc., a conditionally automated vehicle drives in fully automated mode. Meanwhile, the driver can engage in other tasks, such as writing emails, reading a book, etc. In case of a predictable situation (e.g., exiting a highway) or an unpredictable situation (sensor failure, traffic collision, road works, extreme weather conditions, etc.) that the vehicle cannot safely manage, it alerts the driver and the driver must take over. Normally, a takeover (TO) process starts with a takeover request (TOR) issued by the vehicle.

The timing of the TOR is one of the crucial parameters of every TO process [8]. The takeover lead time (TORlt) is defined as the time interval between the TOR and the anticipated situation requiring a TO, e.g., a collision or the end of the road. Researchers around the globe use different TORlts in their research. Sanghavi et al. showed that drivers' reactions were fastest when using three seconds as the TORlt, but the safest and least demanding reactions were provoked with a TORlt of seven seconds [21]. Shi et al. similarly showed that drivers elicited the overall best reactions when using a TORlt of six seconds [22], while Tan and Zhang concluded that drivers' situational awareness was best when using a TORlt between 16 and 30 s [23]. Eriksson and Stanton conducted a literature review and found that the most commonly used TORlts among the studies reviewed were 3, 4, 5, 6, 7, and 9 s [24].

After the vehicle initiates a TOR, the driver is responsible for taking over in a decent time, referred to as reaction time (RT). It is most commonly measured from the TOR until the driver turns the steering wheel more than two degrees (steering reaction time) or presses the brake pedal more than 10% (brake reaction time) [25]. For a deeper understanding, RT could be further divided into shorter intervals, e.g., for gaze switching from a secondary task to driving, gathering information (becoming aware of the situation), and deciding on the appropriate action [26]. According to the situation awareness (SA) theory [27], which is often used in vehicular human–machine interaction planning [28,29], achieving SA entails perception of environmental elements (SA level 1), comprehension of their significance (SA level 2), and the ability to anticipate their future status (SA level 3). Therefore, some researchers divide the driver's readiness to take over the vehicle into visual, mental, and physical readiness [9,26].

Regarding the user interfaces (UIs) for issuing the TOR, researchers first explored the appropriate modalities. Auditory interfaces (beeps) are known to provoke the fastest reactions [30], while tactile patterns [31–33] and ambient light [34,35] reduce drivers' effort and increase their situational awareness. Most concluded that multimodal user interfaces should be used when issuing a TOR [36,37]. If researchers' conclusions are consistent in terms of UI modality, we did not find this to be the case for the different types of stimuli. Auditory beeps represent the most often used stimuli [38–40]; Stojmenova et al. reported that a pure tone of 4000 Hz provoked the fastest reactions during driving experiments [40]. Politis et al. recommend abstract visual cues in non-urgent situations and additional auditory warning for increased urgency of the situation [38]. Among tactile interfaces, meaningful stimuli (e.g., different patterns) generally do not improve takeover performance, but could increase the driver's mental load [39,41]. Wu et al. showed that including the recommended steering direction in the TOR increases TO safety by decreasing the reaction time and reducing mental workload [42]. Additionally, Kraut et al. recently reported that assertive TO requests lead to shorter reaction times without finding any other effect on driver performance, stress, and subjective perceptions [43].

While the driver is taking over the demanding driving task, they can already observe the road to become aware of the situation as soon as possible. Various support systems besides mere TOR mechanism would be of additional benefit due to lack of time, possible fatigue, bad weather conditions, and other factors that could occur during the TO [44]. Recent related studies considered extended reality (XR) technologies in the first few moments of TO [44], steering wheel with torque guidance mechanism [45], and additional gradual braking systems [46]. It seems that strategies which monitor the driver's reactions and adapt the user interface accordingly should be used to achieve the best TO performance [46,47].

1.1.2. On Post-Takeover Stabilization

Overall, a proper takeover that ensures a safe and efficient transition of control between the automated system and the human driver involves a combination of timely response, attentiveness, readiness, smooth transition, correct actions, and awareness of the system's limitations. However, this process does not end with taking over the vehicle, i.e., by grabbing the steering wheel or applying the brakes, as the driver must resolve the critical situation that led to the initiation of a TO procedure and continue driving the vehicle manually. We refer to this phase of adaptation back to manual driving as achieving post-takeover stabilization or simply stabilization.

To account for stabilization, Shull et al. [48] and Ma et al. [49] recommended multi-stage TO requests. However, Butmee et al. [50] and Pipkorn et al. [51] showed that stabilization is almost impossible to achieve in some cases and therefore recommended automatically stopping the vehicle without even trying to issue a TOR. Gruden et al. noted in their study [39] that some drivers only took over the vehicle (i.e., applied the brake pedal) but were unable to perform any other action to prevent collisions. Therefore, they proposed a TO UI that helps the driver to take over the vehicle by providing additional warnings to minimize the stabilization time [46].

Determining the required stabilization time is also necessary for planning how often a driver can perform a TO; stated otherwise, this is how long a manual or automated driving must last before initiating a transition back. Feldhütter et al. explored how the duration of automated driving affects TO performance and concluded that there was no difference in performance when drivers had 5 or 20 min between consecutive TOs [52]. This could be understood as a hint that the stabilization phase after the transition of control might have already concluded prior to new TORs in both cases. Bourrelly et al. showed that longer periods of automated driving lead to poorer performance [53]. However, this is more probably a consequence of fatigue or some other phenomenon rather than stabilization after TO. On the other hand, Zhang et al. observed a degraded driving performance in terms of lane control for more than five minutes after TO [54]. Kim et al. measured the stabilization time after TO as reported by the drivers, i.e., the driver had to say "stable" after taking over and safely handling all driving functions [55]. They reported mean stabilization times of 11.5 s, 22.7 s, and 28.7 s for three consequent studies, found that stabilization time was longer when the accident occurred on the road in front of the vehicle compared to accidents in the oncoming lane, and observed some differences in age and gender. Gaspar et al. proposed adapting the TORlt based on whether drivers had achieved sufficient stabilization in previous TO attempts [56]. Riahi Samani and Mishra examined how long the TOR effect lasts and concluded that the first ten seconds after the TOR carry the most significant information, while the effects significantly reduce after 20 s [57]. Choi et al. showed that cognitive and visual load due to secondary tasks have different effects on stabilization after the TO [58]. Cognitive load increased the time from TO until resolving the situation (i.e., lane change), while visual load increased the steering wheel angle variability after the situation was already resolved, alluding to different effects of secondary tasks on stabilization time after a TO.

1.1.3. On Driving-Related and Physiological Predictors of Stabilization

As stabilization could only be achieved after the TO itself, not all common metrics for assessing TO performance could be used (e.g., reaction time or minimal time to collision do not contain any information on post-takeover stabilization). On the other hand, metrics that evaluate overall driving style could be compared within short intervals following the TO. The most obvious would therefore be to measure lane deviation/winding and determine when it stops. This could be achieved by observing steering wheel angle variability, lateral accelerations, or lane position deviation [59–65]. In the previously referred study, Zhang et al. measured the duration of how long the driver was out of their supposed lane per minute and recorded more than 10 s per minute for more than five minutes following the TO in about 25% of drivers [54]. Riahi Samani and Mishra calculated the driving-behavior-related parameters (maximum acceleration/deceleration, standard deviation of lane position, headway, maximum/minimum speed) in time windows of 10 s [57]. Their Multilevel Mixed-Effect Parametric Survival Models analysis showed that the first ten seconds after the takeover request contained the majority of information, while the probability of unsafe behavior significantly reduced only 20 s after the TOR. Choi et al. measured steering angle variability during two consecutive intervals: between TO and lane crossing and in the first five seconds after lane crossing [58]. Their conclusions differed depending on the type of secondary task. Kim et al. recorded stabilization time by instructing the drivers to say "stable" when they were fully capable of driving [55]. The stabilization times they measured in three experiments varied between 10 and 30 s, with longer stabilization times in scenarios involving an accident on the road or multiple vehicles in the vicinity.

Although stabilization in terms of driving-related predictors was almost always measured using lane deviation, it was not often compared to drivers' physiological stabilization or arousal. The duration of physiological responses to TO, such as pupil size, gaze dispersion, heart rate, and skin conductance stabilization was rarely measured. In general, the most commonly measured physiological signals during driving include (1) eye-tracking data such as pupil size, blink rate, horizontal gaze dispersion [25,65–70]; (2) heart rate or heart-rate variability [71–74]; and (3) phasic skin conductance [60,70,72,74]. It is also worth noting that the range of physiological responses during TO vary significantly between individuals [74]. Feldhütter et al. reported slower gaze responses in TOs after 20 min of automated driving than in TOs after five minutes of automated driving [52]. Kerautret et al. reported a long-lasting increase in heart rate after a TO in an emergency situation [75]. Gruden et al. reported on delays and durations of physiological responses to TORs, revealing that pupil diameter had the fastest response with an average duration of about 10 s, while the phasic skin conductance response lasted about 20 s and heart rate almost 60 s [76].

1.2. Our Contribution

Achieving post-takeover stabilization seems to be an often-overlooked aspect of TO scenarios. Understanding stabilization time is vital for optimizing the design and implementation of conditionally automated systems. By exploring how drivers adapt to and recover from takeover events, we can inform the development of more intuitive and user-friendly automation interfaces, as well as tailor training programs to enhance drivers' preparedness for takeover situations. Previous research studies have inadequately addressed the complexity of stabilization time, primarily due to the following pitfalls:

- Limited scope: Many studies have focused narrowly on the time it takes a driver to physically regain control of the vehicle, such as braking reaction time or steering wheel movement, neglecting the broader cognitive and behavioral aspects that influence stabilization time. Zeeb et al. [77], Kim et al. [78], and Radlmayr et al. [79] have shown that analyzing reaction time alone does not provide sufficient insight into the takeover quality. Moreover, Gold et al. concluded that although some interfaces led to faster reaction times, the drivers' actions were of poorer quality [7]. This narrow focus fails to capture the full extent of the transition process and its implications for safety.
- Lack of generalization: Many studies were conducted exclusively in controlled laboratory environments, using a predetermined TO user interface. For example, Zhang et al. issued a takeover request with only an audible, spoken warning [54]. Kim et al. did not even report how a TOR was issued in their study [80]. Additionally, almost all of the presented studies issued a TOR as a one-time event, while Gruden et al. [46] showed that TO is a process that should be monitored and that warnings should be adapted to the driver's reactions. This limits the generalizability of findings and may not accurately reflect the challenges drivers face when driving different vehicles.
- Insufficient consideration of physiological factors: Previous research has often overlooked the role of physiological factors, such as stress, cognitive load, and fatigue, in influencing stabilization time. For example, Riahi Samani and Mishra analyzed driving behavior by measuring only vehicle acceleration, speed, and position [57]. Choi et al. reported numerous driving-related parameters (speed, reaction times, maximal wheel angle, etc.) before and after TO, but only measured drivers' subjective perception with a single visual analog scale at the end of the driving trials [58]. Similarly, Zhang et al. performed a thorough analysis of driver behavior with driving-related parameters and a questionnaire at the end of the trial, but also included heart-rate variability as the only physiological measurement [54]. Therefore, it is possible that some long-lasting effects on the driver's state after the TO were overlooked by measuring only vehicle parameters.

In this paper, we aim to address these shortcomings by conducting comprehensive investigations into stabilization time in conditionally automated driving environments. We expand the current scope of knowledge by determining the post-takeover stabilization time with multiple variables using data from studies with different TO user interfaces. Furthermore, our investigation of physiological signals, such as heart rate variability and electrodermal activity, offers novel insights into the cognitive and emotional states underlying post-takeover performance, enriching our understanding of driver behavior in dynamic driving environments. To summarize, the research questions of our study are:

1. How long after the takeover could stabilization be achieved? Could this be before reaching the system limit (e.g., impact), i.e., is it more or less than the provided TOR lead time (6 s on average in the reviewed literature)?
2. Do physiological signals elicit a similar stabilization time as driving-related parameters? Does a driver remain stressed or aroused longer after the TO than could be predicted from vehicle parameters?

The reminder of this article is organized as follows: Section 2 presents the datasets used and our analysis procedure. Section 3 presents the results. The discussion is presented in Section 4, and a brief conclusion can be found in Section 5.

2. Materials and Methods

Datasets from two driving simulator user studies conducted as part of the European Union's Horizon 2020 research and innovation program HADRIAN (Holistic Approach for Driver Role Integration and Automation–Allocation for European Mobility Needs) were combined and used for the analysis. In the first user study, Stojmenova Pečečnik et al. explored four types of head-up displays (HUD) to assist drivers in conditionally automated vehicles [81,82]. In the second user study, Strle et al. evaluated the proposed HUD against a baseline condition in terms of cognitive load (physiological signals analysis) induced by the HUD [83]. Both studies involved similar conditions: a within-subject design, the driving scenarios featured a city road, the drivers were asked to take over the vehicle four times per scenario, and the same driving simulator setup was used. The only difference was the interfaces used for driving assistance and takeover requests. As we seek general results that are valid for any type of takeover user interface, we joined the datasets for analysis.

2.1. Technical Set-Up

2.1.1. Driving Simulator

Both studies were conducted in a NERVtech™ motion-based driving simulator (Nervtech d.o.o., Ljubljana, Slovenia) [84] that consists of three 49″ FullHD curved displays covering the driver's viewing angle of about 145°, a steering wheel, pedals, a 4-DoF (degrees of freedom) motion platform, and a physical dashboard display; see Figure 1. Using a high-fidelity driving simulator offered several advantages for studying takeover performance (a controlled, safe, risk-free, and standardized environment). Some critical driving scenarios that provide valuable insights into stabilization time could never be safely performed in the real world. In simulators, they can be systematically manipulated and repeated. This level of control also allows for precise measurement of key variables. In addition, driving simulators have been widely used in previous research to study driver behavior, cognitive processes, and performance in various driving tasks. Some studies showed comparable results between studies using real vehicles and motion-based driving simulators, both in terms of physical validity (e.g., vehicle dynamics [85]) and behavioral validity (e.g., car sickness [86]). However, Bellem et al. recommended a motion scaling factor of approximately 50% to 60%, as speed may be underestimated in virtual environments [87]. While we acknowledge that simulators cannot perfectly replicate the complexity of real driving, we believe that our studies have minimized any significant discrepancies between simulated and real vehicles.

The software used to create and play the scenario was AVSimulation's SCANeR Studio 1.7 [88]. The scenario was developed internally to mimic an urban journey for a person driving from home to work through different parts of the city during the day. It is 13 km long and takes about 16 min. At the beginning, the vehicle was parked on the road with two lanes (one for each direction), and the speed limit was 50 km/h. The surrounding traffic was included to simulate a busy small-town road. After about 3 km there were some crosswalks where pedestrians wanted to cross the road. About 6 km from the starting point, the scenario included a complicated intersection where the road widened to 5 lanes and the driver had to move to the appropriate lane based on the navigation system's instructions. The journey then continued on a four-lane road (two lanes in each direction).

About 10 km from the starting point there was a school area where children crossed the road at crosswalks without traffic lights. After driving 13 km in total, the scenario ended with the driver being asked to park in the street parking lot on the right side of the road, which simulated arrival at the office. During each journey, the vehicle requested four takeover attempts. First, to handle some of the crosswalks; second, to change lanes at the complicated intersection; third, to drive through the school grounds; and fourth, to park the vehicle at the end. Before and after this, the driver was asked to engage the automated driving function. The driver could take over the vehicle by turning the steering wheel or pressing the brake or accelerator pedal.

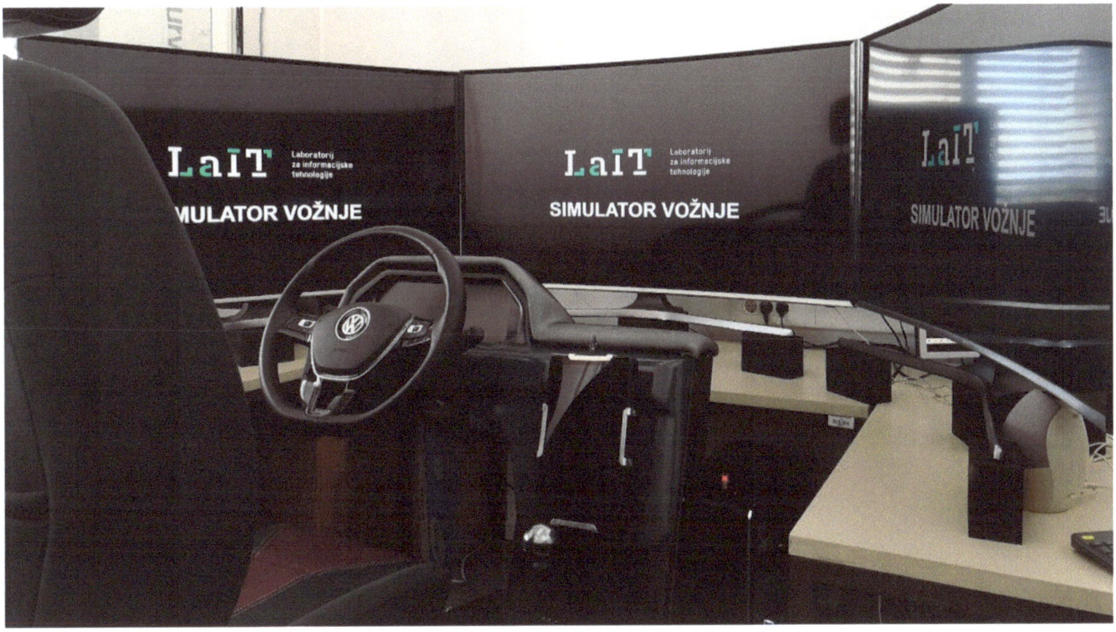

Figure 1. NERVtech^{TM} motion-based driving simulator located at the University of Ljubljana, Faculty of Electrical Engineering.

2.1.2. Head-Up Display and Takeover Requests

The simulated vehicle featured a head-up display (HUD) designed to assist the driver in monitoring the environment by presenting driving-related information such as current speed, speed limit, safety distance, and the status of assistance systems (see Figure 2a). The amount and location of the presented information varied between trial conditions. For example, some conditions also included navigation information projected directly onto the road (see Figure 2b).

When the vehicle in the automated mode approached its system limits, it issued a TOR five seconds before automated driving would become unavailable. The request was presented with an auditory signal (sine wave, 4000 Hz) and a visual notification with countdown on the HUD (see Figure 2c). At some TORs (when approaching a school area), the TO countdown was already displayed five seconds before the acoustic warning. A period of manual driving followed every TO. When automated driving became available, a synthesized female voice asked the driver to turn on the automated driving function by pressing a dedicated button on the steering wheel lever.

(a) (b) (c)

Figure 2. Head-up display (HUD) in the simulated vehicle: (**a**) represents an example of the information displayed on the HUD; (**b**) shows navigation instructions, projected directly onto the road; and (**c**) shows the takeover request notification.

2.1.3. Wearable Sensor Devices

The physiological data were collected using two wearable sensor devices. The eye-tracking data (pupil diameter, gaze) were recorded using Tobii pro glasses 2 [89] with a sampling frequency of 50 Hz. The Empatica E4 wristband [90] was used to record blood volume pulse (BVP) and extract heart rate using the E4's internal algorithm, galvanic skin response (GSR), and skin temperature. The current heart rate was calculated once per second, while GSR and skin temperature were captured with a sampling frequency of 4 Hz.

2.2. Participants and Their Tasks

A total of 30 drivers participated in the first study [82] (16 male, 14 female; aged 23 to 55). As the study had a within-subject design with four trials (types of HUD) and there were four TOs in each trial, this resulted in a total number of 480 TO attempts. 28 drivers participated in the second study [83] (14 male, 14 female; aged 21 to 57). As each participant in the study completed two trials (types of HUD) and there were four TOs in each trial, the study provided 224 TO attempts. In total, we analyzed 704 TOs.

Participation in the study was completely voluntary. Participants were informed that they could stop the experiment at any time without providing a reason. They received a gift voucher of 10 € as compensation for their time. Informed consent was obtained from each participant. The study was conducted in accordance with the code of ethics of the University of Ljubljana which is consistent with the Declaration of Helsinki.

The driver's primary responsibility was to ensure safe driving continuity. They were asked to drive the vehicle according to the navigation system and get to the destination parking lot. When the automatic driving mode was available, drivers were asked to activate it and engage in a secondary task—a 2048 puzzle game [91] on a smartphone.

Before participating in the study, the detailed procedure was explained to each participant and the conductor collected written informed consent and a demographic questionnaire. Participants began with a test drive to familiarize themselves with the driving simulator and ask questions. Each trial began and ended with manual driving. During the whole trial (four TOs) the experiment was not interrupted. After the trial, participants completed four questionnaires (a custom made Likert scale about the system, a System Usability Scale—SUS [92], a User Experience Questionnaire—UEQ [93], and an Acceptance of Advanced Transport Telematics questionnaire [94]). However, these questionnaires focused on the HUD being tested and not on the takeover procedure [95]. As they served other research purposes, they do not provide information on our research questions (stabilization time) and were therefore omitted from our analysis. More detailed descriptions of the study procedures can be found in Stojmenova Pečečnik et al. [82] and Strle et al. [83].

2.3. Variables of Interest

According to related work, driving-related stabilization could be assessed by measurements of lane deviation [54,58]. Among physiological signals, GSR, pupil diameter, and heart rate seem to be good indicators of drivers' mental or physiological state [74,96,97]. Similar to the way alterations in the GSR signal can indicate the intensity of emotional state, such as happiness or stress [98], pupil diameter directly reflects the driver's mental load [66], while heart rate and heart rate variability reflect the autonomic nervous system which might indicate a stress response [99,100].

To determine and compare the post-takeover stabilization time we collected the following driving-related and physiological data:

1. Driving-related variables:
    - Winding (standard deviation of steering wheel angle);
    - Speed;
    - Deceleration.
2. Physiological variables:
    - Eyes off-road ratio (E-OFF);
    - Pupil diameter (PD);
    - Heart rate (HR);
    - Phasic skin conductance (SC).

2.4. Analysis Procedure

Steering wheel angle, vehicle speed, deceleration, eye gaze, pupil diameter, heart rate, and skin conductance were measured continuously throughout the driving trial. To study post-takeover stabilization, we applied two time windows: one of two seconds and one of five seconds, and we performed a separate analysis for both cases. The time windows (data bins) started at the moment the driver took over the vehicle, i.e., started driving manually, and were calculated every other second (starting at 0 s, 2 s, 4 s, 6 s, etc. after TO). In this way, we obtained non-overlapping windows in the first case, when the window length was two seconds, and overlapping windows in the second case, when the window length was five seconds.

Winding was calculated as the standard deviation of the steering wheel angle for each time window. Additionally, we used the mean speed and deceleration of each time window. Among the physiological variables, we observed eye gaze and determined whether the gaze was directed toward or away from the road. We then calculated the ratio of samples with eye gaze off the road (E-OFF). The mean pupil diameter (PD) and heart rate (HR) for each time window were also calculated. To account for individual differences in PD, we subtracted the value at TO from each subsequent sample of the same driver (shifted the data) so that each PD starts at zero at the TO. Phasic skin conductance (SC) was calculated from the raw GSR signal using the cvxEDA algorithm (convex optimization approach) by Greco et al. [101]. We used the mean phasic skin components in each time window.

The means of the calculated variables over all TO attempts were plotted in the time range from the TO to the time window starting 12 s after the TO (7 windows). We performed repeated measures analysis of variance (RM ANOVA) with Greenhouse-Geisser correction when the sphericity assumption was violated for each variable of interest with the window start time as the factor. If differences between time windows were confirmed, we performed paired-samples *t*-tests for every pair of time windows with Bonferroni adjustment for multiple comparisons. We considered the variable stabilized in a given time window if no differences were observed between that time window and all subsequent windows. An alpha level of 0.05 was used if not stated otherwise.

Data were processed with Python 3.9.12 (anaconda distribution) [102] and analyzed with IBM SPSS Statistics 22 [103].

3. Results

The following subsections present the plots of the measured variables for each time window and the statistical analysis of the differences between the adjacent time windows.

3.1. Driving-Related Variables

Repeated measures ANOVA confirmed that differences in winding exist between time windows: $F(3.1, 1957.4) = 251.1$, $p < 0.001$ for time windows of two seconds, and $F(2.1, 1335.6) = 185.8$, $p < 0.001$ for time windows of five seconds. The mean values are presented in Figure 3, and the pairwise comparisons can be found in Appendix A, Tables A1 and A2. Winding stabilized in the time interval starting eight seconds after the TO for both window lengths.

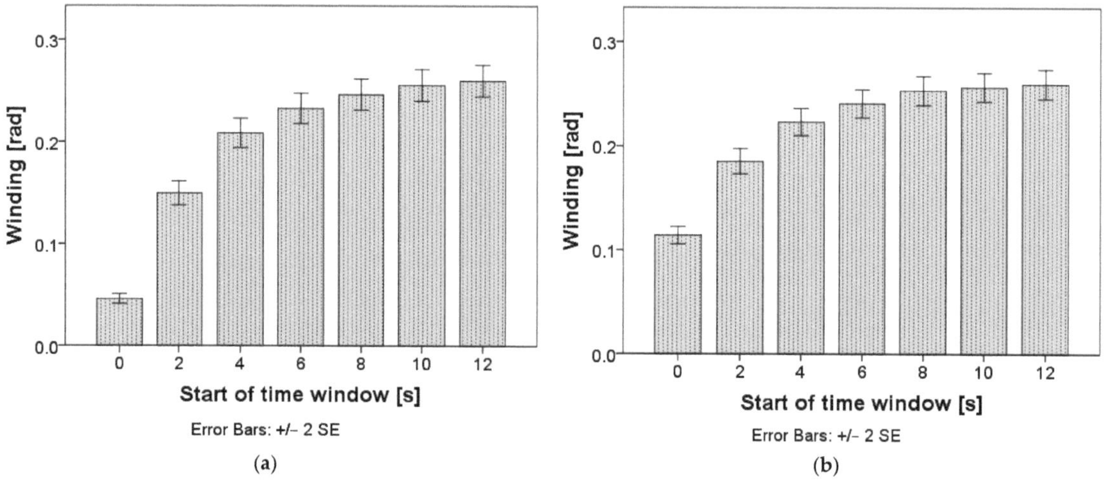

Figure 3. Winding (standard deviation of steering wheel angle) after the takeover. Chart (**a**) represents calculations with a window length of two seconds; chart (**b**) represents calculations with a window length of five seconds.

Repeated measures ANOVA confirmed that differences in speed exist between time windows: $F(2.1, 1338.0) = 34.9$, $p < 0.001$ for time windows of two seconds, and $F(1.5, 989.7) = 13.5$, $p < 0.001$ for time windows of five seconds. The mean values are presented in Figure 4, and the pairwise comparisons can be found in Appendix A, Tables A1 and A2. When using a window length of two seconds, speed stabilized only in the time interval starting ten seconds after the TO. The drivers slowed down immediately after the TO and then accelerated until the speed stabilized. When using a window length of five seconds, speed already first stabilized in the time interval starting two seconds after the TO, when the drivers slowed down. Afterwards, the drivers accelerated and the speed stabilized again in the time interval starting eight seconds after the TO.

Repeated measures ANOVA revealed no differences in deceleration between time windows when using a window length of two seconds $F(4.6, 317.9) = 1.01$, $p = 0.410$, but revealed statistically significant differences between time windows when using a window length of five seconds $F(3.4, 684.4) = 5.46$, $p = 0.001$. The mean values are presented in Figure 5, and the pairwise comparisons for the window length of five seconds can be found in Appendix A, Table A2. When using a window length of five seconds, deceleration stabilized in the time interval starting two seconds after the TO.

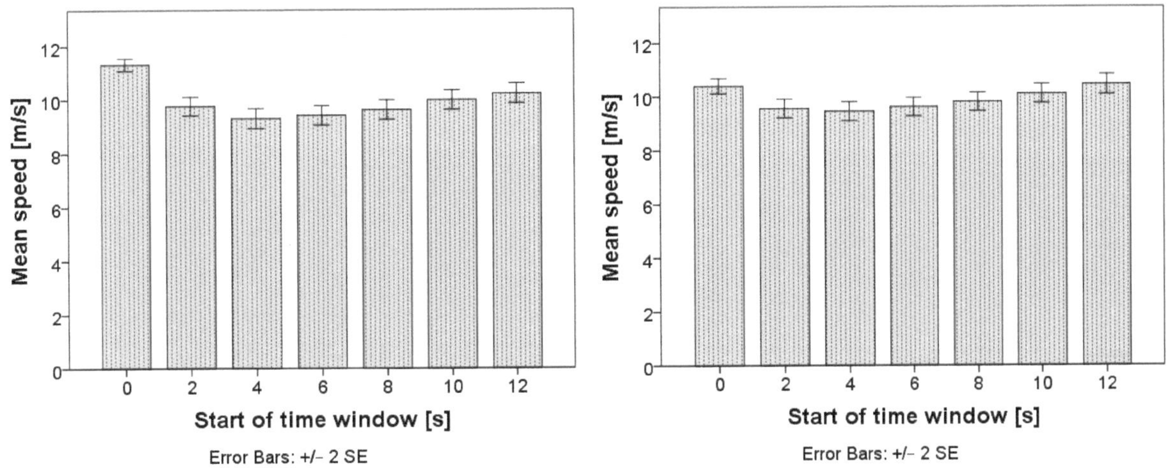

(a) (b)

Figure 4. Mean speed after the takeover. Chart (**a**) represents calculations with a window length of two seconds; chart (**b**) represents calculations with a window length of five seconds.

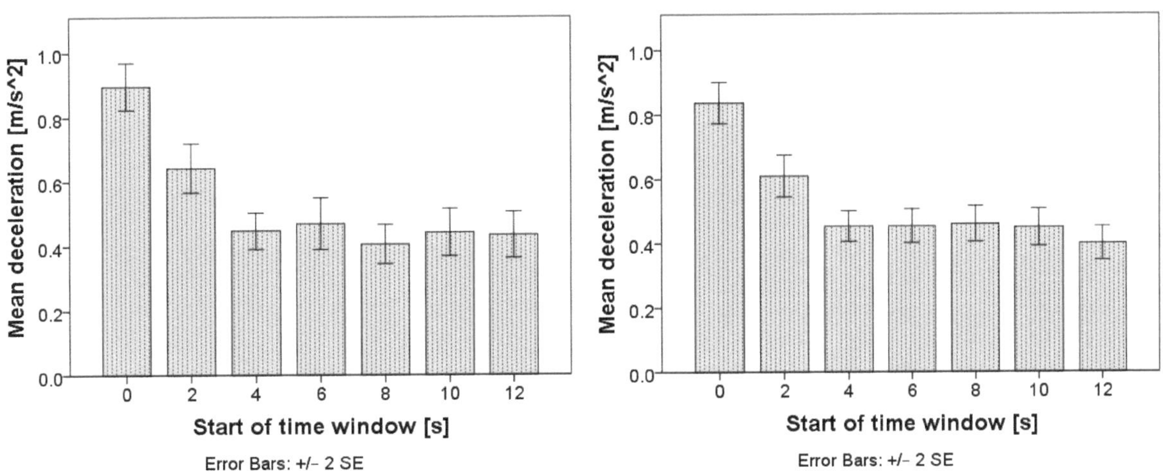

(a) (b)

Figure 5. Mean deceleration after the takeover. Chart (**a**) represents calculations with a window length of two seconds; chart (**b**) represents calculations with a window length of five seconds.

3.2. Physiological Variables

Repeated measures ANOVA confirmed that differences in eyes off-road ratio exist between time windows: $F(5.4, 3506.1) = 6.12$, $p < 0.001$ for time windows of two seconds, and $F(2.8, 1797.8) = 4.27$, $p = 0.006$ for time windows of five seconds. The mean values are presented in Figure 6, and the pairwise comparisons can be found in Appendix A, Tables A1 and A2. E-OFF stabilized in the time interval starting two seconds after the TO for both window lengths.

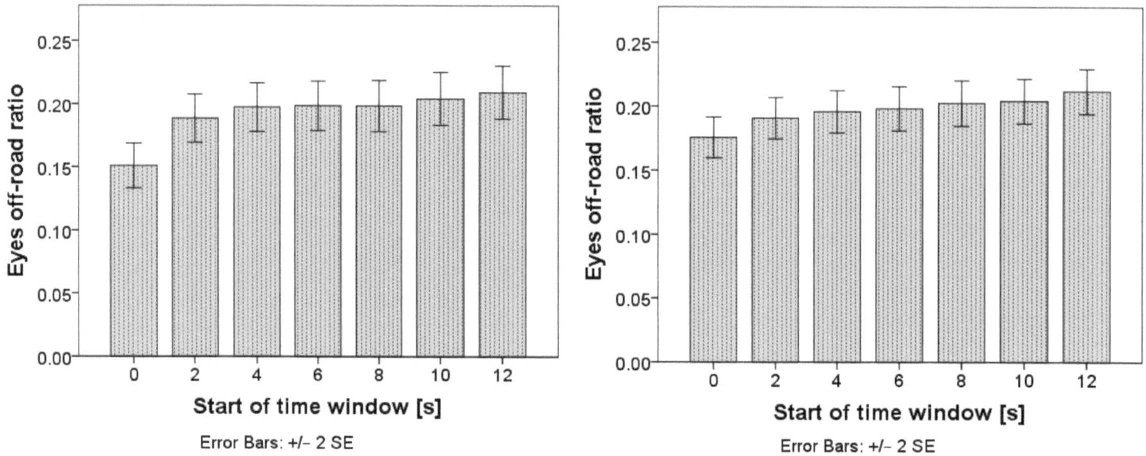

(**a**) (**b**)

Figure 6. Eyes off-road ratio after the takeover. Chart (**a**) represents calculations with a window length of two seconds; chart (**b**) represents calculations with a window length of five seconds.

Repeated measures ANOVA confirmed that differences in pupil diameter exist between time windows: F(3.5, 2150.0) = 143.9, $p < 0.001$ for time windows of two seconds, and F(2.1, 1311.5) = 129.8, $p < 0.001$ for time windows of five seconds. The mean values are presented in Figure 7, and the pairwise comparisons can be found in Appendix A, Tables A1 and A2. Pupil diameter stabilized in the time interval starting six seconds after the TO for both window lengths. When using a window length of five seconds, the pupil diameter started dropping again in the interval starting twelve seconds after the TO.

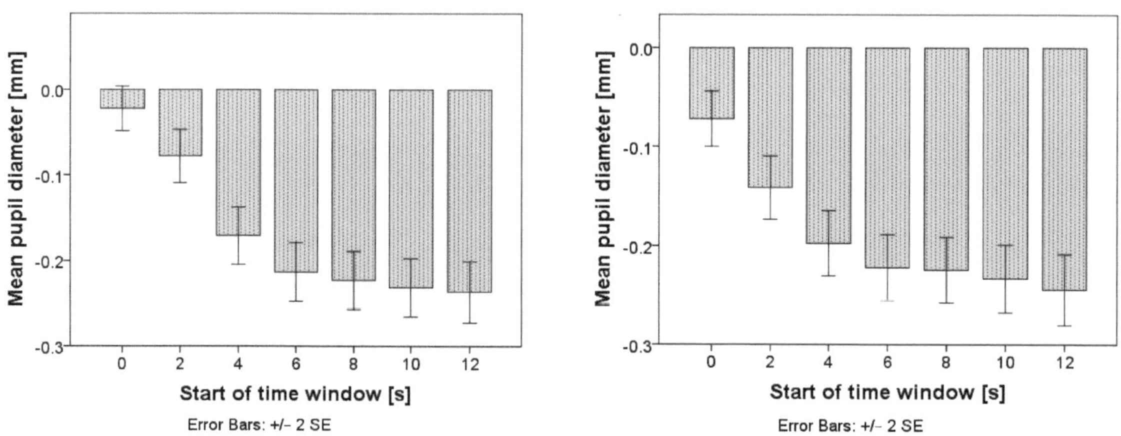

(**a**) (**b**)

Figure 7. Mean pupil diameter after the takeover. Chart (**a**) represents calculations with a window length of two seconds; chart (**b**) represents calculations with a window length of five seconds.

Repeated measures ANOVA confirmed that differences in heart rate exist between time windows: F(3.6, 150.1) = 2.76, $p = 0.035$ for time windows of two seconds, and F(2.6, 263.7) = 15.8, $p < 0.001$ for time windows of five seconds. The mean values are presented in Figure 8, and the pairwise comparisons can be found in Appendix A, Tables A1 and A2. When using a window length of two seconds, no statistically significant differences between data in adjacent time intervals were observed. When using a window length of five seconds, the HR only started dropping in the interval starting eight seconds after the TO and stabilized again in the interval starting ten seconds after the TO.

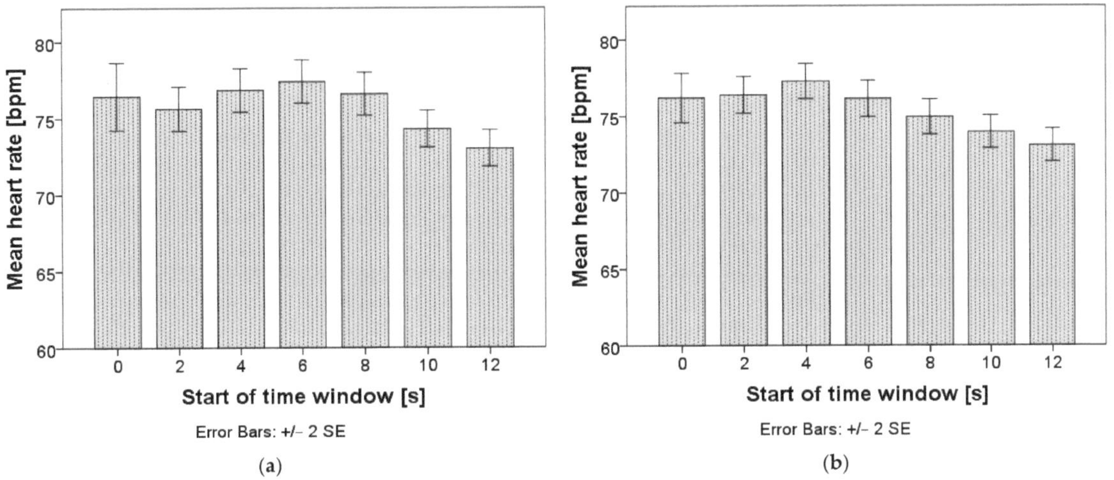

Figure 8. Mean heart rate after the takeover. Chart (**a**) represents calculations with a window length of two seconds; chart (**b**) represents calculations with a window length of five seconds.

Repeated measures ANOVA confirmed that differences in phasic skin conductance exist between time windows: F(1.2, 708.6) = 9.90, $p = 0.001$ for time windows of two seconds, and F(1.1, 653.7) = 12.6, $p < 0.001$ for time windows of five seconds. The mean values are presented in Figure 9, and the pairwise comparisons can be found in Appendix A, Tables A1 and A2. Phasic skin conductance only started dropping in the time interval starting six seconds after the TO and did not stabilize while measuring for both window lengths.

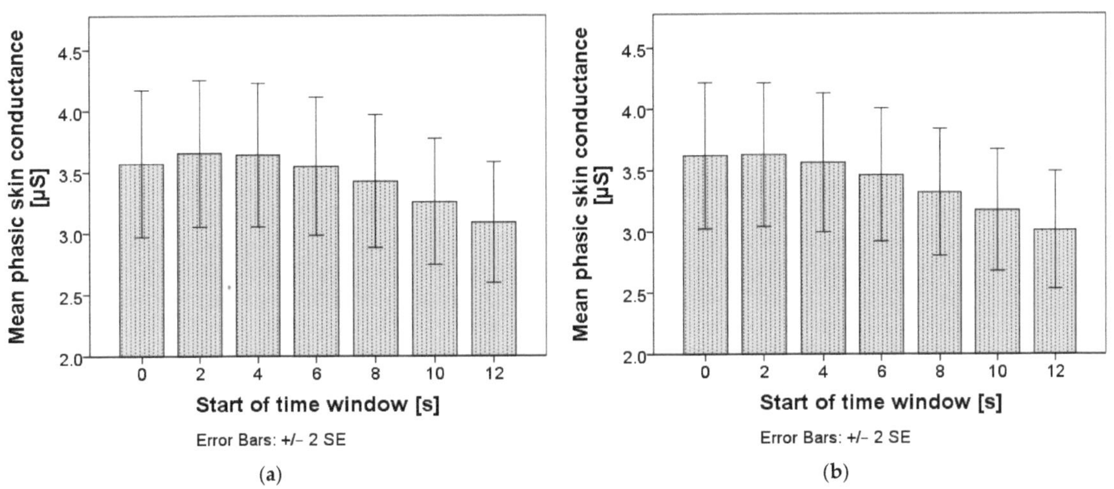

Figure 9. Mean phasic skin conductance after the takeover. Chart (**a**) represents calculations with a window length of two seconds; chart (**b**) represents calculations with a window length of five seconds.

4. Discussion

This study aimed to investigate two primary research questions: (1) How long after takeover could stabilization be achieved, and could it occur before reaching the system limit (e.g., impact); and (2) Do physiological signals exhibit a similar stabilization time to driving-related parameters, and does a driver remain stressed or aroused longer after takeover than predicted from vehicle parameters? Our findings provide insights into these research questions, shedding light on the temporal and cognitive dynamics underlying the stabilization process in takeover scenarios in conditionally automated driving environments.

Regarding the first research question, our results indicated that the stabilization time following takeover (TO) varied for different parameters. Notably, deceleration and eyes off-road ratio stabilized approximately two seconds after TO, then pupil diameter stabilized approximately six seconds after TO, winding stabilized approximately eight seconds after TO, speed stabilized eight to ten seconds after TO, heart rate stabilized more than ten seconds after TO, and phasic skin conductance did not stabilize at all in both the two-second and five-second intervals.

Deceleration being one of the first parameters to stabilize indicates that the drivers probably applied the brake pedal to take over the vehicle, slow down a bit, and then stopped braking, which is in line with the findings of Gruden et al. [47]. The E-OFF ratio also stabilized approximately two seconds after TO, indicating a rapid adjustment of visual attention after the transition, in line with Stephenson et al. [69]. Pupil diameter, often considered a direct indicator of driver cognitive load [66], showed stabilization starting several seconds after TO. It should be noted that pupil diameter stabilizing at a lower value than at TO could reflect lower cognitive load after TO or adaptation to changes in light conditions [75,104]. As the driver was performing a secondary task before TO, the driver's gaze was directed off the driving simulator screen, which could have different illumination than when looking at the screen (road). However, as pupil constriction due to sudden illumination changes occurs in fractions of a second [105], we believe that our measurements reliably reflect the decrease in mental load that stabilized six seconds after TO. After the decrease in mental load, the stabilization times of winding and speed show that driving-related parameters stabilize about eight to ten seconds after TO. This is less than the stabilization reported by Kim et al. [55], but is consistent with Riahi Samani and Mishra [57], who found that most information can be obtained by observing the first ten seconds following TO. It should be noted that they measured stabilization based on a different method. As Kim et al. instructed drivers to say "stable" when they felt stable, it is reasonable that this self-reported stabilization is longer, as the vehicle should already be stable for some time before the driver considers it stable. However, this is still more than the usually adopted TORlt of six seconds [24], implying that drivers might not be able to stabilize the vehicle in time, e.g., before impact with an obstacle. Therefore, using longer TORlt, as proposed by Tan and Zhang [23], might be beneficial. As the driver should have enough time to react and stabilize to perform a qualitative TO, the sum of reaction and stabilization times should be lower than or at least similar to the TORlt. Heart rate and phasic skin conductance did not begin to decrease until approximately six to eight seconds after TO, indicating a prolonged arousal state following TO and vehicle-related stabilization.

Addressing the second research question, our findings suggest that physiological signals do not necessarily exhibit a similar stabilization time as driving-related parameters. While driving-related parameters such as winding and speed stabilized within a few seconds after TO, physiological signals, especially HR and SC, demonstrated more prolonged stabilization periods, as proposed by Gruden et al. [76]. HR and SC exhibited variations between adjacent time intervals, indicating a longer period of arousal following TO than predicted from vehicle parameters alone. HR stabilized only in the last observed time window when using the window length of five seconds, in line with Kerautret et al. [75], while SC did not stabilize during the observed duration. Therefore, we propose longer observations of physiological parameters in future studies to reliably detect drivers' arousal stabilization. We suspect that the drivers remain aroused while being able to reasonably drive the vehicle, i.e., after achieving vehicle-related stabilization. However, if another unforeseen situation occurs while being stressed, drivers might react worse due to still being under the influence of the last TO. This discrepancy highlights the importance of considering both physiological and driving-related parameters when assessing drivers' arousal and stress levels during takeover scenarios.

Additionally, vehicle speed exhibited a different profile than other parameters, which monotonously stabilized after TO. It initially dropped and showed an earlier stabilization with a five-second interval occurring two seconds after TO, followed by an increasing trend and a subsequent stabilization eight seconds after TO. The initial drop could be explained by the immediate deceleration at TO, which settled afterwards. Speed, deceleration, and heart rate also exhibited variations between the lengths of the time windows. It should be noted that the deceleration at the interval length of two seconds was probably not found significant, as there were not many TO attempts (only about 10%) where drivers gradually decelerated and therefore deceleration data could be calculated. The longer size, on the other hand, resulted in overlapping time windows that contained more data samples. Something similar could be stated for HR data measured with the E4 wristband. As the driver is moving during a TO, many HR samples were not available due to motion artifacts [71]. Since heartbeats are produced about once per second, there is a high probability that data were not available in most of the two-second intervals. According to our results, this could, however, not be stated for the five-second intervals.

Therefore, it is essential to acknowledge some limitations of the study. (1) As previously mentioned, the different illumination of the environment and the driving simulator screen may have affected pupil diameter measurements [104]. (2) There were many missing samples for deceleration and heart rate. Future studies could interpolate the missing values or use some other tools for analysis. (3) The experiments were conducted in a controlled high-fidelity driving simulator environment. While driving simulators offer several advantages, including controlled experimental conditions and enhanced safety, they may not fully replicate the stress, distraction, and other real-world driving conditions. However, user studies with potentially dangerous scenarios could never be conducted on the road without exposing participants to actual risk. Moreover, simulators allow us to systematically manipulate variables, assess driver responses, and collect detailed data that are difficult to obtain in on-road studies. Therefore, despite their limitations, driving simulators are a valuable tool for studying takeover performance. (4) Additionally, the study focused on a limited set of physiological parameters, and further investigation of additional measures of cognitive workload and emotional state could provide a more comprehensive understanding of stabilization time.

Next, we open up some interesting discussion topics related to stabilization time. These considerations could be further investigated by readers, as a deeper analysis would exceed the sole purpose of this manuscript. Our analysis could be further generalized by drawing on and comparing the results of other researchers' TO studies or tailored to individual technological systems. While our study focused primarily on identifying factors that reveal stabilization time, future research could explore interventions or strategies aimed at shortening the time it takes for drivers to regain full control after a takeover, similar to studies conducted to find the optimal TO user interface [35,37,46]. For example, Petermeijer et al. showed that auditory TOR stimuli provoked the fastest response [30]. The guidelines by Naujoks et al. [106] also provide a way to design an optimal user interface and show which features should be included. Possible approaches include the development of advanced automation interfaces (integration of real-time feedback mechanisms, adaptive displays, and ergonomic design to support rapid and effective TO), driver training programs, or predictive algorithms to improve driver readiness and responsiveness to takeover events. The datasets we used consisted of takeover attempts with the same lead time of five seconds, as this is considered optimal for takeover requests in the literature [21]. However, future research could dive further into the relationship between takeover lead time and stabilization time and how different lead times affect driver adaptation and performance during takeover events. The effects of the urgency of the takeover scenario and the driver's secondary task on overall TO performance have been studied in depth, but their effects on stabilization time have not yet been adequately addressed. The aim of this paper is to determine the stabilization time in a general takeover scenario. In the dataset used, participants were allowed to perform any type of secondary task at their own discretion. Although the investigation of the influence of scenario complexity is beyond the scope of the present manuscript, it could be useful for the development of adaptive, context-aware, and robust automated driving systems.

5. Conclusions

Our study contributes to the understanding of stabilization time in takeover scenarios in conditionally automated driving environments. We have shown that driving-related stabilization can be achieved approximately eight to ten seconds after the TO, which is more than the commonly assumed TOR lead time. We also demonstrated that physiological signals, particularly heart rate and phasic skin conductance, exhibited prolonged stabilization periods, indicating that drivers remain aroused even after driving-related stabilization is achieved.

Future studies should extend the observation period after TO to reliably determine the stabilization time of phasic skin conductance, which did not stabilize during our observations. In addition, a more thorough analysis of the missing values for heart rate and deceleration should be performed. Ultimately, a model of drivers' mental states throughout the TO process could be derived from this and similar studies to better understand the process. The diversity of drivers could be taken into account by clustering drivers based on demographic data and analyzing different driver profiles. To improve the generalizability of the results, future analysis could combine data from many TO studies conducted by research groups around the world to obtain valid comparisons under many different conditions and using different technologies. Using artificial intelligence and machine learning techniques, predictive algorithms could anticipate upcoming takeover events and proactively assist drivers to reduce stabilization time. By addressing these future research directions, we can further improve our understanding of stabilization time and accelerate progress towards a future where automated vehicles seamlessly coexist with human drivers, ushering in a new era of mobility characterized by enhanced safety, efficiency, and accessibility.

Author Contributions: Conceptualization, T.G. and G.J.; methodology, G.J.; software, T.G. and G.J.; validation, S.T. and G.J.; formal analysis, S.T.; investigation, T.G. and G.J.; resources, S.T.; data curation, T.G.; writing—original draft preparation, T.G.; writing—review and editing, S.T. and G.J.; visualization, T.G. and G.J.; supervision, S.T. and G.J.; project administration, G.J.; funding acquisition, S.T. All authors have read and agreed to the published version of the manuscript.

Funding: This research was funded in part by the Slovenian Research Agency within the research program ICT4QoL—Information and Communications Technologies for Quality of Life, grant number P2-0246.

Institutional Review Board Statement: Not applicable.

Informed Consent Statement: Informed consent was obtained from all subjects involved in the study.

Data Availability Statement: The data are available upon request to the corresponding author.

Conflicts of Interest: The authors declare no conflicts of interest. The funders had no role in the design of the study; in the collection, analyses, or interpretation of data; in the writing of the manuscript; or in the decision to publish the results.

Appendix A

Pairwise comparisons for the dependent variables in the analysis.

Table A1. Pairwise comparisons for window length of two seconds.

Start of Time w. [s]		Winding [rad]		Speed [m/s]		Eyes Off-Road Ratio		Pupil Diameter [mm]		Heart Rate [bpm]		Phasic Skin Conductance [μS]	
(I)	(II)	Mean Diff. (II–I)	p-Value	Mean Diff. (II–I)	p-Value	Mean Diff. (II–I)	p-Value	Mean Diff. (II–I)	p-Value	Mean Diff. (II–I)	p-Value	Mean Diff. (II–I)	p-Value
0	2	0.093 *	<0.001	−1.579 *	<0.001	0.038 *	0.001	−0.062 *	<0.001	0.21	1.000	0.080	0.071
	4	0.148 *	<0.001	−1.875 *	<0.001	0.049 *	<0.001	−0.155 *	<0.001	−0.16	1.000	0.067	1.000
	6	0.178 *	<0.001	−1.585 *	<0.001	0.047 *	<0.001	−0.197 *	<0.001	0.96	1.000	−0.019	1.000
	8	0.195 *	<0.001	−1.273 *	<0.001	0.044 *	0.002	−0.204 *	<0.001	−0.40	1.000	−0.124	1.000
	10	0.209 *	<0.001	−1.035 *	<0.001	0.051 *	<0.001	−0.216 *	<0.001	−1.26	1.000	−0.267	0.492
	12	0.213 *	<0.001	−0.962 *	<0.001	0.055 *	<0.001	−0.221 *	<0.001	−2.11	1.000	−0.391	0.153
2	4	0.055 *	<0.001	−0.296 *	0.002	0.010	1.000	−0.093 *	<0.001	−0.37	1.000	−0.013	1.000
	6	0.084 *	<0.001	−0.006	1.000	0.008	1.000	−0.134 *	<0.001	0.75	1.000	−0.099	0.963
	8	0.102 *	<0.001	0.306	1.000	0.006	1.000	−0.142 *	<0.001	−0.61	1.000	−0.204	0.165
	10	0.115 *	<0.001	0.544	0.090	0.013	1.000	−0.153 *	<0.001	−1.47	1.000	−0.347 *	0.021
	12	0.119 *	<0.001	0.617	0.114	0.017	1.000	−0.159 *	<0.001	−2.32	0.788	−0.471 *	0.010
4	6	0.030 *	<0.001	0.290 *	0.001	−0.002	1.000	−0.041 *	<0.001	1.12	1.000	−0.086 *	0.031
	8	0.047 *	<0.001	0.602 *	<0.001	−0.004	1.000	−0.049 *	<0.001	−0.23	1.000	−0.191 *	0.015
	10	0.061 *	<0.001	0.840 *	<0.001	0.003	1.000	−0.061 *	<0.001	−1.09	1.000	−0.333 *	0.002
	12	0.065 *	<0.001	0.913 *	0.001	0.006	1.000	−0.066 *	<0.001	−1.95	1.000	−0.457 *	0.002
6	8	0.018 *	0.008	0.311 *	0.001	−0.002	1.000	−0.008	1.000	−1.36	0.713	−0.105 *	0.017
	10	0.031 *	<0.001	0.550 *	0.002	0.005	1.000	−0.019	0.926	−2.21	0.123	−0.248 *	0.002
	12	0.035 *	0.001	0.623 *	0.025	0.008	1.000	−0.024	0.601	−3.07 *	0.026	−0.372 *	0.002
8	10	0.014	0.153	0.238 *	0.049	0.007	1.000	−0.012	1.000	−0.86	1.000	−0.143 *	0.001
	12	0.018	0.349	0.311	0.664	0.011	1.000	−0.017	1.000	−1.72	1.000	−0.267 *	0.001
10	12	0.004	1.000	0.073	1.000	0.004	1.000	−0.005	1.000	−0.86	1.000	−0.124 *	0.007

* The mean difference is significant at the 0.05 level.

Table A2. Pairwise comparisons for window length of five seconds.

Start of Time w. [s]		Winding [rad]		Speed [m/s]		Deceleration [m/s^2]		Eyes Off-Road Ratio		Pupil Diameter [mm]		Heart Rate [bpm]		Phasic Skin Conductance [μS]	
(I)	(II)	Mean Diff. (II–I)	p-Value	Mean Diff. (II–I)	p-Value	Mean Diff. (II–I)	p-Value	Mean Diff. (II–I)	p-Value	Mean Diff. (II–I)	p-Value	Mean Diff. (II–I)	p-Value	Mean Diff. (II–I)	p-Value
0	2	0.067 *	<0.001	−0.763 *	<0.001	−0.155 *	<0.001	0.016 *	0.006	−0.072 *	<0.001	−0.29	1.000	0.012	1.000
	4	0.105 *	<0.001	−0.709 *	<0.001	−0.238 *	<0.001	0.021 *	0.049	−0.127 *	<0.001	−0.60	1.000	−0.048	1.000
	6	0.128 *	<0.001	−0.417 *	0.014	−0.202 *	0.001	0.020	0.331	−0.150 *	<0.001	−1.25	1.000	−0.150	0.757
	8	0.141 *	<0.001	−0.186	1.000	−0.126	0.714	0.023	0.152	−0.158 *	<0.001	−2.40 *	0.014	−0.277	0.109
	10	0.149 *	<0.001	−0.016	1.000	−0.115	1.000	0.025	0.072	−0.166 *	<0.001	−3.65 *	<0.001	−0.400 *	0.035
	12	0.152 *	<0.001	0.173	1.000	−0.143	0.317	0.032 *	0.007	−0.179 *	<0.001	−3.80 *	<0.001	−0.515 *	0.016

Table A2. *Cont.*

Start of Time w. [s]		Winding [rad]		Speed [m/s]		Deceleration [m/s²]		Eyes Off-Road Ratio		Pupil Diameter [mm]		Heart Rate [bpm]		Phasic Skin Conductance [µS]	
(I)	(II)	Mean Diff. (II–I)	*p*-Value	Mean Diff. (II–I)	*p*-Value	Mean Diff. (II–I)	*p*-Value	Mean Diff. (II–I)	*p*-Value	Mean Diff. (II–I)	*p*-Value	Mean Diff. (II–I)	*p*-Value	Mean Diff. (II–I)	*p*-Value
2	4	0.038 *	<0.001	0.054	1.000	−0.083	0.422	0.005	1.000	−0.056 *	<0.001	−0.31	1.000	−0.059	0.380
	6	0.061 *	<0.001	0.346 *	0.029	−0.047	1.000	0.004	1.000	−0.079 *	<0.001	−0.96	0.751	−0.162 *	0.044
	8	0.074 *	<0.001	0.577 *	0.005	0.023	1.000	0.007	1.000	−0.086 *	<0.001	−2.11 *	0.001	−0.289 *	0.009
	10	0.081 *	<0.001	0.748 *	0.003	0.040	1.000	0.010	1.000	−0.094 *	<0.001	−3.36 *	<0.001	−0.412 *	0.005
	12	0.084 *	<0.001	0.936 *	0.001	0.012	1.000	0.017	1.000	−0.108 *	<0.001	−3.51 *	<0.001	−0.527 *	0.003
4	6	0.023 *	<0.001	0.292 *	<0.001	0.036	1.000	−0.001	1.000	−0.023 *	<0.001	−0.65	0.663	−0.102 *	0.007
	8	0.036 *	<0.001	0.523 *	<0.001	0.112	0.258	0.002	1.000	−0.031 *	<0.001	−1.80 *	<0.001	−0.230 *	0.002
	10	0.044 *	<0.001	0.693 *	0.001	0.123	0.245	0.004	1.000	−0.039 *	0.001	−3.05 *	<0.001	−0.352 *	0.002
	12	0.046 *	<0.001	0.882 *	0.001	0.095	0.927	0.011	1.000	−0.052 *	<0.001	−3.20 *	<0.001	−0.468 *	0.001
6	8	0.013 *	0.001	0.231 *	0.012	0.076	0.847	0.003	1.000	−0.008	1.000	−1.15 *	<0.001	−0.127 *	0.001
	10	0.020 *	0.004	0.402 *	0.031	0.087	0.832	0.006	1.000	−0.016	1.000	−2.39 *	<0.001	−0.250 *	0.001
	12	0.023 *	0.019	0.590 *	0.013	0.059	1.000	0.012	1.000	−0.029 *	0.018	−2.55 *	0.002	−0.365 *	0.001
8	10	0.008	0.303	0.170	0.223	0.011	1.000	0.002	1.000	−0.008	1.000	−1.25 *	0.007	−0.123 *	0.002
	12	0.011	1.000	0.359	0.062	−0.17	1.000	0.009	1.000	−0.021 *	0.021	−1.40	0.191	−0.238 *	0.002
10	12	0.003	1.000	0.188 *	0.048	−0.028	1.000	0.007	1.000	−0.013 *	0.003	−0.16	1.000	−0.115 *	0.002

* The mean difference is significant at the 0.05 level.

References

1. SAE International J3016_202104: Taxonomy and Definitions for Terms Related to Driving Automation Systems for On-Road Motor Vehicles. Available online: https://www.sae.org/standards/content/j3016_202104/ (accessed on 28 January 2023).
2. Jiang, L.; Chen, H.; Chen, Z. City Readiness for Connected and Autonomous Vehicles: A Multi-Stakeholder and Multi-Criteria Analysis through Analytic Hierarchy Process. *Transp. Policy* **2022**, *128*, 13–24. [CrossRef]
3. Miller, K.; Chng, S.; Cheah, L. Understanding Acceptance of Shared Autonomous Vehicles among People with Different Mobility and Communication Needs. *Travel Behav. Soc.* **2022**, *29*, 200–210. [CrossRef]
4. Susilawati, S.; Wong, W.J.; Pang, Z.J. Safety Effectiveness of Autonomous Vehicles and Connected Autonomous Vehicles in Reducing Pedestrian Crashes. *Transp. Res. Rec.* **2022**, *2677*, 1605–1618. [CrossRef]
5. Tao, T.; Cao, J. Examining Motivations for Owning Autonomous Vehicles: Implications for Land Use and Transportation. *J. Transp. Geogr.* **2022**, *102*, 103361. [CrossRef]
6. Maggi, D.; Romano, R.; Carsten, O.; Winter, J.C.F.D. When Terminology Hinders Research: The Colloquialisms of Transitions of Control in Automated Driving. *Cogn. Technol. Work.* **2022**, *24*, 509–520. [CrossRef]
7. Gold, C.; Damböck, D.; Lorenz, L.; Bengler, K. "Take over!" How Long Does It Take to Get the Driver Back into the Loop? *Proc. Hum. Factors Ergon. Soc. Annu. Meet.* **2013**, *57*, 1938–1942. [CrossRef]
8. Melcher, V.; Rauh, S.; Diederichs, F.; Widlroither, H.; Bauer, W. Take-Over Requests for Automated Driving. *Procedia Manuf.* **2015**, *3*, 2867–2873. [CrossRef]
9. McDonald, A.D.; Alambeigi, H.; Engström, J.; Markkula, G.; Vogelpohl, T.; Dunne, J.; Yuma, N. Toward Computational Simulations of Behavior During Automated Driving Takeovers: A Review of the Empirical and Modeling Literatures. *Hum. Factors* **2019**, *61*, 642–688. [CrossRef]
10. Agrawal, S.; Peeta, S. Evaluating the Impacts of Situational Awareness and Mental Stress on Takeover Performance under Conditional Automation. *Transp. Res. Part F Traffic Psychol. Behav.* **2021**, *83*, 210–225. [CrossRef]
11. Yi, B.; Cao, H.; Song, X.; Zhao, S.; Guo, W.; Li, M. How to Identify the Take-over Criticality in Conditionally Automated Driving? An Examination Using Drivers' Physiological Parameters and Situational Factors. *Transp. Res. Part F Traffic Psychol. Behav.* **2022**, *85*, 161–178. [CrossRef]
12. Parseh, M.; Asplund, F. New Needs to Consider during Accident Analysis: Implications of Autonomous Vehicles with Collision Reconfiguration Systems. *Accid. Anal. Prev.* **2022**, *173*, 106704. [CrossRef] [PubMed]
13. Xing, Y.; Lv, C.; Cao, D.; Hang, P. Toward Human-Vehicle Collaboration: Review and Perspectives on Human-Centered Collaborative Automated Driving. *Transp. Res. Part C Emerg. Technol.* **2021**, *128*, 103199. [CrossRef]
14. Hamadneh, J.; Esztergár-Kiss, D. Multitasking Onboard of Conventional Transport Modes and Shared Autonomous Vehicles. *Transp. Res. Interdiscip. Perspect.* **2021**, *12*, 100505. [CrossRef]
15. Zou, X.; Logan, D.B.; Vu, H.L. Modeling Public Acceptance of Private Autonomous Vehicles: Value of Time and Motion Sickness Viewpoints. *Transp. Res. Part C Emerg. Technol.* **2022**, *137*, 103548. [CrossRef]
16. Ayiku, J. *Economic Commission for Europe, Inland Transport Committee, Working Party on Road Traffic Safety Report of the Sixty-Eighth Session of the Working Party on Road Traffic Safety*; United Nations Economic Commission for Europe: Geneva, Switzerland, 2014.
17. Working Party on Automated/Autonomous and Connected Vehicles UNECE. Available online: https://unece.org/transport/vehicle-regulations/working-party-automatedautonomous-and-connected-vehicles-introduction (accessed on 27 September 2023).
18. Ivanov, A.M.; Shadrin, S.S.; Makarova, D.A. The Analysis of International Standards in the Field of Safety Regulation of Highly Automated and Autonomous Vehicles. In Proceedings of the 2022 Systems of Signals Generating and Processing in the Field of on Board Communications, Moscow, Russia, 15–17 March 2022; pp. 1–6.
19. *ISO 26262-1:2018*; Road Vehicles — Functional Safety. ISO: Geneva, Switzerland, 2018. Available online: https://www.iso.org/standard/68383.html (accessed on 27 September 2023).
20. *ISO 21448:2022*; Road Vehicles — Safety of the Intended Functionality. ISO: Geneva, Switzerland, 2022. Available online: https://www.iso.org/standard/77490.html (accessed on 27 September 2023).
21. Sanghavi, H.; Jeon, M.; Nadri, C.; Ko, S.; Sodnik, J.; Stojmenova, K. Multimodal Takeover Request Displays for Semi-Automated Vehicles: Focused on Spatiality and Lead Time. In *HCI in Mobility, Transport, and Automotive Systems*; Krömker, H., Ed.; Springer International Publishing: Cham, Switzerland, 2021; pp. 315–334.
22. Shi, D.; Wang, T.; Chen, Y.; Liu, C. Exploring the Effects of Request Time, Secondary Task, and Take-Over Mode on Take-Over Performance. In Proceedings of the 2021 6th International Conference on Intelligent Transportation Engineering (ICITE 2021), Beijing, China, 29–31 October 2021; Zhang, Z., Ed.; Springer Nature: Singapore, 2022; pp. 993–1006.
23. Tan, X.; Zhang, Y. The Effects of Takeover Request Lead Time on Drivers' Situation Awareness for Manually Exiting from Freeways: A Web-Based Study on Level 3 Automated Vehicles. *Accid. Anal. Prev.* **2022**, *168*, 106593. [CrossRef] [PubMed]
24. Eriksson, A.; Stanton, N.A. Takeover Time in Highly Automated Vehicles: Noncritical Transitions to and From Manual Control. *Hum. Factors* **2017**, *59*, 689–705. [CrossRef]
25. Jing, C.; Shang, C.; Yu, D.; Chen, Y.; Zhi, J. The Impact of Different AR-HUD Virtual Warning Interfaces on the Takeover Performance and Visual Characteristics of Autonomous Vehicles. *Traffic Inj. Prev.* **2022**, *23*, 277–282. [CrossRef] [PubMed]
26. Zeeb, K.; Buchner, A.; Schrauf, M. What Determines the Take-over Time? An Integrated Model Approach of Driver Take-over after Automated Driving. *Accid. Anal. Prev.* **2015**, *78*, 212–221. [CrossRef] [PubMed]

27. Endsley, M.R. Toward a Theory of Situation Awareness in Dynamic Systems. *Hum. Factors* **1995**, *37*, 32–64. [CrossRef]
28. Winner, H. ADAS, Quo Vadis? In *Handbook of Driver Assistance Systems: Basic Information, Components and Systems for Active Safety and Comfort*; Winner, H., Hakuli, S., Lotz, F., Singer, C., Eds.; Springer International Publishing: Cham, Switzerland, 2016; pp. 1557–1584, ISBN 978-3-319-12352-3.
29. Kridalukmana, R.; Lu, H.Y.; Naderpour, M. A Supportive Situation Awareness Model for Human-Autonomy Teaming in Collaborative Driving. *Theor. Issues Ergon. Sci.* **2020**, *21*, 658–683. [CrossRef]
30. Petermeijer, S.; Doubek, F.; Winter, J. de Driver Response Times to Auditory, Visual, and Tactile Take-over Requests: A Simulator Study with 101 Participants. In Proceedings of the 2017 IEEE International Conference on Systems, Man, and Cybernetics (SMC), Banff, AB, Canada, 5–8 October 2017; pp. 1505–1510.
31. Petermeijer, S.M.; Cieler, S.; de Winter, J.C.F. Comparing Spatially Static and Dynamic Vibrotactile Take-over Requests in the Driver Seat. *Accid. Anal. Prev.* **2017**, *99*, 218–227. [CrossRef] [PubMed]
32. Wan, J.; Wu, C. The Effects of Vibration Patterns of Take-Over Request and Non-Driving Tasks on Taking-Over Control of Automated Vehicles. *Int. J. Hum.–Comput. Interact.* **2018**, *34*, 987–998. [CrossRef]
33. Martinez, K.D.; Huang, G. In-Vehicle Human Machine Interface: Investigating the Effects of Tactile Displays on Information Presentation in Automated Vehicles. *IEEE Access* **2022**, *10*, 94668–94676. [CrossRef]
34. Mirnig, A.G.; Gärtner, M.; Wallner, V.; Demir, C.; Özkan, Y.D.; Sypniewski, J.; Meschtscherjakov, A. Enlightening Mode Awareness. *Pers. Ubiquitous Comput.* **2023**, *27*, 2307–2320. [CrossRef]
35. Borojeni, S.S.; Chuang, L.; Heuten, W.; Boll, S. Assisting Drivers with Ambient Take-Over Requests in Highly Automated Driving. In Proceedings of the 8th International Conference on Automotive User Interfaces and Interactive Vehicular Applications, Ann Arbor, MI, USA, 24–26 October 2016; ACM: New York, NY, USA, 2016; pp. 237–244.
36. Naujoks, F.; Mai, C.; Neukum, A. The Effect of Urgency of Take-Over Requests During Highly Automated Driving under Distraction Conditions. In *Advances in Human Aspects of Transportation Part I: Proceedings of AHFE International Conference (AHFE 2021), New York, NY, USA, 25–29 July 2021*; AHFE International: Louisville, KY, USA, 2021; pp. 431–438.
37. Petermeijer, S.; Bazilinskyy, P.; Bengler, K.; de Winter, J. Take-over Again: Investigating Multimodal and Directional TORs to Get the Driver Back into the Loop. *Appl. Ergon.* **2017**, *62*, 204–215. [CrossRef] [PubMed]
38. Politis, I.; Brewster, S.; Pollick, F. To Beep or Not to Beep? Comparing Abstract versus Language-Based Multimodal Driver Displays. In Proceedings of the 33rd Annual ACM Conference on Human Factors in Computing Systems, Seoul, Republic of Korea, 18–23 April 2015; Association for Computing Machinery: New York, NY, USA, 2015; pp. 3971–3980.
39. Gruden, T.; Tomažič, S.; Sodnik, J.; Jakus, G. A User Study of Directional Tactile and Auditory User Interfaces for Take-over Requests in Conditionally Automated Vehicles. *Accid. Anal. Prev.* **2022**, *174*, 106766. [CrossRef]
40. Stojmenova, K.; Policardi, F.; Sodnik, J. On the Selection of Stimulus for the Auditory Variant of the Detection Response Task Method for Driving Experiments. *Traffic Inj. Prev.* **2018**, *19*, 23–27. [CrossRef]
41. Huang, G.; Pitts, B.J. To Inform or to Instruct? An Evaluation of Meaningful Vibrotactile Patterns to Support Automated Vehicle Takeover Performance. *IEEE Trans. Hum.-Mach. Syst.* **2022**, *53*, 678–687. [CrossRef]
42. Wu, C.; Wu, H.; Lyu, N.; Zheng, M. Take-Over Performance and Safety Analysis Under Different Scenarios and Secondary Tasks in Conditionally Automated Driving. *IEEE Access* **2019**, *7*, 136924–136933. [CrossRef]
43. Kraut, V.; Motnikar, L.; Mirnig, A.; Fröhlich, P. Assertive Takeover Requests: Immediate and Sustained Effects on Stress and Performance. In Proceedings of the 15th International Conference on Automotive User Interfaces and Interactive Vehicular Applications, Ingolstadt, Germany, 17–21 September 2023; Association for Computing Machinery: New York, NY, USA, 2023; pp. 103–111.
44. Mukhopadhyay, A.; Sharma, V.K.; Gaikwad, P.T.; Sandula, A.K.; Biswas, P. Exploring the Use of XR Interfaces for Driver Assistance in Take Over Request. In *AutomotiveUI '22: Proceedings of the 14th International Conference on Automotive User Interfaces and Interactive Vehicular Applications, Seoul, Republic of Korea, 17–20 September 2022*; Association for Computing Machinery: New York, NY, USA, 2022; pp. 58–61.
45. Morales-Alvarez, W.; Certad, N.; Tadjine, H.H.; Olaverri-Monreal, C. Automated Driving Systems: Impact of Haptic Guidance on Driving Performance after a Take Over Request. In Proceedings of the 2022 IEEE Intelligent Vehicles Symposium (IV), Aachen, Germany, 4–9 June 2022; pp. 1817–1823.
46. Gruden, T.; Sodnik, J.; Jakus, G. Assisted Partial Take-Over in Conditionally Automated Driving: A User Study. *IEEE Access* **2023**, *11*, 107940–107957. [CrossRef]
47. Gruden, T.; Jakus, G. Determining Key Parameters with Data-Assisted Analysis of Conditionally Automated Driving. *Appl. Sci.* **2023**, *13*, 6649. [CrossRef]
48. Shull, E.M.; Gaspar, J.G.; McGehee, D.V.; Schmitt, R. Using Human–Machine Interfaces to Convey Feedback in Automated Driving. *J. Cogn. Eng. Decis. Mak.* **2022**, *16*, 29–42. [CrossRef]
49. Ma, S.; Zhang, W.; Yang, Z.; Kang, C.; Wu, C.; Chai, C.; Shi, J.; Zeng, Y.; Li, H. Take over Gradually in Conditional Automated Driving: The Effect of Two-Stage Warning Systems on Situation Awareness, Driving Stress, Takeover Performance, and Acceptance. *Int. J. Hum.–Comput. Interact.* **2021**, *37*, 352–362. [CrossRef]
50. Butmee, T.; Lansdown, T.C.; Walker, G.H. Alternative Options for Dealing with Automation Failures: Automated Stopping vs. Taking over Manual Control. *Transp. Res. Part F Traffic Psychol. Behav.* **2022**, *88*, 248–257. [CrossRef]

51. Pipkorn, L.; Dozza, M.; Tivesten, E. Driver Visual Attention Before and After Take-Over Requests During Automated Driving on Public Roads. *Hum. Factors* **2022**, *66*, 336–347. [CrossRef] [PubMed]
52. Feldhütter, A.; Gold, C.; Schneider, S.; Bengler, K. How the Duration of Automated Driving Influences Take-Over Performance and Gaze Behavior. In *Advances in Ergonomic Design of Systems, Products and Processes*; Schlick, C.M., Duckwitz, S., Flemisch, F., Frenz, M., Kuz, S., Mertens, A., Mütze-Niewöhner, S., Eds.; Springer: Berlin/Heidelberg, Germany, 2017; pp. 309–318.
53. Bourrelly, A.; de Naurois, C.J.; Zran, A.; Rampillon, F.; Vercher, J.-L.; Bourdin, C. Long Automated Driving Phase Affects Take-over Performance. *IET Intell. Transp. Syst.* **2019**, *13*, 1249–1255. [CrossRef]
54. Zhang, N.; Fard, M.; Xu, J.; Davy, J.L.; Robinson, S.R. Influence of Non-Driving Related Tasks on Driving Performance after Takeover Transition in Conditionally Automated Driving. *Transp. Res. Part F Traffic Psychol. Behav.* **2023**, *96*, 248–264. [CrossRef]
55. Kim, H.; Kim, W.; Kim, J.; Lee, S.-J.; Yoon, D.; Jo, J. A Study on Re-Engagement and Stabilization Time on Take-Over Transition in a Highly Automated Driving System. *Electronics* **2021**, *10*, 344. [CrossRef]
56. Gaspar, J.G.; Schwarz, C.W.; Ahmad, O.; Fincannon, T. Measuring the Transition Window in Conditional Automation. In Proceedings of the 27th International Technical Conference on the Enhanced Safety of Vehicles (ESV), Yokohama, Japan, 3–6 April 2023; National Highway Traffic Safety Administration: Washington, DC, USA, 2023.
57. Riahi Samani, A.; Mishra, S. How Long the Effect of Take-Over Conditions Lasts? A Survival Analysis of Commercial Motor Vehicle Drivers' Reaction Time and Driving Behavior in Highly Automated Vehicles. *Surviv. Anal. Commer. Mot. Veh. Drivers' React. Time Driv. Behav. Highly Autom. Veh.* **2023**, 4592428. [CrossRef]
58. Choi, D.; Sato, T.; Ando, T.; Abe, T.; Akamatsu, M.; Kitazaki, S. Effects of Cognitive and Visual Loads on Driving Performance after Take-over Request (TOR) in Automated Driving. *Appl. Ergon.* **2020**, *85*, 103074. [CrossRef]
59. Wu, Y.; Kihara, K.; Hasegawa, K.; Takeda, Y.; Sato, T.; Akamatsu, M.; Kitazaki, S. Age-Related Differences in Effects of Non-Driving Related Tasks on Takeover Performance in Automated Driving. *J. Saf. Res.* **2020**, *72*, 231–238. [CrossRef] [PubMed]
60. Yun, H.; Yang, J.H. Multimodal Warning Design for Take-over Request in Conditionally Automated Driving. *Eur. Transp. Res. Rev.* **2020**, *12*, 34. [CrossRef]
61. Wang, Y.; Zhang, W.; Zhou, R. Speech-Based Takeover Requests in Conditionally Automated Driving: Effects of Different Voices on the Driver Takeover Performance. *Appl. Ergon.* **2022**, *101*, 103695. [CrossRef]
62. Gold, C.; Körber, M.; Lechner, D.; Bengler, K. Taking Over Control From Highly Automated Vehicles in Complex Traffic Situations: The Role of Traffic Density. *Hum. Factors* **2016**, *58*, 642–652. [CrossRef] [PubMed]
63. Wintersberger, P.; Schartmüller, C.; Shadeghian-Borojeni, S.; Frison, A.-K.; Riener, A. Evaluation of Imminent Take-Over Requests With Real Automation on a Test Track. *Hum. Factors* **2021**, *65*, 1776–1792. [CrossRef] [PubMed]
64. Shahini, F.; Park, J.; Welch, K.; Zahabi, M. Effects of Unreliable Automation, Non-Driving Related Task, and Takeover Time Budget on Drivers' Takeover Performance and Workload. *Ergonomics* **2022**, *66*, 182–197. [CrossRef] [PubMed]
65. Dillmann, J.; den Hartigh, R.J.R.; Kurpiers, C.M.; Pelzer, J.; Raisch, F.K.; Cox, R.F.A.; de Waard, D. Keeping the Driver in the Loop through Semi-Automated or Manual Lane Changes in Conditionally Automated Driving. *Accid. Anal. Prev.* **2021**, *162*, 106397. [CrossRef] [PubMed]
66. Čegovnik, T.; Stojmenova, K.; Jakus, G.; Sodnik, J. An Analysis of the Suitability of a Low-Cost Eye Tracker for Assessing the Cognitive Load of Drivers. *Appl. Ergon.* **2018**, *68*, 1–11. [CrossRef] [PubMed]
67. Zhou, F.; Yang, X.J.; de Winter, J.C.F. Using Eye-Tracking Data to Predict Situation Awareness in Real Time During Takeover Transitions in Conditionally Automated Driving. *IEEE Trans. Intell. Transp. Syst.* **2021**, *23*, 2284–2295. [CrossRef]
68. Marti, P.; Jallais, C.; Koustanaï, A.; Guillaume, A.; Mars, F. Effect of Driver's Distraction on Gaze Behaviour, Mental Workload and Takeover Quality. In Proceedings of the Driving Simulation Conference Europe 2021 VR, Munich, Germany, 14–17 September 2021; Driving Simulation Association: Boulogne-Billancourt, France, 2021; pp. 99–106.
69. Stephenson, A.C.; Eimontaite, I.; Caleb-Solly, P.; Morgan, P.L.; Khatun, T.; Davis, J.; Alford, C. Effects of an Unexpected and Expected Event on Older Adults' Autonomic Arousal and Eye Fixations During Autonomous Driving. *Front. Psychol.* **2020**, *11*, 571961. [CrossRef]
70. Du, N.; Zhou, F.; Tilbury, D.M.; Robert, L.P.; Jessie Yang, X. Behavioral and Physiological Responses to Takeovers in Different Scenarios during Conditionally Automated Driving. *Transp. Res. Part F Traffic Psychol. Behav.* **2024**, *101*, 320–331. [CrossRef]
71. Gruden, T.; Stojmenova, K.; Sodnik, J.; Jakus, G. Assessing Drivers' Physiological Responses Using Consumer Grade Devices. *Appl. Sci.* **2019**, *9*, 5353. [CrossRef]
72. Perello-March, J.R.; Burns, C.G.; Woodman, R.; Elliott, M.T.; Birrell, S.A. Driver State Monitoring: Manipulating Reliability Expectations in Simulated Automated Driving Scenarios. *IEEE Trans. Intell. Transp. Syst.* **2021**, *23*, 5187–5197. [CrossRef]
73. Müller, A.L.; Fernandes-Estrela, N.; Hetfleisch, R.; Zecha, L.; Abendroth, B. Effects of Non-Driving Related Tasks on Mental Workload and Take-over Times during Conditional Automated Driving. *Eur. Transp. Res. Rev.* **2021**, *13*, 16. [CrossRef]
74. Deng, M.; Gluck, A.; Zhao, Y.; Li, D.; Menassa, C.C.; Kamat, V.R.; Brinkley, J. An Analysis of Physiological Responses as Indicators of Driver Takeover Readiness in Conditionally Automated Driving. *Accid. Anal. Prev.* **2024**, *195*, 107372. [CrossRef] [PubMed]
75. Kerautret, L.; Dabic, S.; Navarro, J. Exploration of Driver Stress When Resuming Control from Highly Automated Driving in an Emergency Situation. *Transp. Res. Part F Traffic Psychol. Behav.* **2023**, *93*, 222–234. [CrossRef]
76. Gruden, T.; Stojmenova Pečečnik, K.; Jakus, G.; Sodnik, J. Quantifying Drivers' Physiological Responses to Take-Over Requests in Conditionally Automated Vehicles. In Proceedings of the 7th Human-Computer Interaction Slovenia Conference 2022, Ljubljana, Slovenia, 29 November 2022; Stojmenova Pečečnik, K., Jakus, G., Eds.; CEUR: Aachen, Germany, 2022; Volume 3300.

77. Zeeb, K.; Buchner, A.; Schrauf, M. Is Take-over Time All That Matters? The Impact of Visual-Cognitive Load on Driver Take-over Quality after Conditionally Automated Driving. *Accid. Anal. Prev.* **2016**, *92*, 230–239. [CrossRef] [PubMed]
78. Kim, J.W.; Yang, J.H. Understanding Metrics of Vehicle Control Take-Over Requests in Simulated Automated Vehicles. *Int. J. Automot. Technol.* **2020**, *21*, 757–770. [CrossRef]
79. Radlmayr, J.; Weinbeer, V.; Löber, C.; Farid, M.; Bengler, K. How Automation Level and System Reliability Influence Driver Performance in a Cut-In Situation. In *Advances in Human Aspects of Transportation*; Stanton, N.A., Ed.; Springer International Publishing: Berlin/Heidelberg, Germany, 2018; pp. 684–694.
80. Kim, Y.W.; Kim, D.Y.; Yoon, S.H. Understanding Driver's Situation Awareness in Highly Automated Driving. In *Adjunct Proceedings of the 14th International Conference on Automotive User Interfaces and Interactive Vehicular Applications, Seoul, Republic of Korea, 17–20 September 2022*; Association for Computing Machinery: New York, NY, USA, 2022; pp. 132–136.
81. Stojmenova Pečečnik, K.; Sodnik, J. Evalvacija projekcijskega zaslona za pogojno avtomatizirana vozila. *Elektrotehniški Vestn.* **2022**, *89*, 269–274.
82. Stojmenova Pečečnik, K.; Tomažič, S.; Sodnik, J. Design of Head-up Display Interfaces for Automated Vehicles. *Int. J. Hum. -Comput. Stud.* **2023**, *177*, 103060. [CrossRef]
83. Strle, G.; Košir, A.; Sodnik, J.; Stojmenova Pečečnik, K. Physiological Signals as Predictors of Cognitive Load Induced by the Type of Automotive Head-Up Display. *IEEE Access* **2023**, *11*, 87835–87848. [CrossRef]
84. Vengust, M.; Kaluža, B.; Stojmenova, K.; Sodnik, J. NERVteh Compact Motion Based Driving Simulator. In Proceedings of the 9th International Conference on Automotive User Interfaces and Interactive Vehicular Applications Adjunct, Oldenburg, Germany, 24–27 September 2017; Association for Computing Machinery: New York, NY, USA, 2017; pp. 242–243.
85. Sekar, R.; Jacome, O.; Chrstos, J.; Stockar, S. *Assessment of Driving Simulator for Longitudinal Vehicle Dynamics Evaluation*; SAE International: Warrendale, PA, USA, 2022.
86. Kuiper, O.X.; Bos, J.E.; Diels, C.; Cammaerts, K. Moving Base Driving Simulators' Potential for Carsickness Research. *Appl. Ergon.* **2019**, *81*, 102889. [CrossRef] [PubMed]
87. Bellem, H.; Klüver, M.; Schrauf, M.; Schöner, H.-P.; Hecht, H.; Krems, J.F. Can We Study Autonomous Driving Comfort in Moving-Base Driving Simulators? A Validation Study. *Hum. Factors* **2017**, *59*, 442–456. [CrossRef] [PubMed]
88. AVSimulation SCANeR Studio. Available online: https://www.avsimulation.com/scanerstudio/ (accessed on 12 August 2020).
89. Tobii Pro Glasses 2 Wearable Eye Tracker. Available online: https://www.tobiipro.com/product-listing/tobii-pro-glasses-2/ (accessed on 6 February 2020).
90. E4 Wristband | Real-Time Physiological Signals | Wearable PPG, EDA, Temperature, Motion Sensors. Available online: https://www.empatica.com/research/e4 (accessed on 12 August 2020).
91. Cirulli, G. 2048. Available online: https://play2048.co/ (accessed on 23 March 2024).
92. Brooke, J. SUS: A "Quick and Dirty" Usability Scale. In *Usability Evaluation In Industry*; CRC Press: Boca Raton, FL, USA, 1996; ISBN 978-0-429-15701-1.
93. UEQ—User Experience Questionnaire. Available online: https://www.ueq-online.org/ (accessed on 24 March 2024).
94. Van Der Laan, J.D.; Heino, A.; De Waard, D. A Simple Procedure for the Assessment of Acceptance of Advanced Transport Telematics. *Transp. Res. Part C Emerg. Technol.* **1997**, *5*, 1–10. [CrossRef]
95. Pečečnik, K.S.; Mirnig, A.; Meschtscherjakov, A.; Sodnik, J. Evaluation of Head-Up Display for Conditionally Automated Vehicles. In *Disruptive Information Technologies for a Smart Society*; Trajanovic, M., Filipovic, N., Zdravkovic, M., Eds.; Springer Nature: Cham, Switzerland, 2024; pp. 174–182.
96. Du, N.; Yang, X.J.; Zhou, F. Psychophysiological Responses to Takeover Requests in Conditionally Automated Driving. *Accid. Anal. Prev.* **2020**, *148*, 105804. [CrossRef] [PubMed]
97. Weigl, K.; Schartmüller, C.; Wintersberger, P.; Steinhauser, M.; Riener, A. The Influence of Experienced Severe Road Traffic Accidents on Take-over Reactions and Non-Driving-Related Tasks in an Automated Driving Simulator Study. *Accid. Anal. Prev.* **2021**, *162*, 106408. [CrossRef] [PubMed]
98. Dawson, M.E.; Schell, A.M.; Filion, D.L. The Electrodermal System. In *Handbook of Psychophysiology*, 3rd ed.; Cambridge University Press: New York, NY, USA, 2007; pp. 159–181, ISBN 978-0-521-84471-0.
99. Appelhans, B.M.; Luecken, L.J. Heart Rate Variability as an Index of Regulated Emotional Responding. *Rev. Gen. Psychol.* **2006**, *10*, 229–240. [CrossRef]
100. Laborde, S.; Mosley, E.; Thayer, J.F. Heart Rate Variability and Cardiac Vagal Tone in Psychophysiological Research—Recommendations for Experiment Planning, Data Analysis, and Data Reporting. *Front. Psychol.* **2017**, *8*, 213. [CrossRef] [PubMed]
101. Greco, A.; Valenza, G.; Lanata, A.; Scilingo, E.P.; Citi, L. cvxEDA: A Convex Optimization Approach to Electrodermal Activity Processing. *IEEE Trans. Biomed. Eng.* **2016**, *63*, 797–804. [CrossRef]
102. Anaconda Python. Available online: https://www.anaconda.com/ (accessed on 24 March 2024).
103. IBM Corp. *IBM SPSS Statistics for Windows, Version 23.0*; IBM Corp: Armonk, NY, USA, 2015.
104. Pfleging, B.; Fekety, D.K.; Schmidt, A.; Kun, A.L. A Model Relating Pupil Diameter to Mental Workload and Lighting Conditions. In Proceedings of the 2016 CHI Conference on Human Factors in Computing Systems, San Jose, CA, USA, 7–12 May 2016; Association for Computing Machinery: New York, NY, USA, 2016; pp. 5776–5788.

105. Belliveau, A.P.; Somani, A.N.; Dossani, R.H. Pupillary Light Reflex. In *StatPearls*; StatPearls Publishing: Treasure Island, FL, USA, 2024.
106. Naujoks, F.; Wiedemann, K.; Schömig, N.; Hergeth, S.; Keinath, A. Towards Guidelines and Verification Methods for Automated Vehicle HMIs. *Transp. Res. Part F Traffic Psychol. Behav.* **2019**, *60*, 121–136. [CrossRef]

sensors

MDPI

Article

Effect of Weight Distribution and Active Safety Systems on Electric Vehicle Performance

Valerio Gori [1], Will Hendrix [2], Amritam Das [2] and Zhiyong Sun [2,*]

[1] Faculty of Mechanical Engineering, Delft University of Technology, 2628 CD Delft, The Netherlands; v.gori@student.tudelft.nl
[2] Department of Electrical Engineering, Eindhoven University of Technology, 5612 AZ Eindhoven, The Netherlands; w.h.a.hendrix@tue.nl (W.H.); am.das@tue.nl (A.D.)
* Correspondence: z.sun@tue.nl

Abstract: This paper describes control methods to improve electric vehicle performance in terms of handling, stability and cornering by adjusting the weight distribution and implementing control systems (e.g., wheel slip control, and yaw rate control). The vehicle is first simulated using the bicycle model to capture the dynamics. Then, a study on the effect of weight distribution on the driving behavior is conducted. The study is performed for three different weight configurations. Moreover, a yaw rate controller and a wheel slip controller are designed and implemented to improve the vehicle's performance for cornering and longitudinal motion under the different loading conditions. The simulation through the bicycle model is compared to the experiments conducted on a rear-wheel driven radio-controlled (RC) electric vehicle. The paper shows how the wheel slip controller contributes to the stabilization of the vehicle, how the yaw rate controller reduces understeering, and how the location of the center of gravity (CoG) affects steering behavior. Lastly, an analysis of the combination of control systems for each weight transfer is conducted to determine the configuration with the highest performance regarding acceleration time, braking distance, and steering behavior.

Keywords: electric vehicle; weight distribution; active safety systems; wheel slip controller; torque vectoring

Citation: Gori, V.; Hendrix, W.; Das, A.; Sun, Z. Effect of Weight Distribution and Active Safety Systems on Electric Vehicle Performance. *Sensors* **2024**, *24*, 3557. https://doi.org/10.3390/s24113557

Academic Editors: Guillermo Villanueva, Yafei Wang, Chao Huang, Hailong Huang, Zhiqiang Zuo, Bo Leng and Peng Hang

Received: 20 December 2023
Revised: 26 April 2024
Accepted: 22 May 2024
Published: 31 May 2024

1. Introduction

Electric vehicles (EVs) have increased in popularity due to their influence on reducing greenhouse gas emissions, reducing the impact of pollution on human health and hence contributing to a cleaner environment [1,2]. According to the market outlook, 58% of new car sales will be of electric vehicles by 2040 [3]. Due to the electrification in the automotive world, the possibility of the functionality of a vehicle has significantly been affected. Research aimed at improving the safety of vehicles to reduce car accident fatalities has increased substantially. An example is an active stability control system that prevents vehicles from spinning, drifting, and rolling over. The most commercialized stability control systems are based on differential braking and torque vectoring which apply a different braking or driving torque to each driving wheel to achieve the desired yaw moment, respectively. This can be achieved when the wheels are driven separately by two electric motors [4,5]. Moreover, electric vehicles, due to their architecture, offer greater potential compared to conventional vehicles regarding longitudinal motion.

EVs with individually driven wheels allow the development of control algorithms that can significantly improve vehicle performance through anti-lock braking system (ABS) and traction control (TC) [6]. Such systems are essential in vehicles as they assist the driver to keep the vehicle stable and follow the desired trajectory. The systems are based on the feedback control of the lateral dynamics parameters, such as the side-slip angle and yaw rate responses. Some researchers have focused solely either on the control of the yaw rate response to increase responsiveness to steering inputs or on the feedback control of the side slip angles to enhance stability; others attempted to combine both feedback controls to further increase vehicle stability performance [7]. To achieve this, various control algorithms have been developed and presented in the literature: for example, regarding the electronic stability control, PID, state feedback control, optimal control and sliding mode-based controllers have been used [8]. Das et al. proposed a modular hierarchical control architecture for multi-wheeled vehicles [9]. M. K. Aripin et al. evaluated a non-linear feedback algorithm and sliding mode for yaw rate control [10]. Benoit Lacroix et al. conducted a study to compare different methods on direct yaw moment control (i.e., PID and sliding mode) using a 2-DOF vehicle model [7]. Similar methods were implemented by Andoni Medina et al. who compared typical control methods used for ensuring vehicle stability and improving lap time for electric racing cars using PID and sliding controllers [11] as well as Leonardo De Novellis, et al. who analyzed and compared different PID and sliding mode-based control techniques (e.g., SOSM controllers) [12]. Gökhan Tekin et al. developed a fuzzy logic control scheme for active yaw rate and side slip angles feedback control [13]. Haiping Du et al. analyzed the yaw rate and side slip angle responses of a vehicle when applying a controller based on a finite numbers of linear matrix inequalities (LMIs) [14]. Alberto Parra et al. presented a study on nonlinear model predictive controller on EV with multiple drive trains to enhance energy efficiency through the control of the cornering performance [15]. Last, A. Parra et al. and Q. Lu et al. proposed intelligent and H_∞ controllers, respectively [16,17].

Furthermore, weight distribution is a key parameter in road vehicle design as different loading conditions may lead to more aggressive under- or oversteering behavior aggravating the stability of a vehicle. Weight distribution also influences the maximum force that can be transmitted to the wheels. Research has been conducted to analyze the dependency of weight distribution on the driving behavior. Ekalak Prompakdee et al. conducted a research aimed at studying the relation of the understeer gradient with the weight distributions on intercity buses under steady state conditions [18]. The driving performances under various loading conditions have also been analyzed in [19] to study the effect on the braking distance in road freight transport. An analysis on weight distribution aimed at maximizing the cornering speed of formula cars has also been conducted by H. Nozaki [20]. Lastly, a lot of research has been ongoing for developing anti-lock braking systems and traction controllers: often, such systems are based on wheel slip controllers; however, various methods have been investigated. Regarding the control schemes, PID is most often used [21]. For example, Min et al. show the performance of PID and fuzzy controller on ABS development [22]. Taketoshi Kawabe et al. developed a wheel slip controller based on sliding mode for commercial vehicles on low friction roads [23]. Ma et al. evaluated the performance of wheel slip controller based on model predictive control considering road roughness and low adhesion surfaces [24]. Dzmitry Savitski et al. compared PI, first-order sliding mode, integral sliding mode and continuous twisting algorithms applied to a wheel slip controller on fully electric vehicles [25].

This research combines the implementation of a yaw rate controller and a wheel slip controller to improve longitudinal and cornering performance to different weight distributions in order to determine the configuration with the highest performance in terms of safety, handling and stability. To conclude, the literature presents studies on the implementation of effective electronic stability controls, on wheel slip controllers for ABS or TC and also on the effect of loading condition on driving behavior for different vehicles; however, a study combining the weight distribution analysis and the implementation of control system algorithms to fully improve performance on rear-wheel driving electric vehicles is missing. In other words, the article is aimed at showing how weight distribution and control systems can be designed and combined to improve the driving behavior.

This paper is organized as follows: Section 2 shows the experimental set up and the methods used to model the vehicle and implement the control schemes. Then the results are shown and discussed in Sections 3 and 4, respectively. Finally, some concluding remarks are made in Section 5. A video on the results obtained can be seen at the following url: https://youtu.be/wRzeLYJABbQ (accessed on 25 May 2024).

2. Materials and Methods

2.1. Experimental Set Up

The RC car used to perform the experiments is an FG Competition EVO 08-510 (FG Modellsport GmbH, Winterbach, Germany). The vehicle is in scale 1:5; about five times smaller than a real car. The main components can be seen in Figure 1 and an overview on the processor boards and their connections is shown in Figure 2. The vehicle originally had an internal combustion engine which had been replaced by two electric motors, each driving the rear wheel through a gearbox designed in-house at the Eindhoven University of Technology. The vehicle is driven by a remote controller which sends the steering and throttle percentage to the receiver on board. The steering percentage is directly sent to the front tires servos, while the throttle percentage is first sent to the DSP. By implementing control schemes on the DSP it is possible to actively control the torque that is delivered to each driving wheel. The car originally had friction brakes; however, they have been disconnected and the vehicle is braked through the electric motors on the rear tires. The motors are three-phase synchronous machines with a maximum output power of 419 W corresponding to a maximum torque of 5.29 Nm delivered to each driving tire. In case of emergency, the motor can be short-circuited through a resistive circuit. To power the vehicle, a Makita 40 V, 4 Ah Li-ion battery pack is used. For safety reason a battery management system is implemented: if the battery voltage drops below 3 Volts, the motors are deactivated and short-circuited.

The torque setpoints for the left and right rear motor (from the control actions) are sent using PWM signals to the power board electronics. The full mechanical range of the throttle handle on the RC transmitter is mapped to a maximum of 0–100% value in the forward driving direction (pulling the handle) and 0–100% in the backward driving direction (pushing the handle). The reading on the display shows the current value. This mapping of the mechanical range is set by the End Point Adjustment (EPA) value. The throttle handle and torque generation are shown in Figure 3.

The DSP is equipped with a Texas Instruments (TI)eZDSP F28355 board (Dallas, TX, USA) and AMBER wireless data transfer system (Trier, Germany) that allows the data to be transferred to a PC with a frequency of 200 Hz through the data logger unit. The board is supported by Matlab embedded encoder. Hence, the control schemes can be directly implemented on Matlab and Simulink 2021b [26]. The program is implemented and uploaded on the vehicle through the USB programming cable. Last, the vehicle is equipped with the following sensors:

- wheel speed sensors on each tire
- gyroscope

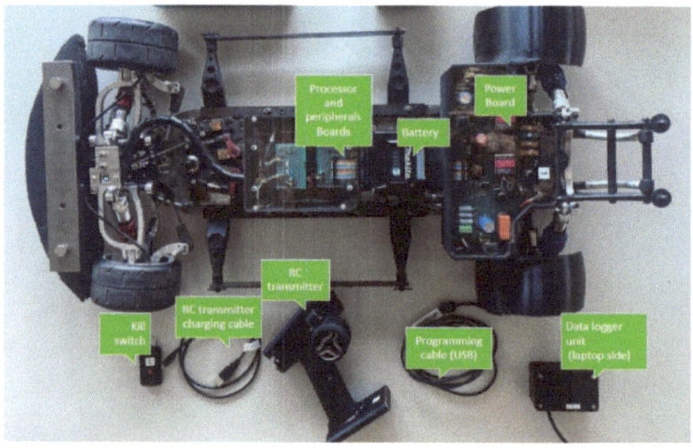

Figure 1. Experimental set up: vehicle system.

The wheel speed sensors are located on the inside of each wheel. They consist of six black and six white stripes of film located along the circumference in the inside of the wheel. As the wheel rotates, the transition of the black and white regions are detected by CNY70 optical sensors located on a PCB inside the wheel. The velocity is calculated from the time elapsed between each transition. The gyroscope is an MPU-6000 (InvenSense Inc., Sunnyvale, CA, USA) that calculates the rotation around the z-axis (i.e., the yaw rate). To conclude, the data gathered from the gyroscope and speed sensor allow for the implementation of feedback to individually control the torque in each motor.

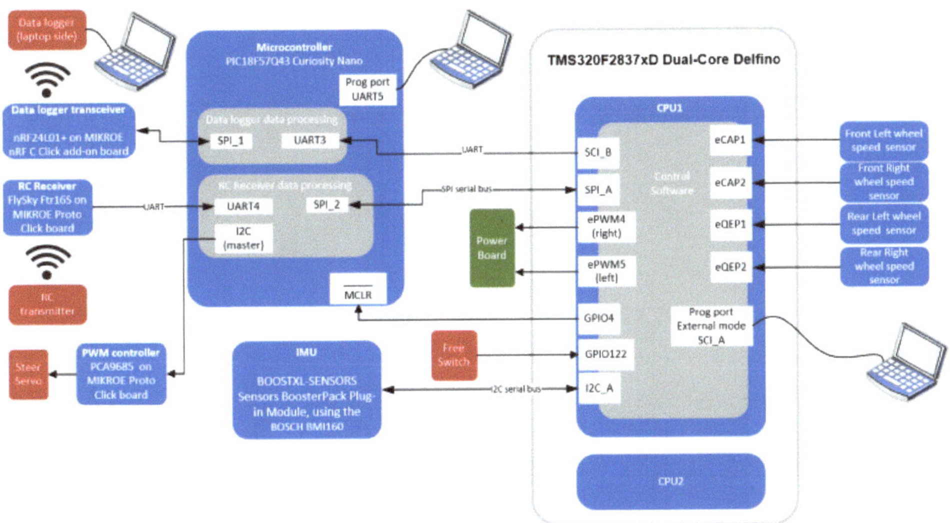

Figure 2. Experimental set up: software system overview.

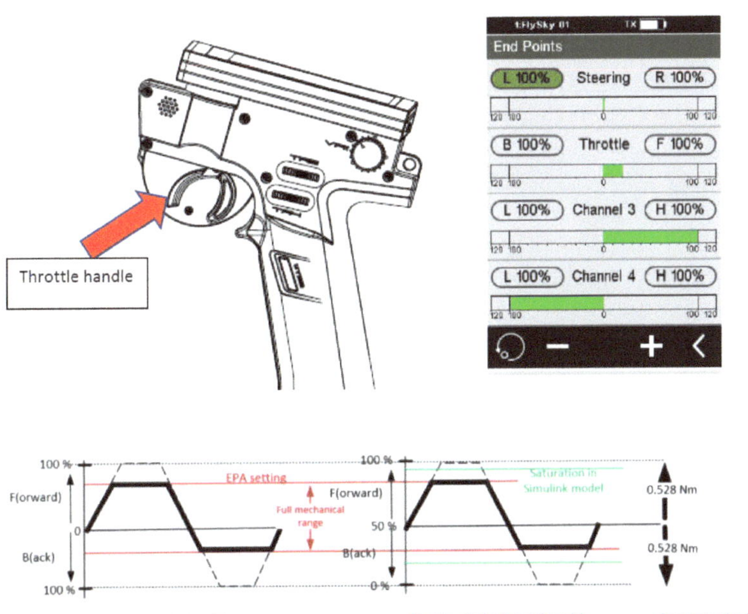

Figure 3. Experimental set up: throttle handle and torque generation.

2.2. Vehicle Dynamics Modeling

The main factor for analyzing the weight distribution of a vehicle is the position of its center of gravity (CoG). Determining the CoG involves weighing the car. As depicted in Figure 4, each tire is precisely positioned on the center point of a scale to measure its weight. Furthermore, the vehicle configuration ensures that the Center of Gravity (CoG) is equidistant from the left and right wheels. It results in equal forces being experienced by the left and right tires.

Figure 4. Vehicle on scales for weight measurement.

Next, given that the car is standstill, the following equations of motion are derived:

$$\begin{cases} I_{yy}\dot{\omega}_y = F_{z1}a - F_{z2}b = 0 \\ l = a + b \\ mg = F_{z1} + F_{z2} \end{cases} \tag{1}$$

where F_{z1} and F_{z2} are the normal forces acting on the front and rear tire, respectively, obtained by the measurements conducted as in Figure 4. I_{yy} is the pitch mass moment of inertia, $\dot{\omega}_y$ is the pitch acceleration, l is the wheelbase and m is the mass. Solving the equations above for each weight configuration allows to derive the corresponding a and b values representing the distance between the CoG and the front and rear tires, respectively, as shown below:

$$\begin{cases} a = \frac{F_{z2}}{mg} l \\ b = \frac{F_{z1}}{mg} l. \end{cases} \tag{2}$$

In order to investigate various loading conditions, weights were placed on the front and on the rear of the vehicle to achieve different CoG locations. More specifically, the weights have been attached to the tail (Figure 5a) and to the front bumper (Figure 5b) to maximize the change in weight. The set up for the testing vehicle with front- and rear-loaded weights is illustrated in Figure 5. The experiments were conducted on these two configurations as they represent the most extreme loading condition possible for the vehicle in consideration. In order to validate the yaw rate and wheel slip controllers, the tests were conducted on the unloaded vehicle shown in Figure 4.

(**a**) Rear loaded vehicle

(**b**) Front loaded vehicle

Figure 5. Vehicle loading condition.

2.3. Vehicle Model

This section deals with the differential equation used to describe and model the vehicle behavior. For the scope of this research, the bicycle model has been used due to its simplicity, computational efficiency, and accuracy related to the control objective.

The bicycle model consists of the longitudinal (u), lateral (v) and yaw motion (r) as shown in ([27] Figure 1.9) which shows the global coordinates denoted as X and Y, the lateral and longitudinal directions, u and v, together with the yaw rate (r) moments denoted as M_z and the parameters a, b, l. Moreover, the model is based on the following assumptions:

- The left and right tires are lumped together in one equivalent tire.
- Pitch and roll are not taken into account: the height of the center of gravity is assumed to be zero.
- The vehicle is assumed to drive on a flat surface.

We remark that these assumptions are reasonable in practice. The first assumption is for deriving a bicycle model; the second assumption is based on the bicycle dynamics (in a 2-D space); the third assumption is according to the actual testing environment. In order to model the vehicle, two reference frames are used, i.e., the global or ground reference frame (X, Y) and the body reference frame frame (i.e., the one relative to the vehicle direction), denoted by (u,v,r) where the u-axis is the the longitudinal axis of the vehicle. The origin of the body frame is given by the center of gravity [28]. The subscript x and y are used to indicate the longitudinal and lateral directions, respectively. Next, the forces and moments acting on the rear-wheel driven vehicle are determined and the following equations are derived to describe the vehicle behavior in the body frame ([27]).

$$\begin{cases} F_u &= F_{x2} - \sin(\delta)F_{y1} - F_d \\ F_v &= F_{y2} + \cos(\delta)F_{y1} \\ M_z &= M_{tv} + a(\cos(\delta))F_{y1} - bF_{y2}. \end{cases} \tag{3}$$

where M_z denotes the yaw moment around the z-axis. For simplicity, this model only considers the yaw moment around the z-axis denoted as M_z rather than the moments around each tire. The model also takes into account the drag forces F_d (F_d is a sum of rolling resistance and air resistance forces.) Moreover, since the test vehicle is RWD, the longitudinal force on the rear tires is equal to zero. The relation between torque and throttle is assumed to be linear, hence the force applied to the driving wheel F_{x2} is related to the input torque through the wheel radius r_w. The vehicle drag forces are given as a combination of rolling resistance, and aerodynamic drag. For the scope of this research, such forces have not been measured individually, but the combined resistance force denoted by F_d is calculated experimentally from a coast-down test. M_{tv} represents the extra moment due to torque vectoring, and is determined by the following equation:

$$M_{tv} = \frac{\Delta T}{r_w} w, \quad (4)$$

where ΔT is the torque difference applied to the wheels by the yaw rate controller, e.g., the output of the controller. w is the vehicle width. Furthermore, the vehicle trajectory in terms of the global coordinate systems is derived as follows:

$$\begin{cases} \dot{X} = u\cos(\psi) - v\sin(\psi) \\ \dot{Y} = u\sin(\psi) + v\cos(\psi) \\ \dot{\psi} = r, \end{cases} \quad (5)$$

where ψ is the angle between the body and the global reference frame, and r denotes the angular velocity. Combining Equations (3) and (5) leads to the equations of motion according to Newton's laws:

$$\begin{cases} m\dot{X} = F_u\cos(\psi) - F_v\sin(\psi) \\ m\dot{Y} = F_u\sin(\psi) + F_v\cos(\psi) \\ I_{zz}\dot{r} = M_z, \end{cases} \quad (6)$$

where I_{zz} is yaw moment of inertia. After taking the derivative with respect to time for Equation (5) and substituting it in Equation (6), the vehicle model below is found:

$$\begin{cases} m(\dot{u} - vr) = F_u \\ m(\dot{v} + ur) = F_v \\ I_{zz}\dot{r} = M_z. \end{cases} \quad (7)$$

2.4. Tire Lateral Dynamics and Steering Behavior

The tire is under the effect of a vertical load and a lateral force when turning which contributes to the vehicle heading angle. The tire lateral forces (F_{y1}, F_{y2}) have a non-linear relation with the side slip angles. However, by keeping the angles small, (α_1 and α_2, respectively) the lateral forces are assumed to be linearly proportional to the side slip angles. Such an angle is defined as the angle between the tire orientation and its velocity vector. The front and rear side slip angles are given in Equation (8), respectively.

$$\begin{cases} \alpha_1 = \delta - \arctan\frac{(v+ar)}{u} \\ \alpha_2 = \arctan\frac{(v-br)}{u}. \end{cases} \quad (8)$$

Assuming linear tire behavior, the side slip angles are related to the lateral forces through the cornering stiffness (C_1 and C_2 for the front and the rear tires, respectively). The lateral forces can therefore be written as [5]:

$$\begin{cases} F_{y1} \approx \alpha_1 C_1 \\ F_{y2} \approx \alpha_2 C_2. \end{cases} \quad (9)$$

Last, an indicator of the vehicle cornering behavior is the understeer gradient η. It indicates the path curvature of the vehicle that results from a given steering angle δ at any speed. Given that the steering angle is expressed as the combination of the kinematic steering angle and the additional angle due to the lateral acceleration (a_y), δ can be written as:

$$\delta = \frac{l}{R} + \eta\frac{a_y}{g} = \frac{l}{R} + \alpha_1 - \alpha_2 \quad (10)$$

where R is the turning radius, a_y the lateral acceleration, g the gravitational acceleration, and η the under-steer gradient. Expressing the side slip angles as a function of the vertical forces allows us to express the steering angle as a function of lateral acceleration and static vertical load as shown below:

$$\delta = \frac{l}{R} + \frac{a_y}{g}(\frac{F_{z1}}{C_1} - \frac{F_{z2}}{C_2}). \tag{11}$$

The under-steer gradient is given as:

$$\eta = \left(\frac{F_{z1}}{C_1} - \frac{F_{z2}}{C_2}\right). \tag{12}$$

Note that dynamic load transfer should not be taken into account in this equations, hence the vertical forces F_{z1} and F_{z2} represent the static weight distribution. The driving behavior is related to the previous equations according to the following relation:

$$\begin{cases} \eta = 0 & \text{"neutral steer" for } \alpha_1 = \alpha_2 \\ \eta > 0 & \text{"under steer" for } \alpha_1 > \alpha_2 \\ \eta < 0 & \text{"over steer" for } \alpha_1 < \alpha_2 \end{cases} \tag{13}$$

In other words, to maintain a constant cornering radius R the steering angle has to increase for an understeered vehicle, decrease for an oversteered one and remain the same for a neutral steered vehicle [27].

The equations described in this section have been implemented in Matlab and Simulink to model the vehicle, given a torque and steering percentage as inputs. The data to validate the model have been gathered from a constant cornering experiment. In other words, the vehicle was driven along a constant radius with slowly increasing throttle. Then, the model was fitted to the measurement data by tuning the cornering stiffness values.

2.5. Yaw Rate Controller Design

The following paragraph deals with the implementation of a torque vectoring algorithm. In this research, a PID-type controller is applied due to its practicality, simplicity and effectiveness compared to other control methods [8,29]. However, due to the high frequency noise in the sensors, a PI controller is chosen. The controller designed in this paper is designed to impose a certain yaw rate to the vehicle. In other words, it is designed to minimize the error between the vehicle measured yaw rate and the reference (i.e., corresponding yaw rate for neutral steering condition) by redistributing the torque to the driving wheels. The reference yaw rate is calculated according to the small slip angle approximation. Given that at steady state the expression for path curvature under constant speed and steering angle is given below [27]:

$$\frac{1}{R} = \frac{r}{V} \approx \frac{r}{u}, \tag{14}$$

and that under kinematic steering, the steering angle is related to the turning radius through the following equation:

$$\delta = \frac{l}{R}, \tag{15}$$

substituting Equation (15) into Equation (14), yields the following reference:

$$r_{ref} = u(\frac{\delta}{a+b}) \tag{16}$$

The longitudinal speed u is assumed to be the average velocity of the rear wheels assuming that they are not spinning nor they are locked. Hence, through feedback, the error is reduced by the PI action as follows:

$$\Delta T = K_p e + K_I \int e, \tag{17}$$

where the error e is given as the difference between the reference r_{ref} and measured yaw rate r as follows:

$$e = r_{ref} - r, \tag{18}$$

Next, the torque (i.e., the output of the controller) is distributed between the left (T_{left}) and right motor (T_{right}) as follows:

$$\begin{cases} T_{right} = T - \frac{\Delta T}{2} \\ T_{left} = T + \frac{\Delta T}{2} \end{cases} \tag{19}$$

where T is the input torque. Equation (19) is valid according to the following sign convention: turning clockwise is positive and anti-clockwise negative. The schematic of the controller is shown in Figure 6.

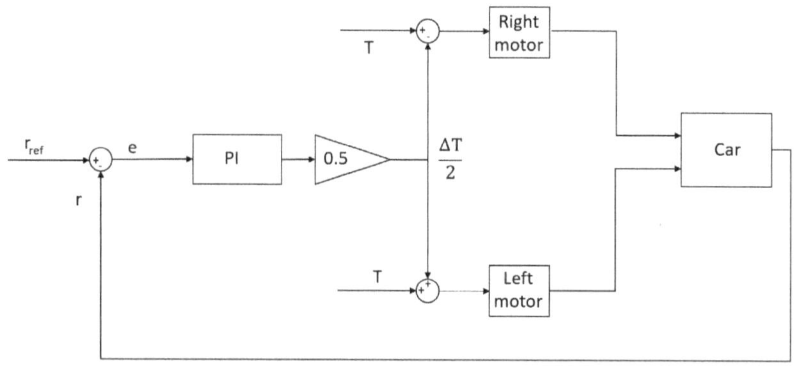

Figure 6. Yaw rate controller scheme.

The parameters are obtained by manual tuning of the proportional K_P and integral K_I gains. The resulting values are displayed in Table 1.

Table 1. Yaw rate controller parameters.

K_P	K_I
2	0.6

2.6. Longitudinal Dynamics: Slip Controller

In order to improve the longitudinal performance of the vehicle, a wheel slip controller is implemented using a PI controller. The PI-type controller is implemented due to its effectiveness and simplicity [21]. A quarter car model consisting of a single wheel attached to a mass is used. According to this model, only longitudinal dynamics are considered; moreover, one of the limitations is the assumption of a fixed load on the wheel. The model and the underlying equations are shown in Equation (20).

$$\begin{cases} m\dot{v} & = -F_x \\ J\dot{\omega} & = r_w F_x + T \\ F_x & = F_z \mu(\kappa), \end{cases} \tag{20}$$

where m is the mass, $\dot{v}$ is the rate of change of velocity, J the wheel inertia, F_x is the longitudinal tire force, F_z is the normal force, T is the torque, $\dot{\omega}$ is the rate of change of the angular velocity and μ is the tyre road friction coefficients dependent on the slip ratio κ. By looking at the equation above it follows that, given a fixed vertical load, the value of the slip that leads to the highest friction coefficient must be found such that the maximum braking/traction force can be developed. The slip ratio is defined as the normalized difference between the vehicle velocity and the wheel velocity as follows [30]:

$$\kappa = -\frac{v - \omega r}{v}. \tag{21}$$

Hence, the following relation holds

$$\begin{cases} \kappa = 0 & \text{free rolling tyre} \\ \kappa > 0 & \text{driving tyre} \\ \kappa < 0 & \text{braking tyre} \\ \kappa = -1 & \text{locked wheel} \end{cases} \tag{22}$$

Since the slip of a free rolling wheel (i.e., non-driving wheel), is equal to zero, the wheel slip of a rear-wheel driven car can be approximated as follows:

$$\kappa \approx -\frac{\omega_f - \omega_r}{\omega_f}, \tag{23}$$

which can be rewritten as:

$$\omega_r \approx \omega_f(1 + \kappa). \tag{24}$$

Since at very low speed ($\omega_r \approx 0$) the slip value is not defined, the slip is achieved by upper bounding the speed of the rear wheel as a function of the front tire as shown in Equation (24). Next, a PI controller is designed such that the wheel slip is kept constant at the point where the maximum force can be developed. The controller takes the difference between the measured rear wheel speed and the reference ($\omega_f(1 + \kappa)$) and minimizes the error according to Equation (18). In this case, the error is given as the difference between the rear tire velocity and the reference. The torque (e.g., output of the PI controller) is added to the input torque such that the error is minimized. Figure 7 shows the control scheme for the left tire only as the scheme is equivalent for both tires. T indicates the input torque (e.g., output of the controller), τ is the output torque (e.g., torque applied to the tire).

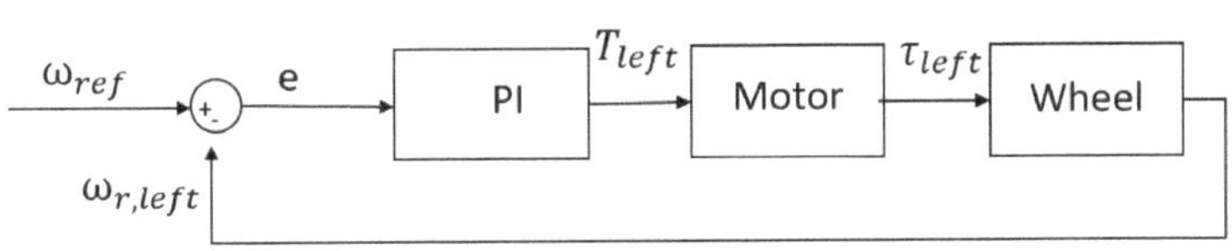

Figure 7. The control loop in the wheel model.

Last, the value of the slip is set to be positive when the driver is accelerating, and negative when braking according to Equation (22), such that both anti-lock braking system (ABS) and traction control (TC) systems are obtained. Given that no data on the tire characteristics are provided, the ideal slip value has been determined experimentally through a braking and acceleration test. Regarding controller design, the gains are obtained manually tuning the controller. The resulting values are displayed in Table 2.

Table 2. Wheel slip controller parameters.

K_P	K_I
4	8

3. Results

The following section deals with the results gathered throughout the research. First, the vehicle driving performance is analyzed and the model is validated. Next, the yaw rate and the wheel slip controllers are evaluated. Then, the weight transfer analysis is conducted. At last, the combination of control algorithms and weight transfer is analyzed.

3.1. Steering Behavior and Model Validation

In order to determine the unknown values of the cornering stiffness, a constant cornering test is conducted. This consists of steady state driving with constant steering angle for different velocities ranging from 0 to 3 m/s. Higher velocities were not reached due to the tire friction limits. The steering angle used throughout the experiments corresponds to 24.3 degrees. By looking at Figure 8, it is possible to compare the vehicle behavior with respect to the model and the reference values. Initially, the vehicle's yaw rate follows the neutral steer reference, then at approximately 2 m/s, the vehicle starts to under-steer as the yaw rate drops. In the linear region, the discrepancy between the model and the measurement data is negligible; however, as the non-linear region approaches, the error reaches a maximum of about 9% at approximately 1.7 m/s. In the non-linear region, where the yaw rate suddenly drops, the error suddenly increments linearly with the speed: this is due to the non-linearities of the tires that are not taken into account by the bicycle model. In order to validate the model, the cornering stiffness values have been tuned manually to minimize the discrepancy. Given that the car is understeering and approximately 60% of the weight is on the rear, solving Equation (11) for a negative understeer gradient leads to the relation of $C_1 < 0.7\, C_2$. As a result, the modeled value of the front cornering stiffness is equal to $0.55\, C_2$. More specifically the modeled value for the rear cornering stiffness is equal to 350 $\frac{\mathrm{N}}{\mathrm{rad/s}}$, while the front value is equal to 192.5 $\frac{\mathrm{N}}{\mathrm{rad/s}}$. The load configuration for this measurement is displayed by the "unloaded" case in Table 3.

Table 3. Vertical static load on each axle; experiments on launch and cornering performance).

Experiment	F_{z1} [N]	F_{z2} [N]
Loaded front	92.9	69
Loaded rear	40.5	97.4
Unloaded	56.3	76.2

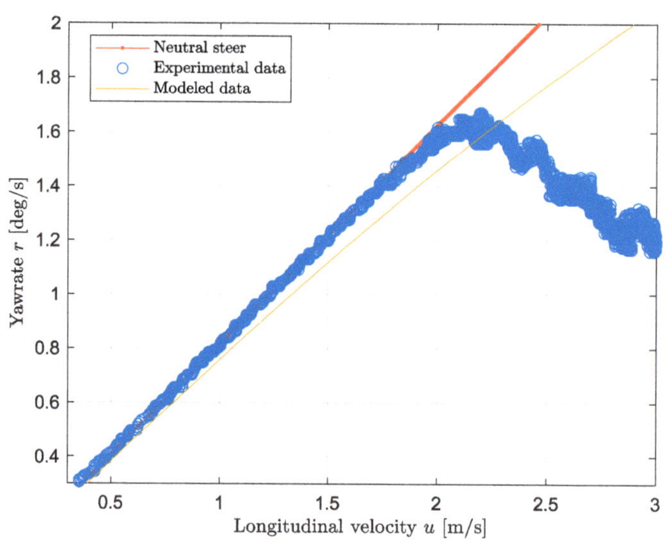

Figure 8. Vehicle steering behavior: modeled and experimental data.

3.2. Yaw Rate Controller

The results in this section are obtained by a steady state cornering test with a steering angle of 24.3 degrees. Figure 9 shows the vehicle behavior with and without the yaw rate controller. It follows that in the linear tire region, the controller does not contribute to the yaw moment as the vehicle is in neutral steering condition; however, in the non-linear tire region, a positive moment is applied to the vehicle improving its performance. In other words, the controller reduces the error by causing an increase in the yaw rate of approximately 15% at 2.5 m/s. Last, it is possible to notice that above a longitudinal speed of 2.5 m/s, the tires exceed their limits, where the input force exceeds the maximum force that can be developed by the tire. As a consequence, the vehicle starts to slide sideways in the direction of the turn. This can be seen by the sudden increase of the yaw rate which instantly reaches a maximum of 1.8 deg/s between 2.5 and 2.6 m/s.

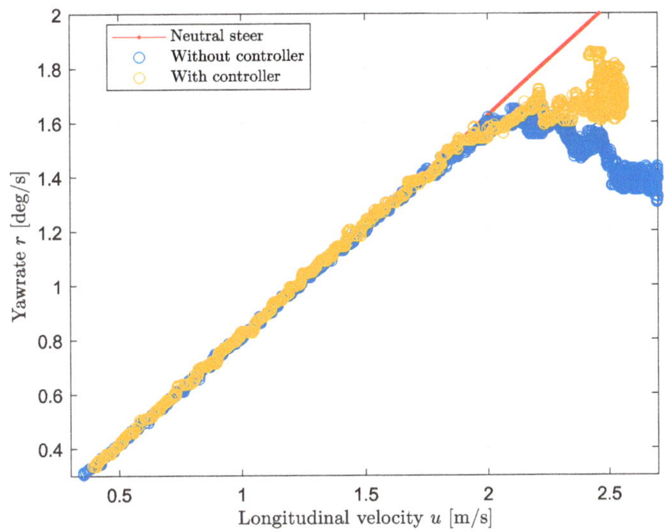

Figure 9. Vehicle steering behavior with yaw rate controller.

3.3. Weight Transfer Analysis

This subsection deals with the weight transfer analysis. Figure 10 shows the yaw rate as a function of velocity for both the modeled and the experimental data together with the reference for three weight configurations without applying any control systems. The loading conditions for this experiment are shown in Table 3, namely the case for $a = 0.37$ corresponds to the "Loaded rear" case and $a = 22$ is the "Loaded front" case. By looking at the figure, it follows that generally, for any weight configuration, the car shows understeering behavior at high speed as the yaw rate drastically drops at approximately 2.2 m/s. Next, both configurations for $a = 0.3$ and $a = 0.37$ show similar behavior at low speed: both cases show neutral steer behavior until the non-linear tire region ($u > 2$ m/s) is entered, whereas the third configuration shows a higher mismatch with the reference. More specifically, when the CoG is moved forward, the vehicle starts to understeer at a lower speed: the error between the reference and the measured data reaches a value greater than 5% already at a velocity of 1 m/s, then it increases linearly with the speed. Regarding the peak values, shifting the CoG forward leads to a slightly higher maximum yaw rate achievable compared to the other cases, namely a 5% increase in the maximum yaw rate is achieved by shifting the CoG to the front by 16 cm. However, the peak value is reached at a slightly higher speed. Last, in the non-linear region, the configuration for $a = 0.37$ m shows the lowest yaw rate for any given velocity, while moving the CoG forward leads to higher yaw rate values. More specifically, moving the CoG rearward by 15 cm leads to a decrease of the yaw rate approximately by 15% in the high speed region. To conclude, loading the front axle leads to a more severe understeering in the linear region while loading the rear axle reduces such behavior. However, the opposite phenomena appears at high speed as the lowest yaw rate is achieved when loading the rear axle. Furthermore, Figure 10 shows that the model matches the observation above; however, a mismatch with regard to the experimental data appears at high speed due to the linear nature of the model.

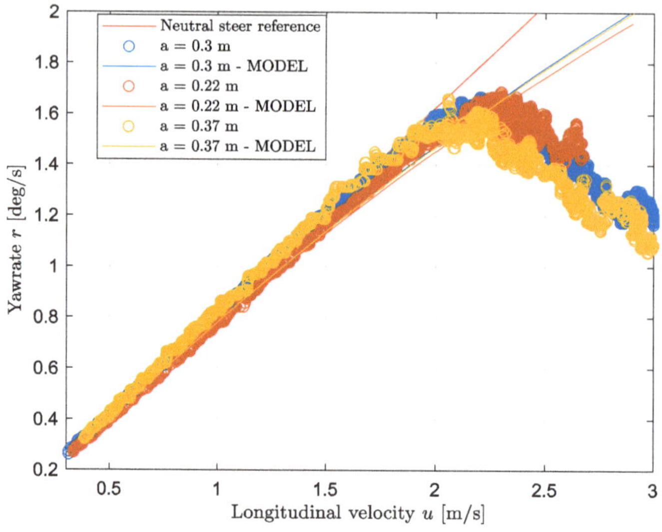

Figure 10. Vehicle steering behavior with yaw rate controller.

3.4. Wheel Slip Controller

This paragraph deals with the validation of the slip controller. Figure 11a,b show the resulting measurements while braking and accelerating when the vehicle is unloaded, respectively. In order to evaluate the controller performance, a step of −2.5 Nm and 2.64 Nm is applied for braking and launching without applying a steering input, respectively. The loading condition corresponds to the "unloaded" case shown in Table 3. By looking at Figure 11a,b, it follows that the wheel slip controller prevents the driving wheels from locking and spinning as in Figure 12a,b. Figure 13 shows the values of the slip as the vehicle accelerates when the controller is applied as a function of velocity. It follows that constant slip of approximately 0.2 is achieved at all speeds. Since the wheels spin when the controller is not applied, it is not possible to calculate a reasonable value of slip using the methods above; however, Figure 12b shows that a very high difference in speed between the front and rear tires is achieved suggesting a high slip value. Hence, the wheel slip controller allows the vehicle to maximize the driving/braking force (see Equation (20)). An average deceleration of 2 m/s^2 is achieved while braking, and an average acceleration of 2.3 m/s^2 is reached when launching. Last, with a slip controller, the stability of the vehicle is highly increased as any unwanted yaw moment is rejected: Figure 14a,b show that when the wheel slip controller is not applied, the vehicle loses stability as a yaw moment is generated. Due to the nature of the experimental set up, the values of velocity and acceleration are given as an indication based on the testing results; they would be used to represent reasonable real-life values (after a scaling) for large-scale vehicles.

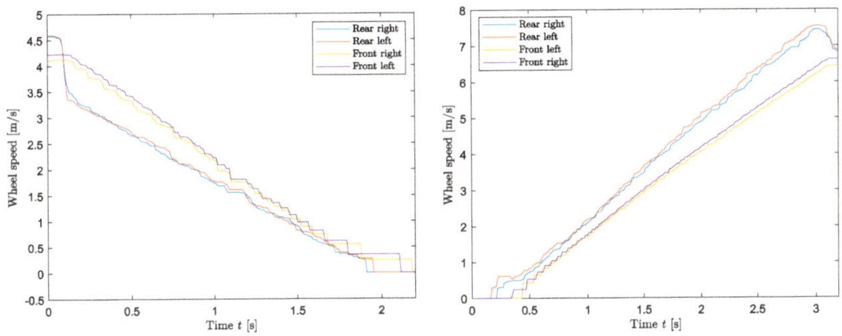

(**a**) (**b**)

Figure 11. Wheel slip controller: wheel velocity against time. (**a**) Wheel slip controller during braking maneuver (ABS). (**b**) Wheel slip controller during acceleration manoeuvre (TC).

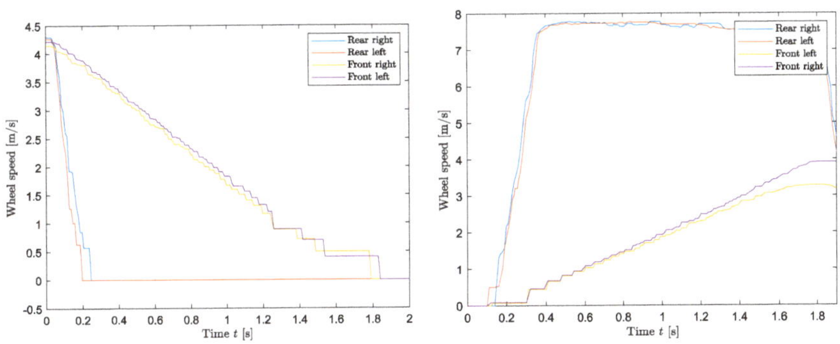

(**a**) Braking (**b**) Acceleration

Figure 12. Wheel speed during a launch and braking maneuver with no slip controller.

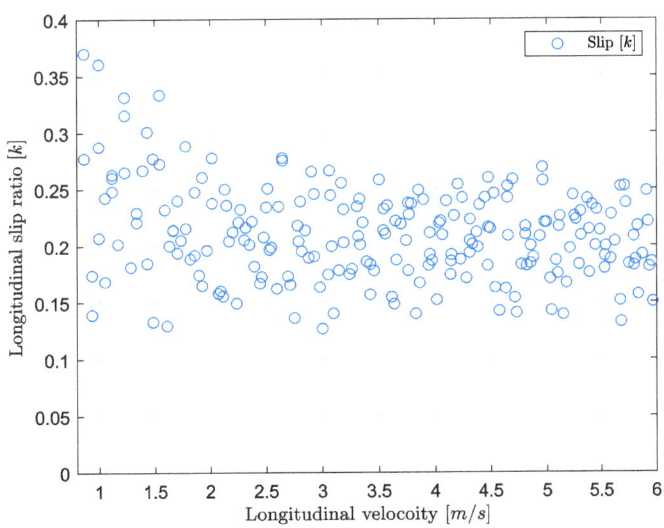

Figure 13. Values of slip: launch control.

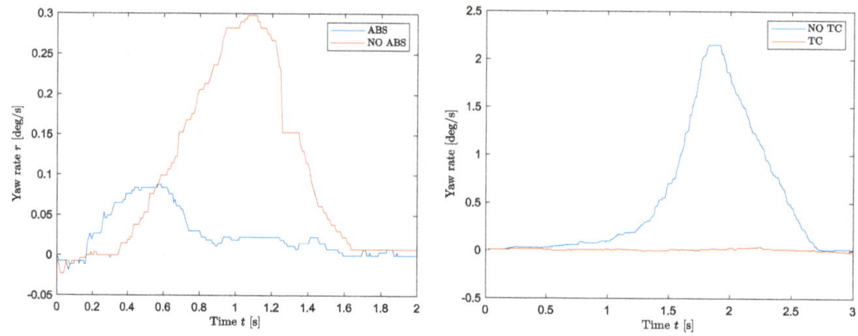

(a) Braking **(b)** Acceleration

Figure 14. Vehicle speed during a launch and braking maneuver with no slip controller.

3.5. Control Algorithms for Different Weight Configurations

The loading conditions shown in Table 3 have been implemented for testing the vehicle performance for different weight configurations. In addition, the experiments have been performed with the yaw rate and wheel slip controllers. Figure 15 shows the yaw rate value as a function of the longitudinal velocity. By looking at the Figure, it follows that the same observation as in Section 3.2 can be gathered. In other words, implementing the controller improves the vehicle performance in the non-linear region of the tires for every loading condition. Moreover, it follows that approximately the same performance can be achieved for each configuration. Therefore, given that in the non-linear region, the configuration with the CoG moved rearward ($a = 0.37$) has the lowest angular velocity, the most significant improvement can be seen in this configuration as the yaw rate is increased by approximately 30% at 2.5 m/s.

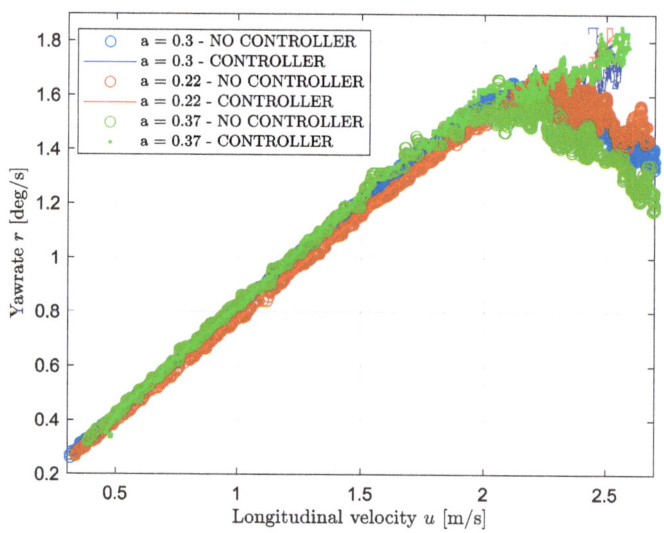

Figure 15. Weight transfer analysis: yaw rate against longitudinal velocity with a yaw rate controller.

Last, an analysis of the vehicle's longitudinal performance with a wheel slip controller for each weight configuration is conducted. Figures 16a and 17a show the longitudinal velocity of the vehicle as a function of time while accelerating and braking for the three loading conditions, respectively. To evaluate the performance while accelerating, the vehicle has been loaded as in Table 3. The loading conditions throughout the braking tests are displayed in Table 4. The same inputs as in Section 3.4 are applied. From Figure 16, it follows that by shifting the CoG rearward, hence increasing the static vertical load on the rear axle, the vehicle shows a higher acceleration. Namely, for the braking case, increasing the rear weight distribution from 42% to 70%, leads to an increase of 48% in the average acceleration. According to Figure 17a, there is a linear relation between the maximum force developed by the tire and the vertical static force on the driving axle. Figure 16b shows the vehicle velocity for a launch with different weight configurations. The same conclusion as for Figure 17a can be drawn. The acceleration linearly increases with the vertical load on the rear axle. In this case, increasing the rear vertical load by 43% leads to an increase in acceleration from 1.78 m/s^2 to 3.2 m/s^2. As a consequence the braking and the acceleration distance are highly reduced when the CoG is shifted to the rear or the rear axle is loaded. Namely for the acceleration, increasing the rear mass percentage from 42% to 70% leads to a 50% decrease in the acceleration distance, while the braking distance is reduced by approximately 34% when the mass percentage on rear is shifted from 44% to 63.5%

Table 4. Vertical static load on each axle braking experiment.

Experiment	F_{z1} [N]	F_{z2} [N]
Loaded front	89.9	71.4
Loaded rear	54.6	95.6
Unloaded	56.3	76.4

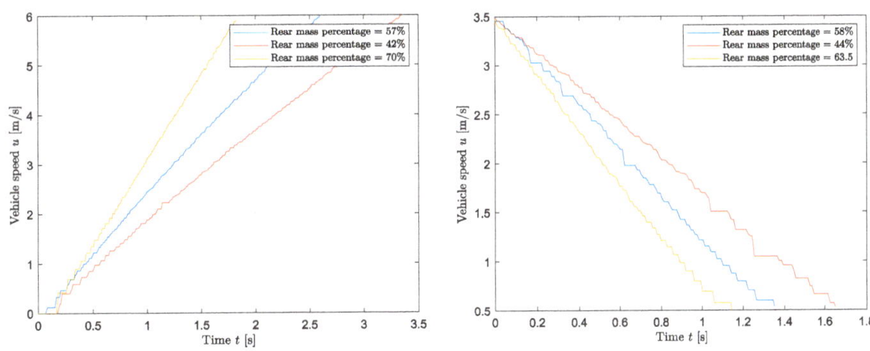

(**a**) Weight transfer analysis: TC. (**b**) Weight transfer analysis: ABS.

Figure 16. Weight transfer analysis: wheel slip controller.

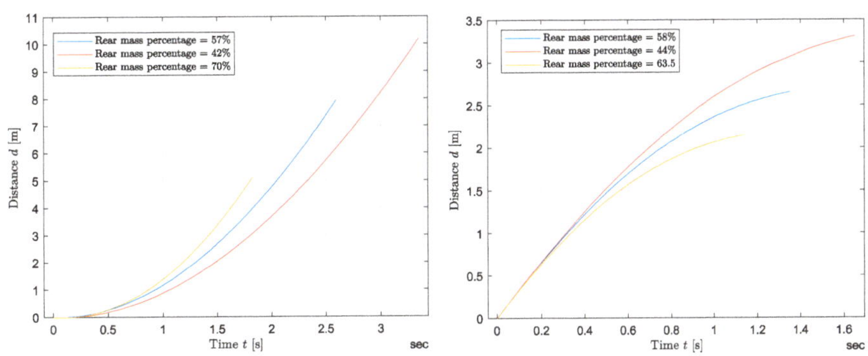

(**a**) Weight transfer analysis: acceleration distance (**b**) Weight transfer analysis: braking distance

Figure 17. Braking and accelerating distance for different weight distribution.

4. Discussion

First, the research shows that, given the assumptions in Section 2.2, the bicycle model captures the essential dynamics of the vehicle especially in the linear tire region, hence it can be considered an accurate model in mimicking the cornering behavior of the vehicle. The bicycle model allows to design and implement an effective yaw rate controller. However, the bicycle model also presents some limitations as it is based on several assumptions as detailed in Section 2. Hence, a different model may be used to examine the impact of the driver or of passengers. Moreover, the model assumes the height of the center of gravity to be zero. In general, EVs are heavier compared to the corresponding vehicles with ICE [31], so this is a reasonable assumption. Moreover, the load transfer is neglected which ensures the accuracy of the model at low lateral and longitudinal acceleration. As for severe maneuvers that produce large lateral accelerations, the bicycle model does not represent the vehicle response accurately due to nonlinear tire forces and associated dynamic load transfer. In other words, a different approach is suggested to model such maneuvers.

As shown in Figure 9, the cornering performance has improved as the yaw rate has increased up to 30% when the CoG is shifted to the rear. However, the effect of the controller is limited by the vehicle physical limits. When the force applied to the wheel exceeds the friction circle, the tires start spinning, hence the vehicle loses stability. Implementing a slip controller to prevent the wheel from spinning or locking while cornering will further improve the vehicle's performance. Regarding the longitudinal motion, applying a wheel slip controller prevents the wheel from spinning and from locking, enhancing performance with regard to the braking distance and acceleration time and it greatly improves vehicle stability as any unwanted moment is completely rejected. However, throughout the experiments the same tires have been fitted to the vehicle and the tests were performed on a flat surface with a constant tyre road coefficient. The paper [32] suggest that to ensure the effectiveness of the control system, information of the peak tyre–road friction coefficient and adjustment of the slip ratios are fundamental. The implementation of a control algorithm based on the estimation of the tyre road friction may be beneficial in practice.

Figure 10 shows the importance of the weight distribution in vehicles regarding cornering performance. The relation between the weight distribution and the driving behavior can be analyzed by observing by the equations of motion in Section 2. Adjusting the CoG location changes the parameters, namely the static vertical load and the values of a and b. Such parameters influence the lateral side slip angles, and hence the lateral forces. By looking at Equation (8), it follows that shifting the CoG towards the front leads to an increase in the front side slip angle, which may lead to understeering as shown in Equation (13). Furthermore, Equation (3) leads to the same conclusion as at constant lateral forces and steering angle, increasing the value of a leads to a higher turning moment (i.e., oversteer). The same reasoning can be applied by looking at the understeer gradient in Equation (11), as it expresses the cornering behavior as a function of the static vertical load. Increasing the vertical front load (i.e., reducing the parameter a) leads to understeering, whereas decreasing it leads to oversteering. Such observations match with the results shown in Figure 10 corresponding to the linear region. Moreover, by applying Equation (12), the modeled value of the understeer gradient is calculated: the gradient has a value of approximately zero when loading the vehicle on the rear while it reaches a value of about 0.3 when loading the vehicle on the front. To model the dynamics at a higher speed more precisely, a more advanced model may be implemented; however, such observations show that the main dynamics of the vehicle are well represented by the bicycle model.

Regarding the longitudinal motion, Equation (20) shows that the force is dependent on the longitudinal slip, and linearly to the vertical force. Hence increasing the tire load leads to an increase in the maximum force assuming that the tire-road coefficient is constant. Furthermore, the tire model assumes no pitch motion, hence no load transfer while accelerating or decelerating. Given that load transfer increases at high acceleration and when the center of gravity is high, it is neglected for the scope of this research. However, implementing load transfer in the tire model may further improve the performance of the controller and of the vehicle for very high acceleration values [30]. Given the results shown in Section 3, it follows that the optimal location of the CoG is highly dependent on the application.

5. Conclusions

From this research, it follows that vehicle performance can be improved by implementing active safety systems such as a yaw rate controller or wheel slip controller. Such systems considerably increase handling, performance and stability as they allow the driver to maintain control of the vehicle, hence preventing accidents by reducing excessive over and understeer while cornering and increasing the stability while accelerating or braking. Moreover, we conclude that performance is highly related to loading conditions, where the weight distribution should be adjusted to meet the cornering requirements.

For a rear-wheel driven vehicle, locating the CoG rearward leads to a higher understeering in the linear tire region, whereas shifting it to the front reduces such behavior. However, such distribution has also an effect on the longitudinal dynamics with regard to braking and launching, as decreasing the vertical load on the driving axle decreases the maximum force that can be developed, hence decreasing the performance in terms of braking distance and acceleration time. Since excessive over- and understeering is often undesired, for a rear-wheel driving vehicle, a trade-off has to be achieved.

To conclude, weight distribution and control algorithms highly influence driving behavior in terms of stability, handling and hence, safety. This research has provided an insight on how control systems can be implemented on commercial vehicles with different weight distribution to increase their performance. For future research, we will further explore topics such as vehicle stability concerning road friction, saturated tire force, and advanced control (prescribed performance control and sliding mode control [33]) to further improve the vehicle driving performance.

Author Contributions: Methodology, V.G., W.H., A.D. and Z.S.; Software, W.H.; Validation, V.G.; Investigation, V.G., W.H., A.D. and Z.S.; Resources, Z.S.; Data curation, V.G. and W.H.; Writing—original draft, V.G.; Writing—review & editing, W.H., A.D. and Z.S.; Visualization, W.H.; Supervision, W.H., A.D. and Z.S. All authors have read and agreed to the published version of the manuscript.

Funding: This research was supported by the European project COVER under the grant No. 101086228.

Institutional Review Board Statement: Not applicable.

Data Availability Statement: The data presented in this study are available on request from the corresponding author.

Conflicts of Interest: The authors declare no conflict of interest.

References

1. Peters, D.R.; Schnell, J.L.; Kinney, P.L.; Naik, V.; Horton, D.E. Public Health and Climate Benefits and Trade-Offs of U.S. Vehicle Electrification. *GeoHealth* **2020**, *4*, e2020GH000275. [CrossRef] [PubMed]
2. Buekers, J.; Van Holderbeke, M.; Bierkens, J.; Int Panis, L. Health and environmental benefits related to electric vehicle introduction in EU countries. *Transp. Res. Part Transp. Environ.* **2014**, *33*, 26–38. [CrossRef]
3. Dik, A.; Omer, S.; Boukhanouf, R. Electric Vehicles: V2G for Rapid, Safe, and Green EV Penetration. *Energies* **2022**, *15*, 803. [CrossRef]
4. Ghosh, J.; Tonoli, A.; Amati, N. A Torque vectoring strategy for improving the performance of a rear wheel drive electric vehicle. In Proceedings of the 2015 IEEE Vehicle Power and Propulsion Conference, VPPC 2015—Proceedings, Montreal, QC, Canada, 19–22 October 2015; Volume 12. [CrossRef]
5. Ling, F.F.; Leckie, F.A.; Gross, D.; Yang, H.T.; Bryant, D.; Wang, P.K.K.; Systems, P.; Klutke, G.A.; Science, T.; Bergles, A.E. *Vehicle Dynamics and Control*, 2 ed.; Springer: Berlin/Heidelberg, Germany, 2012.
6. Knauder, B.; Savitski, D.; Theunissen, J.; De Novellis, L. Electric torque vectoring for electric vehicles. *ATZelektronik Worldw.* **2014**, *9*, 50–55. [CrossRef]
7. Lacroix, B.; Liu, Z.; Seers, P. A comparison of two control methods for vehicle stability control by direct yaw moment. *Proc. Appl. Mech. Mater.* **2012**, *120*, 203–217. [CrossRef]

8. Ahangarnejad, A.H.; Radmehr, A.; Ahmadian, M. A review of vehicle active safety control methods: From antilock brakes to semiautonomy. *J. Vib. Control* **2020**, *27*, 1683–1712. [CrossRef]
9. Das, A.; Kasemsinsup, Y.; Weiland, S. Optimal Trajectory tracking control for automated guided vehicles. *IFAC-PapersOnLine* **2017**, *50*, 303–308. [CrossRef]
10. Aripin, M.K.; Md Sam, Y.; Danapalasingam, K.A.; Peng, K.; Hamzah, N.; Ismail, M.F. A review of active yaw control system for vehicle handling and stability enhancement. *Int. J. Veh. Technol.* **2014**, *2014*, 437515. [CrossRef]
11. Medina, A.; Bistue, G.; Rubio, A. Comparison of Typical Controllers for Direct Yaw Moment Control Applied on an Electric Race Car. *Vehicles* **2021**, *3*, 127–144. [CrossRef]
12. De Novellis, L.; Sorniotti, A.; Gruber, P.; Pennycott, A. Comparison of feedback control techniques for torque-vectoring control of fully electric vehicles. *IEEE Trans. Veh. Technol.* **2014**, *63*, 3612–3623. [CrossRef]
13. Tekin, G.; Ünlüsoy, Y.S. Design and simulation of an integrated active yaw control system for road vehicles. *Int. J. Veh. Des.* **2010**, *52*, 5–19. [CrossRef]
14. Du, H.; Zhang, N.; Naghdy, F. Velocity-dependent robust control for improving vehicle lateral dynamics. *Transp. Res. Part Emerg. Technol.* **2011**, *19*, 454–468. [CrossRef]
15. Parra, A.; Tavernini, D.; Gruber, P.; Sorniotti, A.; Zubizarreta, A.; Perez, J. On Nonlinear Model Predictive Control for Energy-Efficient Torque-Vectoring. *IEEE Trans. Veh. Technol.* **2021**, *70*, 173–188. [CrossRef]
16. Parra, A.; Zubizarreta, A.; Pérez, J.; Dendaluce, M. Intelligent Torque Vectoring Approach for Electric Vehicles with Per-Wheel Motors. *Complexity* **2018**, *2018*, 7030184. [CrossRef]
17. Lu, Q.; Sorniotti, A.; Gruber, P.; Theunissen, J.; De Smet, J. H loop shaping for the torque-vectoring control of electric vehicles: Theoretical design and experimental assessment. *Mechatronics* **2016**, *35*, 32–43. [CrossRef]
18. Prompakdee, E.; Boonporm, P.; Rooppakhun, S. The influence of weight distribution on the handling characteristics of intercity bus under steady state vehicle cornering condition. *MATEC Web Conf.* **2016**, *81*, 04014. [CrossRef]
19. Skrúcaný, T.; Vrábel, J.; Kendra, M.; Kažimír, P. Impact of cargo distribution on the vehicle flatback on braking distance in road freight transport. *MATEC Web Conf.* **2017**, *134*, 00054 [CrossRef]
20. Nozaki, H. Effect of the front and rear weights distribution ratio of a formula car during maximum-speed cornering on a circuit. *Int. J. Automot. Technol.* **2008**, *9*, 307–315. [CrossRef]
21. Park, J.H.; Kim, C.Y. Wheel slip control in traction control system for vehicle stability. *Veh. Syst. Dyn.* **1999**, *31*, 263–278. [CrossRef]
22. Minh, V.T.; Oamen, G.; Vassiljeva, K.; Teder, L. Development of Anti-lock Braking System (ABS) for Vehicles Braking. *Open Eng.* **2016**, *6*, 554–559. [CrossRef]
23. Kawabe, T.; Nakazawa, M.; Notsu, I.; Watanabe, Y. A sliding mode controller for wheel slip ratio control system. *Veh. Syst. Dyn.* **1997**, *27*, 393–408. [CrossRef]
24. Ma, Y.; Zhao, J.; Zhao, H.; Lu, C.; Chen, H. MPC-Based slip ratio control for electric vehicle considering road roughness. *IEEE Access* **2019**, *7*, 52405–52413. [CrossRef]
25. Savitski, D.; Ivanov, V.; Augsburg, K.; Emmei, T.; Fuse, H.; Fujimoto, H.; Fridman, L.M. Wheel Slip Control for the Electric Vehicle with In-Wheel Motors: Variable Structure and Sliding Mode Methods. *IEEE Trans. Ind. Electron.* **2020**, *67*, 8535–8544. [CrossRef]
26. Romijn, T.C.J.; Hendrix, W.H.A.; Donkers, M.C.F. Modeling and Control of a Radio-Controlled Model Racing Car. *IFAC-PapersOnLine* **2017**, *50*, 1962–9167. [CrossRef]
27. Pacejka, H.B. *Tire and Vehicle Dynamics*; Elsevier/BH: Amsterdam, The Netherlands, 2012; p. 632.
28. Ren, H.; Shim, T.; Ryu, J.; Chen, S. *Development of Effective Bicycle Model for Wide Ranges of Vehicle Operations*; SAE Technical Papers; SAE International: Warrendale, PA, USA 2014; Volume 1, p. 4. [CrossRef]
29. Zainal, Z.; Rahiman, W.; Baharom, M.N.R. Yaw Rate and Sideslip Control using PID Controller for Double Lane Changing. *J. Telecommun. Electron. Comput. Eng. (JTEC)* **2017**, *9*, 99–103.
30. Tigelaar, J.A.A. *Design and Vehicle Implementation of an Adaptive ABS*; Technical Report; 2011.
31. Besselink, I.; Nijmeijer, H.; Besselink, I.J.M.; Van Oorschot, P.F.; Meinders, E.; Nijmeijer, H. *Design of an Efficient, Low Weight Battery Electric Vehicle Based on a VW Lupo 3L*; Technical Report; 2010.
32. Zhao, J.; Zhang, J.; Zhu, B. Coordinative traction control of vehicles based on identification of the tyre–road friction coefficient. *Proc. Inst. Mech. Eng. Part D J. Automob. Eng.* **2016**, *230*, 1585–1604. [CrossRef]
33. Liang, Z.; Zhao, J.; Liu, B.; Wang, Y.; Ding, Z. Velocity-based path following control for autonomous vehicles to avoid exceeding road friction limits using sliding mode method. *IEEE Trans. Intell. Transp. Syst.* **2020**, *23*, 1947–1958. [CrossRef]

MDPI AG
Grosspeteranlage 5
4052 Basel
Switzerland
Tel.: +41 61 683 77 34

Sensors Editorial Office
E-mail: sensors@mdpi.com
www.mdpi.com/journal/sensors

www.ingramcontent.com/pod-product-compliance
Lightning Source LLC
Chambersburg PA
CBHW041416210926
51850CB00013B/3282
* 9 7 8 3 7 2 5 8 2 6 5 9 9 *